Foundations of Multiple Regression and Analysis of Variance

This book provides a rigorous development of the foundations of linear models for multiple regression and Analysis of Variance (ANOVA), based on orthogonal projections and relations among linear subspaces. It is appropriate for the linear models course required in most statistics Ph.D. programs.

The presentation is particularly accessible because it is self-contained, general, and taken in logical steps that are linked directly to practicable computations. The broad objective is to provide a path of mastery so that the reader could, if stranded on a desert isle with nothing but pencil, paper, and a computer to perform matrix sums and products, replicate general linear models procedures in extant statistical computing packages.

The primary prerequisite is mathematical maturity, which includes logical thinking and the ability to tell when a proof is a proof. Casual acquaintance with matrices would be helpful but not required. Background in basic statistical theory and methods is assumed, mainly for familiarity with terminology and the purposes of statistics in applications.

The material is developed as a series of propositions, each dependent only on those preceding it. The reader is strongly encouraged to prove each one independently. Mastery requires active involvement.

As part of the broad coverage of the mathematics supporting multiple regression and ANOVA, those propositions also establish several new, key results.

- There is a unique, best numerator sum of squares for testing an estimable function.
- The extra residual sum of squares due to imposing a linear hypothesis tests exclusively the estimable part.
- Models that include exclusively any given set of ANOVA effects can be formulated with contrast coding.
- Tests of any ANOVA effects in any design and model, including unbalanced and empty cells, can be had with extra residual sum of squares due to deleting predictor variables.
- Essential properties of Type III methods are identified and proven.

Lynn Roy LaMotte is Professor Emeritus in the Biostatistics Program, School of Public Health, LSU Health–New Orleans. Elected Fellow of the American Statistical Association, 1985, for "important, innovative, seminal, and diverse contributions to the theory and application of linear statistical models," he is author of about 100 articles in diverse academic journals, cited more than 2,000 times, nearly 500 since 2020.

Foundations of Multiple Regression and Analysis of Variance

Lynn Roy LaMotte

CRC Press
Taylor & Francis Group
Boca Raton London New York

CRC Press is an imprint of the
Taylor & Francis Group, an **informa** business
A CHAPMAN & HALL BOOK

Designed Cover Image: Lynn Roy LaMotte

First edition published 2026
by CRC Press
2385 NW Executive Center Drive, Suite 320, Boca Raton FL 33431

and by CRC Press
4 Park Square, Milton Park, Abingdon, Oxon, OX14 4RN

CRC Press is an imprint of Taylor & Francis Group, LLC

ISBN: 978-1-032-98152-9 (hbk)
ISBN: 978-1-032-98154-3 (pbk)
ISBN: 978-1-003-59732-2 (ebk)

DOI: 10.1201/9781003597322

Typeset in Latin Modern font
by KnowledgeWorks Global Ltd.

Publisher's Note: This book has been prepared from camera-ready copy provided by the author.

Dedicated to
Julia Volaufova

The cover. The figure on the cover is based on data from Alghanim et al. (2017). Its purpose is to illustrate the potential use of percents methylation of DNA at two loci to estimate ages of human subjects. The 91 subjects ranged in age from 6 to 72 years. The percents methylation shown here ($y1$ and $y2$) are from two loci. The upper-left panel shows age by $y1$; the bottom-right, $y2$ by age; and the bottom-left, $y2$ by $y1$. The upper-right panel shows p-values by age for the subject with ID 5, whose true age was 12 years. Red dots and blue lines trace that subject in the other panels. The model fit to the data was bivariate linear regression of the two responses on age. P-values were computed via inverse prediction based on that model. Assuming bivariate normality, the gold spike indicates the maximum p-value, which coincides with the maximum likelihood estimate of this subject's age. The red horizontal line is at $p = 0.05$. The range of ages for which the p-value exceeds 0.05 constitutes a 95% confidence interval on the age of this subject, based on $y1$ and $y2$.

Contents

1

Introduction

In the hands of one who understands the foundations, a program that performs only the most basic computations of multiple regression analysis can be made to perform the most sophisticated analyses. One who doesn't understand the foundations is limited to the items in the program's menu and has scant basis for judging whether those items are relevant to the problem at hand or for adapting them to novel uses.

Textbooks on multiple regression and linear models are roughly of two kinds. Most emphasize models and methods. They might list assumptions and formulate models in terms of matrices, but they don't attempt to establish basic mathematical results. See Kutner et al. (2005), Draper and Smith (1998), and Rawlings et al. (1998).

Others emphasize distributional relations and linear algebra. Shayle Searle's 1971 book and Franklin Graybill's 1976 book, both classics, develop these foundations thoroughly. So also do Hocking (2013), Khuri (2010), Stapleton (1995), Clarke (2008), Seber and Lee (2003), and Christensen (2010). Each of these takes a slightly different approach to the linear algebra. Searle, for example, develops expressions for solutions to the normal equations in terms of generalized inverses. Hocking as well as Seber and Lee reparameterize models to full rank and express estimates in terms of matrix inverses.

Whether the linear algebra is developed in-line (Searle, Graybill) or in appendixes (Christensen, Khuri, Clarke), it is material that the reader must master in order to appreciate the rest of the development. This amounts to a short course in matrix algebra; it is a hurdle that can take up to a third of the course, even when it is assumed that students begin with a good level of mathematical maturity. Most presentations attempt to provide some breadth to the coverage of linear algebra, and they include topics that are potentially useful elsewhere but not necessary for the development of linear models. This is the place where graduate programs in statistics teach their students linear algebra.

Even for students who come with backgrounds in linear algebra more or less equivalent to an upper-level, one-semester course, still it is necessary to establish a common language and approach. For example, when I ask students to describe how they would go about finding solutions to a system of linear equations $A\boldsymbol{x} = \boldsymbol{b}$ when A might be singular, some can describe an approach that would work, but the approaches differ, and most are unnecessarily complicated. Facing this problem many times, I tried to find a simple, direct set of tools that could be mastered quickly and that would provide nearly all the tools required in the rest of the course.

I have written this book assuming that readers are familiar, or have been, with statistical methods for multiple regression and analysis of variance at the level of the standard text by Kutner et al. (2005), and with basic mathematical statistics at the level of Casella and Berger (2002). While the presentation here is mostly self-contained, methods courses help to establish common terminology, conventions, and perspectives of applied statistics. For the most part, the mathematics of linear algebra and matrix algebra that is used here is developed rigorously here. Still, I assume that readers have had some exposure to matrices and vectors and their algebraic operations. That has advantages and disadvantages. On the plus side, expressions like $\mathrm{I} + AB$ do not cause anxiety. On the other side, some habits and ways of thinking might already be set so that they are difficult to displace with different, possibly better ones.

I have assumed that readers have a good level of mathematical maturity. That does not mean that they should already be experts in linear algebra or the theoretical basis of applied statistics. While mathematical maturity is difficult to define, it means at least that they rely on logical thinking and that they can tell when a proof is a proof.

Many authors — Christensen (2010) is an outstanding contemporary example — have exploited the connection between orthogonal projection and least squares: that, given X and $\boldsymbol{y}$, the vector $X\hat{\boldsymbol{b}}$ that minimizes the sum of squares $(\boldsymbol{y} - X\boldsymbol{b})'(\boldsymbol{y} - X\boldsymbol{b})$ is $X\hat{\boldsymbol{b}} = \hat{\boldsymbol{y}} = \mathbf{P}_X\boldsymbol{y}$, which is the orthogonal projection of $\boldsymbol{y}$ in the column space of X. They have differed in how they formulate the orthogonal projection matrix $\mathbf{P}_X$: most common is the formulation $\mathbf{P}_X = X(X'X)^-X'$, in terms of a generalized inverse of $X'X$, from which one choice for a least-squares solution is $\hat{\boldsymbol{b}} = (X'X)^-X'\boldsymbol{y}$. The numerator sum of squares (SS) for testing hypotheses $G'\boldsymbol{\beta} = \mathbf{0}$ based on $\hat{\boldsymbol{b}}$ is then expressed as $(G'\hat{\boldsymbol{b}})'[G'(X'X)^-G]^-(G'\hat{\boldsymbol{b}})$. For those not already comfortable with generalized inverses, this formulation requires additional development and familiarization.

The path taken here is different. The SS is the squared norm of the orthogonal projection of $\boldsymbol{y}$ in the linear subspace specified by the restrictions $G'\boldsymbol{\beta} = \mathbf{0}$ within the full model. That orthogonal projection can be had through the Gram-Schmidt (GS) construction.

In this way, the fundamental structures for inference in these models can be developed rigorously and generally. The need to consider full-rank and less-than-full-rank cases is then unnecessary. It is unnecessary to talk about linear dependence, rank, inverse, and generalized inverse. However, everyone should be conversant with these features, and so bridges are noted to them.

Computations can be accomplished easily with software that is widely available, such as R and R packages, MATLAB®, and the IML procedure in SAS®. With such facility, the reader can obtain numerical results readily. This provides a great opportunity for hands-on experience, to experiment, to construct numerical illustrations and counter-examples where appropriate.

This, too, is a major consideration in choosing a path for developing this methodology. The development and the computations are commensurate, so that the reader can translate the mathematics directly into explicit numerical examples.

Chapters 3 through 7 provide a self-contained development of the foundations in matrix algebra, linear subspaces, and orthogonal projection required for the rest of this book. The chapters in Part II provide a basic, general coverage of methods of statistical inference in linear models.

Part III covers ANOVA models. For factorial treatment arrangements, the objective of these chapters is to provide a general notation, structure, and approach to formulating models with any chosen set of factor effects and covariate-by-factor effects, whether the model is balanced or unbalanced, without or with empty cells. With that facility, any set of factor effects can be examined with a restricted-model-full-model comparison. As an example, it should be clear then that there is a unique, best numerator SS for examining A main effects in a two-factor setting in which the model includes both main effects and interaction effects. The long-running controversy over this topic is summarized and, I hope, resolved satisfactorily. Type III SSs are formulated explicitly, and their properties are detailed and proven.

In the process of writing this book, questions arose that caused me to formulate and prove properties that were until then unknown to me. Among those are the following. Proposition 10.2 establishes the existence and uniqueness of a best numerator sum of squares among F-statistics that test exclusively a given set of linear functions of the mean vector. Proposition 10.3 establishes that the extra sum of squares due to imposing a set of linear conditions tests exclusively the estimable part of those conditions. Chapter 20 shows that widespread beliefs about the "excellence" of proportional subclass numbers are in fact incorrect.

Although those developments were new to me, the theory and mathematics of linear models have been cultivated for a very long time, by many very capable people, and so I doubt that there is any insight possible in this field that is genuinely novel. Still, there is a little thrill and sense of accomplishment that comes from stumbling onto these insights independently. I have tried to write this book in a way that encourages the reader to develop every result from basics, and thereby to feel that satisfaction of discovery.

In order to try to encourage that independent discovery and understanding, for the most part I have not included proofs where propositions first appear. I believe that anyone with the background mentioned above can, given a little time and effort, prove every proposition independently, and that this effort is essential to thoroughly mastering the material. In addition to the formally-stated propositions, I expect the reader to confirm each assertion, explicit or implicit, in the text. Mastering this material requires that kind of active commitment.

As noted above, the approach taken here is different from what has usually been taken in previous books on this subject. The tools and developments are different from formulaic manipulation, and somewhat different habits and reflexes need to be acquired. I hope that the reader will find that the extra initial effort results in unification, illumination, and even simplification of the theory and methodology of inference in linear statistical models.

Mathematical terms, usage, and symbols are conventional and familiar, I hope. "If and only if" is abbreviated "iff." Set inclusion and containment are denoted by $\subset$ and $\supset$, respectively. Some symbols are used for brevity in proofs and solutions: $\Longrightarrow$, $\Longleftarrow$, and $\Longleftrightarrow$ signify "implies," "is implied by," and iff, respectively. Occasionally, $\exists$ and $\forall$ are used for "there exists" and "for each," respectively.

2

Multiple Linear Regression and Analysis of Variance Illustrations

This chapter presumes that the reader has a working familiarity with the setting, terminology, tools, and practice of multiple linear regression (MLR) analysis and Analysis of Variance (ANOVA). The examples illustrate basic methods and techniques, without derivations or proofs, and they use some of the notation, terminology, and structure that are used throughout this book. Some of the techniques and procedures will be familiar, while others may be less so. The purpose is to stimulate the reader's curiosity about the "how" and "why" that make them work, and to begin to justify the usefulness of understanding the undergirding mathematical relations, down to their very foundations.

2.1 Multiple Linear Regression Illustrations

The purpose of this section is to illustrate some of the wide range of tools and special techniques of multiple regression analysis based on least squares. They can be accomplished with any basic procedure that produces sums of squares and coefficient estimates and their standard errors. The best way to follow this discussion is to do the computations yourself.

Table 2.1 shows a small data set with ten records taken from a larger Kentucky Utilities (KU) data set (Thompson and Cady, 1973). The data were compiled by KU as part of an effort to better anticipate peak system load up to 24 hours in advance so that they could adjust their online generating capacity (at that time, coal, gas, nuclear, hydro) accordingly.

The first ten observations shown in Table 2.1 were taken daily at 1 PM. They are used here for illustration only. Variables are: `date`, June 1 is 0601; `dow`, day of week, Monday is 1; x_0 includes ($x_0 = 1$) or excludes ($x_0 = 0$) the intercept term; x_1 is windspeed in knots per hour; x_2 is drybulb temperature in degrees Fahrenheit; x_3 is relative humidity, percent; x_4 is atmospheric pressure, coded as $100\times$(in. Hg - 28); $x_5 = x_4 - x_3$; y is instantaneous load on the KU electrical power generating network, in megawatts. Case $i = 11$

TABLE 2.1: Ten observations extracted from the Kentucky Utilities data set for the first ten weekdays. See the description of the data on p. 5.

i	date	dow	x_0	x_1	x_2	x_3	x_4	x_5	x_6	x_7	y
1	601	2	1	9	79	44	97	53	0	0	1047
2	602	3	1	7	75	64	109	45	0	0	1068
3	603	4	1	8	82	56	119	63	0	0	1115
4	604	5	1	3	84	59	120	61	0	0	1208
5	607	1	1	12	80	69	98	29	0	0	1213
6	608	2	1	7	79	56	99	43	0	0	1165
7	609	3	1	8	73	62	102	40	0	0	1073
8	610	4	1	6	77	48	107	59	0	0	1062
9	611	5	1	3	83	41	111	70	0	0	1200
10	614	1	1	14	81	65	75	10	0	0	1274
11	631	.	1	5	90	80	80	0	-1	0	. $\mid 0^a$
12	632	.	0	0	0	1	1	0	0	-1	. $\mid 0^a$

is included in order to get estimated means and predicted values and their standard errors at $x_1 = 5$, $x_2 = 90$, $x_3 = 80$, and $x_4 = 80$. Case 12 is included to get the estimate of $\beta_3 + \beta_4$ and its standard error. For both, y is coded as missing ($y = .$). Indicator variables x_6 and x_7 are used in Model 2a, for which $y_{11} = y_{12} = 0$ (indicated in the table by 0^a). Although using names of variables (**Load**, **db**, and so on) is preferable when discussing results of statistical analyses, I'll use the symbols y, x_2, and so on in order to simplify notation.

In multiple linear regression, models for the population mean μ of the response variable y generally take the form $\mu(x_1, \ldots, x_k) = x_0\beta_0 + x_1\beta_1 + \cdots + x_k\beta_k$, where x_j is a predictor variable and β_j is its regression coefficient. The common population variance of y is denoted by σ^2.

Estimates are signified by hats ($\hat{\beta}_j$, for example), their variances by $\text{Var}(\hat{\beta}_j)$, and their standard errors by $SE(\hat{\beta}_j) = \sqrt{\hat{\text{V}}\text{ar}(\hat{\beta}_j)}$. Recall that variances of linear estimators are proportional to σ^2. For example, for $\hat{\beta}_j = c_{j1}y_1 + \cdots + c_{jn}y_n$, $\text{Var}(\hat{\beta}_j) = c_j\sigma^2$, and $\hat{\text{V}}\text{ar}(\hat{\beta}_j) = c_j\hat{\sigma}^2$, where $c_j = c_{j1}^2 + \cdots + c_{jn}^2$.

Algorithms that perform multiple linear regression analysis in statistical computing packages specify models by designating the name of one column in the data set as the response and a list of the names of others as predictor variables. Most such programs automatically include the intercept term in the model, with an option to omit it. Output typically consists of an ANOVA table and estimates of the regression coefficients and their standard errors. Most also provide options for other statistics and results, such as predicted values and their standard errors and diagnostic statistics.

Now we'll go through illustrations of a few procedures and techniques. Five models were fit to the data shown in Table 2.1. The response variable is y. Model 0 includes only an intercept term, so $\mu = x_0\beta_0$. Model 1 has an intercept and a single predictor variable, x_2, so the model for the mean μ of

TABLE 2.2: Error SS (SSE) and degrees of freedom (df, ν_E) for five models.

	Model	ν_E	SSE
0	x_0	9	57102.50000
1	x_0, x_2	8	31330.10899
2	x_0, x_1, x_2, x_3, x_4	5	4585.32278
3	x_0, x_1, x_2, x_5	6	4650.85008
2a	$x_0, x_1, x_2, x_3, x_4, x_6, x_7$	5	4585.32278

y at a value x_2 is $\mu = x_0\beta_0 + x_2\beta_2$. The values of the parameters β_0 and β_2 are unknown, but the model assumes that they exist such that this relation holds for all values of x_2, at least in some range of relevance.

In Model 2, population means are modeled in terms of four predictor variables as $\mu = x_0\beta_0 + x_1\beta_1 + x_2\beta_2 + x_3\beta_3 + x_4\beta_4$. Model 2a is the same, but it includes additional terms for the dummy variables x_6 and x_7, and it replaces the missing values of y_{11} and y_{12} by 0. In Model 3, the two predictor variables x_3 and x_4 are replaced by their difference, $x_5 = x_3 - x_4$, so that the model becomes $\mu = x_0\beta_0 + x_1\beta_1 + x_2\beta_2 + x_5\beta_5$, implicitly forcing the coefficients of x_3 and x_4 to sum to zero.

Numerical results of least-squares (LS) computations are shown in the following tables. They were obtained using `proc reg` in SAS version 9.4. Table 2.2 shows SSEs (Error Sums of Squares) and their degrees of freedom for the five models. SSE for Model 0 often is called Total SS, SST. Other statistics related to goodness of fit can be derived from these. For example, for Model 2, $R^2 = SSR/SST = 1 - SSE/SST = 1 - 4585.32278/57102.5000 = 91.97\%$.

Table 2.3 shows estimated regression coefficients, their standard errors, and MSE (Mean Squared Error, $\hat{\sigma}^2$) for Models 1, 2, 2a, and 3. Thus, in Model 1, $\hat{\beta}_2 = 15.29973$ and $\hat{\text{V}}\text{ar}(\hat{\beta}_2) = 5.96406^2$. In Model 2, the t-statistic for testing $\text{H}_0 : \beta_2 = 15$ is

$$t = \frac{19.49258 - 15}{2.99540} = 1.500.$$

In it, $\hat{\sigma}^2$ has $\nu_E = 5$ df, and so the two-sided p-value, from the central Student's t distribution with 5 degrees of freedom, is 0.194. A 95% confidence interval on β_2, based on Model 2, is the range between the two endpoints

$$19.49258 \pm 2.5706 \times 2.99540,$$

the interval from 11.7926 to 27.1926, in units megawatts per degree Fahrenheit.

The proposition that none of the four predictor variables in Model 2 has any effect on μ can be formulated as $\text{H}_0 : \beta_1 = \beta_2 = \beta_3 = \beta_4 = 0$. Unlike inference on just one regression coefficient, like β_2 above, it entails four relations among four regression coefficients. We will call this generally a *conjunctive* hypothesis. The test statistic for it is an F-statistic in which the numerator

TABLE 2.3: Least-squares estimates of regression coefficients and their standard errors in four models for the mean of y. Model 2a is Model 2 plus terms for x_6 and x_7 and with $y = 0$ for the 11-th and 12-th cases. Model 3 replaces x_3 and x_4 by $x_5 = x_3 - x_4$, forcing $\hat{\beta}_3 = -\hat{\beta}_4$. MSEs (degrees of freedom) are shown at the bottom.

	Model 1		Model 2, 2a		Model 3	
Variable	Estimate	StdErr	Estimate	StdErr	Estimate	StdErr
x_0	-70.76839	473.36412	-69.64855	286.12132	-29.47230	223.83491
x_1			-14.95663	6.09869	-15.29290	5.48638
x_2	15.29973	5.96406	19.49258	2.99540	19.44962	2.74991
x_3			5.55286	1.40985		
x_4			-5.12660	1.41500		
x_5					-5.34095	1.07185
x_6			1644.00149[a]	87.35012[a]		
x_7			0.42626[a]	30.32503[a]		
	$\hat{\sigma}^2 = 3916.26362(8)$		$\hat{\sigma}^2 = 917.06456(5)$		$\hat{\sigma}^2 = 775.14168(6)$	

sum of squares is the difference in SSE between the *restricted* model, which incorporates the conditions of H_0, and the full model, which includes all the predictor variables. The denominator SS is SSE for the full model. This form of the numerator SS is called here the *RMFM* SS.

Here, the full model is Model 2, and the restricted model is Model 0. Computation of the F-statistic for H_0 is based on the following table.

Model	df	SSE
Restricted: 0	9	57102.50000
Full: 2	5	4585.32278
Difference	4	52517.17722

The test statistic is $F = \frac{52517.17722/4}{4585.32278/5} = 14.317$. The p-value from a central F distribution with 4 and 5 degrees of freedom is 0.006. In this case, testing that no predictors have any effects, the numerator SS is called *Regression* SS and denoted SSR.

To test the conjunctive hypothesis $H_0 : \beta_1 = \beta_3 = \beta_4 = 0$, the full model is Model 2, and the restricted model is Model 1. The F-statistic is computed from the following table.

Model	df	SSE
Restricted: 1	8	31330.10899
Full: 2	5	4585.32278
Difference	3	26744.78621

The test statistic is $F = \frac{26744.78621/3}{4585.32278/5} = 9.721$, and the p-value from a central F distribution with 3 and 5 df is 0.016.

We turn now to the predicted value of y at the conditions in the 11th case, $\hat{\mu}_{11}$. In Model 2, it can be computed as

$$\begin{aligned}\hat{\mu}_{11} &= -69.64855 + 5 \times (-14.95663) + 90 \times (19.49258) \\ &\quad +80 \times (5.55286) + 80 \times (-5.12660) \\ &= 1644.0013,\end{aligned}$$

with a little truncation error due to using five decimal digits.

We can get the estimate this way, but not its standard error. The SAS procedure `proc reg`, like those in most other packages, includes the option to compute estimated means and their standard errors for all cases in the data set, including those with missing responses. Those statistics are shown in the case-wise output of Tables 2.4 and 2.5. For Model 2, $\hat{\mu}_{11} = 1644.00149$ and $\hat{\text{V}}\text{ar}(\hat{\mu}_{11}) = 81.93227^2$. A 95% confidence interval on the population mean μ_{11} has endpoints given by

$$1644.00149 \pm 2.57058 \times 81.93227,$$

the range from 1433.39 to 1854.61 megawatts.

A 95% prediction interval on an as-yet-unobserved response Y under the same conditions has endpoints given by

$$\begin{aligned}
&\hat{\mu}_{11} \pm 2.57058 \times \sqrt{\hat{\text{V}}\text{ar}(Y - \hat{\mu}_{11})} \\
&= \hat{\mu}_{11} \pm 2.57058 \times \sqrt{\hat{\sigma}^2 + \hat{\text{V}}\text{ar}(\hat{\mu}_{11})} \\
&= 1644.00149 \pm 2.57058 \times \sqrt{917.06456 + 81.93227^2} \\
&= 1644.00149 \pm 2.57058 \times 87.34965,
\end{aligned}$$

the range from $1644.00 - 224.54 = 1419.46$ to $1644.00 + 224.54 = 1868.54$. These are the two values shown for `date` 631 under "LowerPL" and "UpperPL" for Model 2 in Table 2.5.

Residuals are $y_i - \hat{\mu}_i$. For the fifth case, Model 2, the residual is $y_5 - \hat{\mu}_5 = 1213 - 1191.01898 = 21.98102$. It can be shown that $\text{Var}(y_i - \hat{\mu}_i) = \text{Var}(y_i) - \text{Var}(\hat{\mu}_i)$, because $\text{Cov}(y_i, \hat{\mu}_i) = \text{Var}(\hat{\mu}_i)$. Then $\hat{\text{V}}\text{ar}(y_i - \hat{\mu}_i) = \hat{\sigma}^2 - \hat{\text{V}}\text{ar}(\hat{\mu}_i) = 917.06456 - 19.69585^2 = 23.00300^2$, as shown for $i = 5$, Model 2, in Table 2.5.

Next we examine the proposition, $\text{H}_0 : \beta_3 + \beta_4 = 0$, in two ways. Model 3 is formulated specifically as the restricted model for $\text{H}_0 : \beta_3 + \beta_4 = 0$. To see this, note that if H_0 is true, and if $\beta_5 = \beta_3 = -\beta_4$, then $x_3\beta_3 + x_4\beta_4 = (x_3 - x_4)\beta_5$. The model that imposes the restriction H_0 on Model 2 replaces x_3 and x_4 by $x_5 = x_3 - x_4$.

It follows that the sums of squares for an F-statistic for testing H_0 are those in the following table. See Table 2.2 for SSEs and dfs.

Model	df	SSE
Restricted: 3	6	4650.85008
Full: 2	5	4585.32278
Difference	1	65.52730

The test statistic is then $F = 65.52730/917.06546 = 0.07145$. Its p-value, from an F distribution with 1 and 5 df, is 0.7999.

Because $\beta_3 + \beta_4$ is a single linear combination of the regression coefficients, a t-statistic can be constructed to test H_0, too. This requires an estimate of $\beta_3 + \beta_4$ and its standard error. In Model 2, the estimates $\hat{\beta}_3$ and $\hat{\beta}_4$ (and their standard errors) are

$$\hat{\beta}_3 = 5.55286\ (1.40985) \text{ and } \hat{\beta}_4 = -5.12660\ (1.41500).$$

Then the estimate of their sum is $5.55286 - 5.12660 = 0.42626$. An estimate like this, without its estimated variance, is of practically no use, as no inference is possible from it. To be useful at all, we must make the procedure produce together both the estimate and its standard error.

In order to make them appear in output from the regression procedure, the ersatz case in row $i = 12$ was inserted in the data set, with y coded as missing, $x_0 = x_1 = x_2 = 0$, and $x_3 = x_4 = 1$, so that $x_0\beta_0 + x_1\beta_1 + x_2\beta_2 + x_3\beta_3 + x_4\beta_4 = \beta_3 + \beta_4$. In Table 2.5 the predicted value for `date` 632 in Model 2 is $\hat{\beta}_3 + \hat{\beta}_4 = 0.42626$, as we saw above, but now it appears with its standard error, 1.59465. Then the t-statistic for $\mathrm{H_0} : \beta_3 + \beta_4 = 0$ is $t = 0.42626/1.59465 = 0.26731$, from which the two-tailed p-value is $2 \times 0.39995 = 0.7999$ from Student's t distribution with 5 df. Note that $t^2 = 0.07145 = F$. Note, too, that the standard error of $\hat{\beta}_3 + \hat{\beta}_4$ cannot be computed from the two respective standard errors because the two estimates are correlated, as seen by the fact that $1.59456^2 \neq 1.40985^2 + 1.41500^2$.

The responses in the 11th and 12th observations were coded as missing for all models except Model 2a, and so those observations were not used in the LS computations to fit those models. In Model 2a, adding the 11th and 12th observations with $y = 0$ (not missing) and including the dummy variables x_6 and x_7 as predictor variables illustrates a useful device for obtaining so-called *deleted-case* statistics. The dummy variables enable the model to fit $y = 0$ for these two cases perfectly, in effect fitting the model to the first ten observations. It can be shown that, for $i = 11$ or 12, their estimated regression coefficients take the form $-(y_i - \hat{\mu}_i) = -(0 - \hat{\mu}_i) = \hat{\mu}_i$, and that $\hat{\mu}_i$ is based only on the first ten cases. Then, under the customary distributional assumption of independent observations, y_i and $\hat{\mu}_i$ are independent, and $\mathrm{Var}(y_i - \hat{\mu}_i) = \mathrm{Var}(y_i) + \mathrm{Var}(\hat{\mu}_i) = \sigma^2 + \mathrm{Var}(\hat{\mu}_i)$. In this way, predicted values show up as coefficient estimates, as shown in Table 2.3. Then, because x_6 is the indicator of case 11, $\hat{\beta}_6 = \hat{\mu}_{11} = 1644.00149$. The form of its standard error means that a 95% prediction interval on y at the conditions in case 11 is $1644.00149 \pm 2.57058 \times 87.35012 = 1419.46102$ to 1868.54196. This is in agreement with the limits shown for case 11 for Model 2 in Table 2.5 up to truncation error due to calculations with five decimal digits. (It is always best to keep computations internal to the procedure to minimize truncation error.)

For inference on μ_{11} using Table 2.3, we must deduce that $\hat{\mathrm{Var}}(\hat{\mu}_{11}) = \hat{\mathrm{Var}}(\hat{\beta}_6) - \hat{\sigma}^2 = 87.35012^2 - 917.06456 = 81.93277^2$. To test, for example, $\mathrm{H_0} : \mu_{11} = 1500$, the test statistic is $t = (1644.00149 - 1500)/81.93277 = 1.7576$, from which the (two-sided) p-value is 0.14. Similarly, to test $\mathrm{H_0} : \beta_3 + \beta_4 = 0$ with Table 2.3, $\hat{\mathrm{Var}}(\hat{\beta}_3 + \hat{\beta}_4) = 30.32503^2 - 917.06456 = 2.54288 = 1.59464^2$, so that the test statistic is $t = 0.42626/1.59464 = 0.26731$, the same as found above.

2.2 Penalized Least Squares Illustration

This example illustrates a little of the flexibility of linear least squares to fit almost any relation between the response y and a predictor variable x. Figure 2.1 shows results of fitting the data points ("+"s) to a function of x.

TABLE 2.4: Model 1: Least-squares estimates of means and predicted values (they are the same), their standard errors, and standard errors of residuals. LowerPL and UpperPL are lower and upper limits of 95% prediction intervals on the response for the values of the predictors in the same row.

date	i	y	$\hat{\mu}$	$\sqrt{\hat{\text{V}}\text{ar}(\hat{\mu})}$	$\sqrt{\hat{\text{V}}\text{ar}(y-\hat{\mu})}$	LowerPL, UpperPL
Model 1:	$\mu = \beta_0 + x_2\beta_2$					
601	1	1047	1137.91008	19.87027	59.34169	986.50038, 1289.31978
602	2	1068	1076.71117	32.39316	53.54388	914.21427, 1239.20808
603	3	1115	1183.80926	25.51337	57.14308	1027.96713, 1339.65140
604	4	1208	1214.40872	34.31281	52.33445	1049.82992, 1378.98752
607	5	1213	1153.20981	20.22513	59.22168	1001.55047, 1304.86915
608	6	1165	1137.91008	19.87027	59.34169	986.50038, 1289.31978
609	7	1073	1046.11172	42.46648	45.96587	871.71215, 1220.51128
610	8	1062	1107.31063	24.07887	57.76220	952.68698, 1261.93428
611	9	1200	1199.10899	29.64086	55.11518	1039.43017, 1358.78782
614	10	1274	1168.50954	22.23564	58.49649	1015.36087, 1321.65820
631	11	.	1306.20708	66.81349	.	1095.10621, 1517.30796

TABLE 2.5: Model 2. Least-squares estimates of means and predicted values (they are the same), their standard errors, and standard errors of residuals. LowerPL and UpperPL are lower and upper limits of 95% prediction intervals on the response for the values of the predictors in the same row.

date	i	y	$\hat{\mu}$	$\sqrt{\hat{\text{V}}\text{ar}(\hat{\mu})}$	$\sqrt{\hat{\text{V}}\text{ar}(y-\hat{\mu})}$	LowerPL, UpperPL
Model 2 :	$\mu = x_0\beta_0 + x_1\beta_1 + x_2\beta_2 + x_3\beta_3 + x_4\beta_4$					
601	1	1047	1082.70133	21.79597	21.02380	986.78972, 1178.61293
602	2	1068	1084.18233	18.86034	23.69287	992.47425, 1175.89040
603	3	1115	1109.98485	25.62137	16.14341	1008.01601, 1211.95369
604	4	1208	1235.28516	25.03855	17.03337	1134.27753, 1336.29278
607	5	1213	1191.01898	19.69585	23.00300	1098.15758, 1283.88038
608	6	1165	1168.99574	13.93969	26.88400	1083.29933, 1254.69215
609	7	1073	1055.02101	21.11748	21.70522	960.11770, 1149.92432
610	8	1062	1059.53151	15.98504	25.72048	971.50697, 1147.55606
611	9	1200	1161.98045	22.00500	20.80492	1065.75396, 1258.20693
614	10	1274	1276.29864	26.51851	14.62304	1172.82527, 1379.77202
631	11	.	1644.00149	81.93277	.	1419.46085, 1868.54213
632	12	.	0.42626	1.59465		

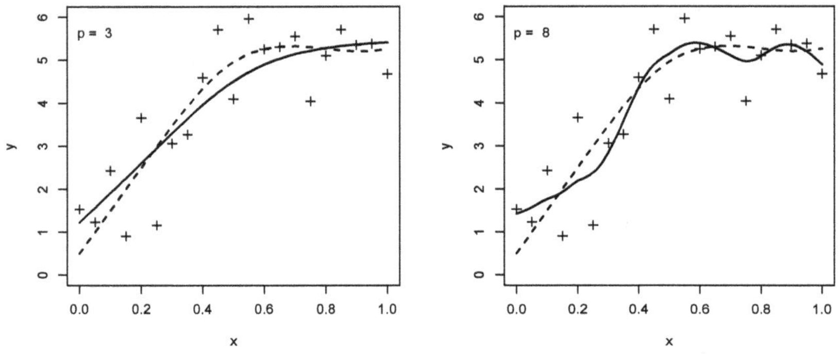

FIGURE 2.1: Responses y $(+)$, from populations with true means along the dashed line, fit (the solid lines) by linear splines, penalized by the sum of squared second divided differences. The smoothing parameter is set so that the model has p "effective" degrees of freedom.

Linear splines in x over a fine grid of knots form the columns of the matrix X. In many computing packages, this is accomplished by one call to a function with a name something like `bspline`. A smoothing constant k was found such that the "effective degrees of freedom" of the model was p, as shown in the figure. The fit is accomplished by regressing the augmented response vector $\binom{y}{0}$ on the matrix $\binom{X}{\sqrt{k}D'}$. $D'\boldsymbol{\beta}$ produces second divided differences, and the sum of their squares times k is the smoothing penalty. Without the smoothing penalty, the model would simply connect the "+"s. The technical part is determining the scalar k. The rest is just plain old least squares. See LaMotte et al. (2020) and Section 11.5.

2.3 ANOVA Illustration

The results shown in Tables 2.7 and 2.8 for the four data sets described in Table 2.6 can be had from most general-purpose statistical computing packages. The SAS package produces the SAS Types III and IV sums of squares (SSs), of course; other packages produce something under the Type III name, but it is not always identical to SAS's Type III SS. See Section 18.3 for detailed descriptions of the items shown in Table 2.8.

TABLE 2.6: Data for the illustration. Levels of A and B for a balanced design and an unbalanced design. Omit cell 3,3 in the unbalanced design for one empty cell, plus cells 1,1 and 2,2 for three empty cells.

Balanced		Unbalanced			Balanced		Unbalanced		
A	B	A	B	y	A	B	A	B	y
1	1	1	1	50.0	2	2	2	2	65.4
1	1	1	2	22.2	2	3	2	3	99.8
1	2	1	2	111.7	2	3	2	3	126.8
1	2	1	3	65.3	3	1	3	1	87.3
1	3	1	3	53.2	3	1	3	1	88.6
1	3	1	3	54.2	3	2	3	1	133.2
2	1	2	1	101.3	3	2	3	2	67.0
2	1	2	1	42.0	3	3	3	2	70.2
2	2	2	1	95.5	3	3	3	3	106.2

TABLE 2.7: SAS Types I-IV A SSs, all based on DV models, all with 2 df. Data are from Table 2.6.

Type	Balanced	Unbal	Unbal-1	Unbal-3
I	3844.074444	3844.074444	3356.083686	3006.556000
II	3844.074444	3944.595077	3369.335045	2150.316056
III	3844.074444	3286.460303	2889.843397	1830.567230
IV	3844.074444	3286.460303	1308.541072	826.404167

In practice, all of these A SSs and degrees of freedom (dfs) have been used, depending on assumptions and the setting, for the purpose of assessing whether there are factor A main effects on the response. Although the coding schemes (DV, RL, EC) are different, they model the nine cell means identically. If everybody agreed on a definition of A effects, then all these SSs and dfs should be identical within each of the four data sets. That clearly is not the case.

These SSs have the mathematical property that, if they differ for the same set of responses, then they test different linear hypotheses on the nine population cell means. Scrutinize these results carefully to observe how the types of SSs and dfs differ within each data set and across coding schemes, and observe how those relations differ across the four data sets. There are lots of moving parts. Much has been written and argued about these differences in the century that has passed since R. A. Fisher's introduction of Analysis of Variance. Discovering the mathematical relations that account for these differences provides a basis for understanding these powerful tools and using them effectively. That is the subject of Part III of this book.

TABLE 2.8: SSs and dfs to test A main effects for three test forms and three coding schemes: dummy variables (DV), reference-level coding (RL), and effect coding (EC). Test forms are General Linear Hypothesis (GLH), Restricted Model - Full Model (RMFM), and Extra SSE due to deleting A effect predictor variables (DelVars). GLH_1 and GLH_2 are based on different least-squares solutions. Data are from Table 2.6.

	DV		RL		EC	
Test	*df*	*SS*	*df*	*SS*	*df*	*SS*
Balanced						
GLH_1	2	3844.074444	2	3844.074444	2	3844.074444
GLH_2	2	3844.074444	2	3844.074444	2	3844.074444
RMFM	2	3844.074444	2	3844.074444	2	3844.074444
DelVars	0	0.000000	2	3581.613333	2	3844.074444
Unbalanced						
GLH_1	2	3286.460303	2	3286.460303	2	3286.460303
GLH_2	2	3286.460303	2	3286.460303	2	3286.460303
RMFM	2	3286.460303	2	3286.460303	2	3286.460303
DelVars	0	0.000000	2	4305.608333	2	3286.460303
Unbalanced, $n_{33} = 0$						
GLH_1	2	3248.714912	2	3248.714912	2	3076.205632
GLH_2	2	2495.982956	2	2495.982956	2	3060.049524
RMFM	1	1914.449167	1	1914.449167	1	1914.449167
DelVars	0	0.000000	1	3727.445333	1	1914.449167
Unbalanced, $n_{11} = n_{22} = n_{33} = 0$						
GLH_1	2	3776.461731	2	3776.461731	2	3776.461731
GLH_2	2	2318.933767	2	1643.109627	2	4174.202931
RMFM	0	0.000000	0	0.000000	0	0.000000
DelVars	0	0.000000	1	3727.445333	0	0.000000

2.4 Summary

The purpose of these examples is to illustrate a little of the range of techniques possible by exploiting the basic computations of multiple regression analysis and to preview perplexing questions that arise in ANOVA.

2.5 Exercises

These questions refer to Tables 2.2–2.5. "Find" means to provide the numerical value or a formula involving parameters. All of these questions can be answered with at most a few simple calculations using only the numerical results shown in the tables.

1. Find $\hat{\sigma}^2$ and its df in Model 2.

2. Find $\hat{\text{V}}\text{ar}(\hat{\beta}_2)$ in Model 2.

3. Recognizing that $\hat{\text{V}}\text{ar}(\hat{\beta}_2) = c_2\hat{\sigma}^2$, find c_2 in Model 2.

4. Find $\text{Var}(\hat{\beta}_2)$ in Model 1 as a constant times σ^2. In which model (1 or 2) is $\text{Var}(\hat{\beta}_2)$ less?

5. Find R^2 for Model 2.

6. Find the t-statistic to test $\text{H}_0 : \beta_1 = 0$ in Model 2. Find a p-value for this proposition.

7. Find SSE for the model formed by deleting x_2 from Model 2, that is, for the regression of y on x_0, x_1, x_3, and x_4.

8. Would it be unreasonable to think that β_2 is the same in Model 2 as it is in Model 1? Describe the computational steps required to address this question. They are straightforward using the GS construction detailed in Chapter 6.

Part I

Basics

3

Matrices and Vectors in $\Re^n$

This chapter summarizes the basic definitions and operations of matrices as they are used in this book. It is assumed that the reader has at least a handshaking familiarity with matrices and their basic operations.

3.1 Definitions and Operations

A ***matrix*** is an rc-tuple of real numbers, arranged as a rectangular array with r rows and c columns. A matrix named A is

$$A = \begin{pmatrix} a_{11} & \cdots & a_{1c} \\ a_{21} & \cdots & a_{2c} \\ \vdots & \vdots & \vdots \\ a_{r1} & \cdots & a_{rc} \end{pmatrix},$$

where the rc entries are denoted a_{ij}, $i = 1, \ldots, r$, $j = 1, \ldots, c$. These entries may be referred to variously as ***components*** or ***elements*** of A. The ij-th entry of A is a_{ij}. Sometimes, a comma is used to distinguish i and j, e.g., $a_{i,j}$.

The number of rows r and the number of columns c of a matrix A are called the ***row dimension*** and ***column dimension*** of A, or sometimes collectively the dimensions or ***order*** of A. A is said to be an $r \times c$ matrix. A is ***square*** if $r = c$.

Two $r \times c$ matrices $A = (a_{ij})$ and $B = (b_{ij})$ are ***equal***, denoted $A = B$, iff $a_{ij} = b_{ij}$, $i = 1, \ldots, r$, $j = 1, \ldots, c$. A and B are equal iff all corresponding elements of A and B are equal.

An $r \times 1$ matrix is called a ***column vector*** or an ***r-vector***, and a $1 \times c$ matrix is called a ***row vector***. When reference is made to a vector, think of it as a column vector unless the context indicates otherwise. Vectors often will be denoted in boldface, like $\boldsymbol{v}$.

Sometimes, the notation $\{A\}_{ij}$ will be used to refer to the ij-th element of a matrix or matrix expression, such as $\{A + B\}_{ij} = a_{ij} + b_{ij}$.

At times, we shall need to refer to specific rows or columns of A. A may be defined in terms of its columns as $A = (\boldsymbol{a}_1, \ldots, \boldsymbol{a}_c)$, where $\boldsymbol{a}_1, \ldots, \boldsymbol{a}_c$ are c r-vectors. Sometimes, $A_{[i,\cdot]}$ or $A_{i\cdot}$ is used to denote the i-th row of A and $A_{[\cdot,j]}$ or $A_{\cdot j}$ to denote the j-th column.

Denote the vector or matrix with all its entries 0 by $\mathbf{0}$ if it is explicitly a vector or by 0 if it is a matrix. The dimension will be clear in context; otherwise, it will be indicated by a subscript, e.g., $\mathbf{0}_r$, $0_{r\times c}$.

As used here, a matrix has at least one row and at least one column. Where it is unambiguous or unnecessary, dimensions of matrices will rarely be stated explicitly, in order to avoid clutter. Assume that dimensions of matrices are commensurate with the operations in the expressions in which they appear. Thus, for example, for matrices A, B, and C in the expression $\mathrm{tr}(AB+C)$, it will be clear (once the trace of a square matrix is defined) that C and AB are square with the same dimensions, and the number of rows in B is the same as the number of columns in A.

The ***identity*** matrix is a special square matrix. An $n \times n$ matrix, it is denoted I_n and defined as having all entries on the ***main diagonal*** equal to 1 and all its other entries equal to 0. The main diagonal is the set of positions along the line from upper-left to lower-right, that is, positions $(1,1),(2,2),\ldots,(n,n)$. For example, the 3×3 identity matrix is

$$\mathrm{I}_3 = \begin{pmatrix} 1 & 0 & 0 \\ 0 & 1 & 0 \\ 0 & 0 & 1 \end{pmatrix}.$$

The subscript often is omitted when the dimension is clear in context. Denote the i-th column of the identity matrix by $\boldsymbol{e}_i$. Its dimension will be clear in context.

For a positive integer n, the set of all n-vectors of real numbers is denoted $\Re^n$. It is sometimes named n-dimensional *Euclidean* space.

In this context, a ***scalar*** is a real number. For a scalar b and an $r\times c$ matrix A, the ***scalar-by-matrix product*** of A and b is defined element-by-element as $\{bA\}_{ij} = ba_{ij}$, $i=1,\ldots,r$, $j=1,\ldots,c$. The order of factors doesn't matter: bA means the same as Ab. The result bA is an $r\times c$ matrix.

The ***matrix sum*** of two $r\times c$ matrices A and B is also defined element-by-element; denoted $A+B$, it is defined by $\{A+B\}_{ij} = a_{ij}+b_{ij}$, $i=1,\ldots,r$, $j=1,\ldots,c$.

Note that matrix sum (also called matrix addition) is defined only for matrices having the same dimensions. The result of the matrix sum of two $r\times c$ matrices is an $r\times c$ matrix.

Putting these two operations together, for scalars b and d and two $r\times c$ matrices A and B, $bA+dB$ is an $r\times c$ matrix with ij-th element $\{bA+dB\}_{ij} = ba_{ij}+db_{ij}$, $i=1,\ldots,r$, $j=1,\ldots,c$. For example,

$$(2)\begin{pmatrix} -1.0 & 2.0 & -1.0 \\ 0.8 & 0.7 & 0.6 \end{pmatrix} + (-1.5)\begin{pmatrix} 2.0 & 0.0 & 0.5 \\ -1.0 & -0.5 & 0.5 \end{pmatrix} = \begin{pmatrix} -5.00 & 4.00 & -2.75 \\ 3.10 & 2.15 & 0.45 \end{pmatrix}.$$

If A and B are $r \times c$ matrices, the expression $A - B$ is the same as $A + (-1)B$: matrix subtraction is defined in terms of matrix addition and scalar-by-matrix product.

For an $r \times c$ matrix A, the ***transpose*** of A, denoted A' here, is the $c \times r$ matrix formed with the rows of A as the columns of A'. Thus $\{A'\}_{ji} = a_{ij}$, $j = 1, \ldots, c$, $i = 1, \ldots, r$. Denoting it by A^T is also common.

Example:

$$A = \begin{pmatrix} 1 & 2 & 3 \\ 4 & 5 & 6 \end{pmatrix}, \quad A' = \begin{pmatrix} 1 & 4 \\ 2 & 5 \\ 3 & 6 \end{pmatrix}.$$

The transpose of a row vector is a column vector, and the transpose of a column vector is a row vector. For example,

$$(1 \ -2 \ 1)' = \begin{pmatrix} 1 \\ -2 \\ 1 \end{pmatrix}$$

and

$$\begin{pmatrix} 1 \\ 1 \\ 1 \end{pmatrix}' = (1\ 1\ 1).$$

Often a column vector is described in text as the transpose of a row vector, as $\boldsymbol{v} = (v_1, \ldots, v_r)'$.

A matrix A is ***symmetric*** iff $A = A'$. This requires that A be square, say $r \times r$, and $a_{ij} = a_{ji}, i, j = 1, \ldots, r$.

Example:

$$S = \begin{pmatrix} 1 & -1 & 0 \\ -1 & 3 & 7 \\ 0 & 7 & 2 \end{pmatrix}$$

is symmetric. Note that I_n is symmetric, and $0_{n \times n}$ is symmetric.

Given an $r \times c$ matrix $A = (\boldsymbol{a}_1, \ldots, \boldsymbol{a}_c)$ (where $\boldsymbol{a}_1, \ldots, \boldsymbol{a}_c$ denote the vectors that form the columns of A) and a c-vector $\boldsymbol{x} = \begin{pmatrix} x_1 \\ \vdots \\ x_c \end{pmatrix}$, the ***matrix product*** $A\boldsymbol{x}$ is defined to be

$$A\boldsymbol{x} = x_1\boldsymbol{a}_1 + \cdots + x_c\boldsymbol{a}_c, \tag{3.1}$$

a ***linear combination*** of the columns of A with coefficients given by the respective elements of $\boldsymbol{x}$. $A\boldsymbol{x}$ is a vector with the same number of rows as A.

For an $r \times c$ matrix A and a $c \times m$ matrix $B = (\boldsymbol{b}_1, \ldots, \boldsymbol{b}_m)$, the ***matrix product*** AB of A and B is defined by applying (3.1) to each column of B:

$$\begin{aligned} A_{(r\times c)}B_{(c\times m)} &= (A\boldsymbol{b}_1, \ldots, A\boldsymbol{b}_m), \\ \{AB\}_{ij} &= \sum_{\ell=1}^{c} a_{i\ell}b_{\ell j},\ i = 1, \ldots, r,\ j = 1, \ldots, m. \end{aligned} \tag{3.2}$$

The matrix product AB is the $r \times m$ matrix with columns $A\boldsymbol{b}_1, \ldots, A\boldsymbol{b}_m$.

The ***trace*** is defined for square matrices A by $\text{tr}(A) = \sum_i a_{ii}$, the sum of the diagonal elements of A. It (can be shown that it) has the ***circulant*** property that $\text{tr}(AB) = \text{tr}(BA)$, which implies that $\text{tr}(ABC) = \text{tr}(CAB) = \text{tr}(BCA)$, and so on for square products of any number of matrices.

A square matrix A is said to be ***idempotent*** iff $A = AA$.

Matrix product (matrix multiplication) is not generally commutative, so that AB is not generally equal to BA even when dimensions are such that both matrix products are defined. For reexpressing matrix formulations, this is the primary characteristic that distinguishes matrix algebra from the algebra of real numbers. In any expression that involves a matrix product, AB cannot automatically be replaced by BA. Otherwise, symbolic expressions involving matrices can be manipulated with much the same relations as for algebraic manipulations with real numbers.

Following is a summary of basic relations for the operations scalar-by-matrix product, matrix addition, transpose, matrix product, and trace. All the algebraic relations and derivations in this book are based on these relations. As you go through any development, practice identifying the relations that justify each step. Let A, B, and C be matrices, and let c be a scalar, such that the expressions are defined.

1. $A + B = B + A$
2. $A + 0 = A$
3. $A + (-1)A = 0$
4. $cA = Ac$
5. If $c = 0$, then $cA = 0$; if $A = 0$ or $B = 0$, then $AB = 0$.
6. $A + (B + C) = (A + B) + C$
7. $A(B + C) = AB + AC$ and $(A + B)C = AC + BC$
8. $c(AB) = (cA)B = A(cB) = (AB)c$
9. $A(BC) = (AB)C$
10. $(cA)' = cA'$
11. $(A')' = A$

12. $(A+B)' = A' + B'$

13. $(AB)' = B'A'$

14. $\text{tr}(cA) = c\,\text{tr}(A)$

15. $\text{tr}(A+B) = \text{tr}(A) + \text{tr}(B)$

16. $\text{tr}(AB) = \text{tr}(BA)$

Note also that the number of rows in the matrix product AB is the number of rows in A, and the number of columns is the number of columns in B.

Our main use of matrix algebra is to show derivations and relations through symbolic reexpression. For that purpose, the properties of the operations are important, but the mechanics aren't. We can manipulate expressions involving matrix products and sums perfectly well without ever actually writing expressions for the individual elements. For example, it's enough to know that $A(B+C) = AB + AC$; we don't need to calculate or express the individual elements of any of the matrix products involved. For derivations, getting down to the individual elements is almost always counter-productive – you can't see the forest for the trees. For actual calculations that produce numerical results, of course, the individual elements need to be calculated. The best way by far to do that is with computer software designed for that purpose, like R or MATLAB or SAS/IML. R is free and available online for download and installation.

Let $\mathbf{1}$ denote a vector of ones, that is,

$$\mathbf{1} = \begin{pmatrix} 1 \\ \vdots \\ 1 \end{pmatrix}. \tag{3.3}$$

The number of rows in $\mathbf{1}$ will usually be clear in context, or it will be specified as $\mathbf{1}_n$ to indicate n rows. Note that in the product $A\mathbf{1}$, the number of rows in $\mathbf{1}$ is the same as the number of columns in the matrix A, and $A\mathbf{1}$ is the sum of the columns of A.

The following notation will be used consistently in this book. For each positive integer n, the matrix U_n is defined to be $(1/n)\mathbf{1}_n\mathbf{1}_n'$, an $n \times n$ matrix with all its entries equal to $1/n$. The matrix S_n is defined to be $\text{I}_n - U_n$.

Here's an example to illustrate algebraic reexpression. Consider the expression $(\text{I} - (1/n)\mathbf{1}\mathbf{1}')\mathbf{1}$, and, for each step, identify the relation from the list above that justifies the step:

$$\begin{aligned}(\text{I} - (1/n)\mathbf{1}\mathbf{1}')\mathbf{1} &= \text{I}\mathbf{1} - (1/n)(\mathbf{1}\mathbf{1}')\mathbf{1} \\ &= \mathbf{1} - (1/n)\mathbf{1}(\mathbf{1}'\mathbf{1}) \\ &= \mathbf{1} - (1/n)\mathbf{1}(n) \\ &= \mathbf{1} - (1/n)(n)\mathbf{1} = \mathbf{1} - \mathbf{1} = \mathbf{0}_n.\end{aligned}$$

Here's another example of reexpression. For an $r \times c$ matrix A, to show that $A\mathrm{I}_c = \mathrm{I}_r A = A$, one way is to express I_c in terms of its columns as $(\boldsymbol{e}_1, \ldots, \boldsymbol{e}_c)$, where

$$\boldsymbol{e}_j = j \rightarrow \begin{pmatrix} 0 \\ \vdots \\ 1 \\ \vdots \\ 0 \end{pmatrix},$$

that is, $\boldsymbol{e}_j$ is a column with 1 in the j-th position and 0s elsewhere. Then

$$A\boldsymbol{e}_j = (0)\boldsymbol{a}_1 + \cdots + (1)\boldsymbol{a}_j + \cdots + (0)\boldsymbol{a}_c = \boldsymbol{a}_j,$$

and hence

$$\begin{aligned} A\mathrm{I}_c &= (A\boldsymbol{e}_1, \ldots, A\boldsymbol{e}_j, \ldots, A\boldsymbol{e}_c) \\ &= (\boldsymbol{a}_1, \ldots, \boldsymbol{a}_j, \ldots, \boldsymbol{a}_c) = A \end{aligned}$$

To show that $\mathrm{I}_r A = A$, a similar argument can be used on the transpose, showing that $(\mathrm{I}_r A)' = A'\mathrm{I}_r = A'$, and so $\mathrm{I}_r A = [(\mathrm{I}_r A)']' = (A')' = A$.

The operations mentioned here – addition and multiplication of scalars, multiplication of a matrix by a scalar, matrix addition, matrix product, and transpose – are not always clearly distinguished in notation; '+' is used to denote both scalar addition and matrix addition, and ab, aB, and AB denote multiplication of scalars a and b, multiplication of the matrix B by the scalar a, and matrix multiplication of the matrices A and B, respectively. In expressions, you must learn to distinguish the different operations. For example, let $\boldsymbol{v}$ be an n-vector, and consider the expression $\boldsymbol{v}/(\boldsymbol{v}'\boldsymbol{v})$. Noting that $\boldsymbol{v}'\boldsymbol{v}$ is a scalar, then provided that it isn't 0, the only reasonable interpretation of this expression is as the scalar $1/(\boldsymbol{v}'\boldsymbol{v})$ multiplied by the vector $\boldsymbol{v}$.

To add further possible confusion, software like MATLAB and SAS/IML provide element-by-element operations that are convenient but not usually used in mathematical (as compared to computational) settings. For example, if A and B are matrices of the same dimensions, in SAS/IML A/B works computationally (provided no entries in B are 0), but it's mathematically ambiguous at best.

3.2 Other Operations.

For matrices A and B, with A $r \times c$ and B $s \times t$, the ***Kronecker product*** of A and B is denoted and defined as

$$A \otimes B = \begin{pmatrix} a_{11}B & a_{12}B & \cdots & a_{1r}B \\ a_{21}B & a_{22}B & \cdots & a_{2r}B \\ \vdots & \vdots & \vdots & \vdots \\ a_{r1}B & a_{r2}B & \cdots & a_{rcB} \end{pmatrix}.$$

$A \otimes B$ has rc $s \times t$ blocks $a_{ij}B$. The dimensions of $A \otimes B$ are $rs \times ct$.

An important relation combining matrix products and Kronecker products is this: if A, B, C, and D are matrices such that the matrix products AC and BD are defined, then

$$(A \otimes B)(C \otimes D) = (AC) \otimes (BD).$$

If $\boldsymbol{x} = (x_1, \ldots, x_r)'$ is an r-vector (an $r \times 1$ matrix), $\mathrm{diag}(\boldsymbol{x})$ denotes the $r \times r$ diagonal matrix with entries of $\boldsymbol{x}$ on its principal diagonal:

$$\mathrm{diag}(\boldsymbol{x}) = \begin{pmatrix} x_1 & 0 & \cdots & 0 \\ 0 & x_2 & \cdots & 0 \\ \vdots & \vdots & \vdots & \vdots \\ 0 & 0 & \cdots & x_r \end{pmatrix}.$$

Sometimes, the same notation is used to denote the diagonal matrix with the diagonal entries of an $r \times r$ matrix $A = (a_{ij})$ on its diagonal:

$$\mathrm{diag}(A) = \begin{pmatrix} a_{11} & 0 & \cdots & 0 \\ 0 & a_{22} & \cdots & 0 \\ \vdots & \vdots & \vdots & \vdots \\ 0 & 0 & \cdots & a_{rr} \end{pmatrix}.$$

Suppose that $A_1, \ldots, A_k$ is a list of matrices with respective dimensions $r_i \times c_i$. By the notation

$$B = \mathrm{Diag}(A_i) = \mathrm{Diag}_{i=1}^{k}(A_i) = \mathrm{Diag}(A_1, \ldots, A_k),$$

we shall denote a matrix that, in partitioned form, has the matrices $A_1, \ldots, A_k$ down the principal "diagonal," that is, B is partitioned into $k \times k$ submatrices B_{ij} such that $B_{ii} = A_i$, $i = 1, \ldots, k$, and $B_{ij} = 0_{r_i \times c_j}$, $j \neq i = 1, \ldots, k$, is a 0-matrix of appropriate dimensions. Such a matrix is said to be ***block diagonal***.

Let A be an $r \times c$ matrix with columns $\boldsymbol{a}_1, \ldots, \boldsymbol{a}_c$, i.e., $A = (\boldsymbol{a}_1, \ldots, \boldsymbol{a}_c)$. Then

$$\mathrm{vec}(A) = \begin{pmatrix} \boldsymbol{a}_1 \\ \boldsymbol{a}_2 \\ \vdots \\ \boldsymbol{a}_c \end{pmatrix}. \tag{3.4}$$

$\mathrm{vec}(A)$ is an rc-vector formed by stacking the columns of A.

Let A, B, and C be matrices such that the matrix product ABC is defined. A fundamental and useful relation is:

$$\mathrm{vec}(ABC) = (C' \otimes A)\mathrm{vec}(B). \tag{3.5}$$

If A and B are $r \times c$ matrices, the ***Hadamard product*** of A and B is defined as the $r \times c$ matrix $A \circ B$ with ij-th entry

$$(A \circ B)_{ij} = a_{ij}b_{ij}, \; i = 1, \ldots, r, \; j = 1, \ldots, c.$$

3.3 Exercises

For Exercises 1–44, let:

$$\begin{aligned} A &= (\boldsymbol{a}_1, \boldsymbol{a}_2) = \begin{pmatrix} 1 & 2 \\ 3 & 4 \\ 5 & 6 \end{pmatrix}, \\ B &= (\boldsymbol{b}_1, \boldsymbol{b}_2, \boldsymbol{b}_3) = \begin{pmatrix} -3 & 0 & 3 \\ 1 & 1 & -2 \\ 1 & -2 & 1 \end{pmatrix}, \\ \boldsymbol{x} &= \begin{pmatrix} 3 \\ 7 \\ 11 \end{pmatrix}, \ \mathbf{1}_3 = \begin{pmatrix} 1 \\ 1 \\ 1 \end{pmatrix}, \ \mathbf{1}_2 = \begin{pmatrix} 1 \\ 1 \end{pmatrix}, \ \boldsymbol{y} = \begin{pmatrix} 2 \\ 3 \end{pmatrix}, \\ U_3 &= (1/3)\mathbf{1}_3\mathbf{1}_3', \ S_3 = \mathrm{I}_3 - U_3. \end{aligned}$$

Evaluate the expressions and prove or justify the statements. Address the first few questions with pencil and paper. There are several good matrix-oriented computing packages, like MATLAB or SAS IML or R, that make manipulating and evaluating such expressions easy. Use one to formulate and evaluate all the expressions in these exercises. The R package is free. R code for Exercises 1–44 is listed beginning on p. 208.

1. $(1/2)\mathbf{1}_2$
2. $(.5)\boldsymbol{y} + (.3)\mathbf{1}_2$
3. A'
4. $B\mathbf{1}_3$
5. $\mathbf{1}_3'\mathbf{1}_3$
6. $B\boldsymbol{x}$
7. $A'\boldsymbol{x}$
8. $B\boldsymbol{a}_1$
9. $\mathbf{1}_2\mathbf{1}_2'$
10. I_3B
11. $A'B$
12. AA'
13. $A'A$
14. BB
15. $\boldsymbol{x} \times 2$
16. $2A$
17. $B/3$
18. S_3
19. S_3S_3
20. S_3U_3
21. U_3U_3
22. $(1/3)\mathbf{1}_3'\boldsymbol{x}$
23. $U_3\boldsymbol{x}$
24. $S_3\boldsymbol{x}$
25. $B\boldsymbol{a}_1$, $B\boldsymbol{a}_2$, BA
26. $\boldsymbol{x} + 2\mathbf{1}_3$
27. $2AA' + .5B$
28. $\mathrm{tr}(B)$
29. $\mathrm{tr}(AA')$
30. $\mathrm{tr}(A'A)$
31. $\mathrm{tr}(S_3)$
32. $\boldsymbol{x}'S_3\boldsymbol{x}$
33. $\mathrm{vec}(A')$
34. $\mathrm{vec}(BA)$
35. $A \otimes \mathbf{1}_2$
36. $A \otimes \boldsymbol{x}$
37. $\mathbf{1}_3 \otimes A$
38. $\mathrm{diag}(\boldsymbol{x})$
39. $\mathrm{Diag}(A, \mathbf{1}_2\mathbf{1}_2', \boldsymbol{y})$

40. Let $C = (A, \boldsymbol{b}_1)$, $D = \begin{pmatrix} A' \\ \mathbf{1}_3' \end{pmatrix}$. Find CD.

41. Is B symmetric? Verify numerically.

42. Is $A'A$ symmetric?

43. Is U_3 idempotent?

44. Is S_3 symmetric? Is S_3 idempotent? Is $S_3 + U_3$ idempotent?

45. Let $A = \begin{pmatrix} 1 & 2 \\ 2 & 1 \end{pmatrix}$. If $B = \mathrm{I}_2$ or $B = 0_{2\times 2}$, then $AB = BA$. Find another 2×2 matrix B such that $AB \neq BA$. Find another 2×2 matrix B such that $AB = BA$.

46. Let A be an $r \times c$ matrix. Prove: If $A\boldsymbol{y} = \mathbf{0}$ for every $\boldsymbol{y} \in \Re^c$, then $A = 0$. [Hint: Note that $A\boldsymbol{e}_j = \boldsymbol{a}_j$, where $\boldsymbol{e}_j$ is the j-th column of I_c and $\boldsymbol{a}_j$ is the j-th column of A.]

47. Give an example of a non-zero 3×2 matrix A and a non-zero 2-vector $\boldsymbol{y}$ such that $A\boldsymbol{y} = \mathbf{0}$. Clearly, then, $A\boldsymbol{y} = \mathbf{0}$ does not imply that either A or $\boldsymbol{y}$ is zero.

48. Suppose A is an $n \times n$ symmetric matrix. If $\boldsymbol{y}'A\boldsymbol{y} = 0$ for all $\boldsymbol{y} \in \Re^n$, then $A = 0$. [Hint: Note that $\boldsymbol{e}_i'A\boldsymbol{e}_i = a_{ii}$ and $(\boldsymbol{e}_i + \boldsymbol{e}_j)'A(\boldsymbol{e}_i + \boldsymbol{e}_j) = a_{ii} + a_{jj} + 2a_{ij}$ for $i \neq j$.]

49. If $\boldsymbol{z}$ is an n-vector, $\boldsymbol{z}'\boldsymbol{z} = 0$ iff $\boldsymbol{z} = \mathbf{0}$.

50. If A is an $r \times c$ matrix, $A'A = 0_{c\times c}$ iff $A = 0_{r\times c}$. [Hint: Suppose $A'A = 0_{c\times c}$. For any $\boldsymbol{y} \in \Re^c$, $\boldsymbol{y}'A'A\boldsymbol{y} = (A\boldsymbol{y})'(A\boldsymbol{y}) = 0 \Longrightarrow A\boldsymbol{y} = \mathbf{0}$.]

51. If A is an $r \times c$ matrix and B is an $r \times c$ matrix, then

$$\mathrm{tr}(AB') = \sum_{i=1}^{r} \sum_{j=1}^{c} a_{ij} b_{ij}.$$

Show: As a consequence, $\mathrm{tr}(AA') = \sum_{i=1}^{r} \sum_{j=1}^{c} a_{ij}^2$.

52. If A is an $r \times c$ matrix, $\mathrm{tr}(A'A) = 0$ iff $A = 0_{r\times c}$.

53. Let $\boldsymbol{y} = (y_1, \dots, y_n)'$ be an n-vector, and define $\bar{y} = (1/n)\sum_{i=1}^{n} y_i$ and $SOS(\boldsymbol{y}) = \sum_{i=1}^{n}(y_i - \bar{y})^2$. Let $\mathbf{1}_n$ be an n-vector of ones, $U_n = \frac{1}{n}\mathbf{1}_n\mathbf{1}_n'$, and $S_n = \mathrm{I}_n - U_n$. Prove the following assertions.

    (a) $\mathbf{1}_n'\boldsymbol{y} = \sum_{i=1}^{n} y_i$.
    (b) $\sum_{i=1}^{n}(y_i - \bar{y}) = 0$.
    (c) U_n is symmetric and idempotent.

(d) S_n is symmetric and idempotent.

(e) $U_n \mathbf{1}_n = \mathbf{1}_n$ and $S_n \mathbf{1}_n = \mathbf{0}$.

(f) $U_n \boldsymbol{y} = \bar{y}\mathbf{1}_n$.

(g) $S_n \boldsymbol{y} = \boldsymbol{y} - \bar{y}\mathbf{1}_n = (y_i - \bar{y})$.

(h) $SOS(\boldsymbol{y}) = \sum_{i=1}^{n} y_i^2 - n\bar{y}^2$.

(i) $SOS(\boldsymbol{y}) = \boldsymbol{y}' S_n \boldsymbol{y}$.

4

Linear Subspaces in $\Re^n$

This chapter presents basic definitions and terminology of linear subspaces in $\Re^n$. This is the general mathematical foundation of all the material in this book.

4.1 Inner Product, Linear Subspaces

The ***inner product*** of two n-vectors $\boldsymbol{u}$ and $\boldsymbol{v}$ is defined to be

$$\boldsymbol{u}'\boldsymbol{v} = u_1v_1 + \cdots + u_nv_n = \sum_{i=1}^{n} u_iv_i.$$

The two vectors are said to be ***orthogonal*** iff $\boldsymbol{u}'\boldsymbol{v} = 0$.

Most of this chapter has to do with sets of vectors in $\Re^n$. Unless indicated otherwise, a set of vectors shall mean a set containing at least one vector, that is, a non-empty set.

Let $\mathcal{S}$ and $\mathcal{T}$ be sets of n-vectors. The ***sum*** of $\mathcal{S}$ and $\mathcal{T}$ is defined to be the set

$$\mathcal{S} + \mathcal{T} = \{\boldsymbol{z} : \boldsymbol{z} = \boldsymbol{s} + \boldsymbol{t} \text{ for some } \boldsymbol{s} \in \mathcal{S} \text{ and } \boldsymbol{t} \in \mathcal{T}\}.$$

A vector $\boldsymbol{z}$ is in the set $\mathcal{S} + \mathcal{T}$ iff there exist vectors $\boldsymbol{s}$ and $\boldsymbol{t}$ in $\mathcal{S}$ and $\mathcal{T}$, respectively, such that $\boldsymbol{z} = \boldsymbol{s} + \boldsymbol{t}$. The sum is a ***direct sum*** iff for each $\boldsymbol{z}$ in $\mathcal{S} + \mathcal{T}$, there are exactly one $\boldsymbol{s} \in \mathcal{S}$ and exactly one $\boldsymbol{t} \in \mathcal{T}$ such that $\boldsymbol{z} = \boldsymbol{s} + \boldsymbol{t}$, that is, the representation of $\boldsymbol{z}$ as $\boldsymbol{s} + \boldsymbol{t}$ is unique. A direct sum is signified by $\mathcal{S} \oplus \mathcal{T}$.

Sometimes, one of the component sets contains only one vector, say $\mathcal{T} = \{\boldsymbol{t}_0\}$. Often this sum is written simply as $\mathcal{S} + \boldsymbol{t}_0$, and it must be surmised from context that $\boldsymbol{t}_0$ stands for the singleton set $\{\boldsymbol{t}_0\}$.

For a matrix A with n columns and a set $\mathcal{S}$ of n-vectors, occasionally as a notational shortcut we will denote the set $\{A\boldsymbol{x} : \boldsymbol{x} \in \mathcal{S}\}$ by $A\{\mathcal{S}\}$.

Let $\mathcal{U}$ be a set of vectors in $\Re^n$. Suppose k is a positive integer, $\{\boldsymbol{u}_1, \ldots, \boldsymbol{u}_k\}$ is a tuple (an ordered list, not necessarily all distinct) of k vectors from $\mathcal{U}$, and $c_1, \ldots, c_k$ are scalars. We shall say that $c_1\boldsymbol{u}_1 + \cdots + c_k\boldsymbol{u}_k$ is a ***finite linear combination*** of vectors in $\mathcal{U}$.

The ***orthogonal complement*** of a set $\mathcal{U}$ in $\Re^n$ is the set consisting of all the vectors in $\Re^n$ that are orthogonal to all the vectors in $\mathcal{U}$. This set is denoted $\mathcal{U}^\perp$. A vector $\boldsymbol{v}$ is in $\mathcal{U}^\perp$ iff for each vector $\boldsymbol{u}$ in $\mathcal{U}$, $\boldsymbol{u}'\boldsymbol{v} = 0$. Note that every non-empty set $\mathcal{U}$ in $\Re^n$ has a non-empty orthogonal complement, because $\mathbf{0} \in \mathcal{U}^\perp$.

The ***span*** of a set $\mathcal{U}$ of vectors in $\Re^n$ is defined to be the set of all finite linear combinations of vectors in $\mathcal{U}$. It is a set denoted by $\text{sp}(\mathcal{U})$.

A set $\mathcal{V}$ of n-vectors is said to be ***orthonormal*** if (1) $\boldsymbol{v}'\boldsymbol{v} = 1$ for all $\boldsymbol{v} \in \mathcal{V}$, and (2) $\boldsymbol{v}_1'\boldsymbol{v}_2 = 0$ for all $\boldsymbol{v}_1 \neq \boldsymbol{v}_2$ in $\mathcal{V}$. Thus an orthonormal set of vectors is a set of pairwise-orthogonal, norm-1 vectors. Show that the set consisting of the columns of $\text{I}_n = (\boldsymbol{e}_1, \ldots, \boldsymbol{e}_n)$ is an orthonormal set.

A subset of $\Re^n$ that is closed under scalar-by-vector multiplication and vector addition is a ***vector space*** or a ***linear subspace*** of $\Re^n$. Thus a set $\mathcal{S}$ in $\Re^n$ is a linear subspace of $\Re^n$ iff (1) for any vector $\boldsymbol{u} \in \mathcal{S}$ and any scalar c, $c\boldsymbol{u}$ is in $\mathcal{S}$, and (2) for any vectors $\boldsymbol{u}_1$ and $\boldsymbol{u}_2$ in $\mathcal{S}$, $\boldsymbol{u}_1 + \boldsymbol{u}_2$ is in $\mathcal{S}$. The set $\{\mathbf{0}\}$ is a linear subspace, and any other linear subspace will be called a ***non-trivial*** linear subspace.

You may show that an equivalent definition is that $\mathcal{S}$ is a linear subspace of $\Re^n$ iff, for any scalars c_1 and c_2 and any vectors $\boldsymbol{u}_1$ and $\boldsymbol{u}_2$ in $\mathcal{S}$, $c_1\boldsymbol{u}_1 + c_2\boldsymbol{u}_2 \in \mathcal{S}$. That in turn is equivalent to $\mathcal{S}$ being closed under finite linear combinations.

Let $U = (\boldsymbol{u}_1, \ldots, \boldsymbol{u}_k)$ be the matrix with columns $\boldsymbol{u}_1, \ldots, \boldsymbol{u}_k$. We shall use $\text{sp}(U)$ to mean the same as $\text{sp}(\{\boldsymbol{u}_1, \ldots, \boldsymbol{u}_k\})$. Note that a linear combination of columns of U can be expressed as a matrix product:

$$c_1\boldsymbol{u}_1 + \cdots + c_k\boldsymbol{u}_k = U\boldsymbol{c},$$

where $\boldsymbol{c}$ is the column vector with components $c_1, \ldots, c_k$. That is,

$$\text{sp}(U) = \{U\boldsymbol{c} : \boldsymbol{c} \in \Re^k\}.$$

Using the shortcut above, this set also can be expressed as $\text{sp}(U) = U\{\Re^k\}$.

We shall say that a linear subspace $\mathcal{S}$ of $\Re^n$ is ***spanned by*** a set $\mathcal{U}$ of n-vectors iff $\text{sp}(\mathcal{U}) = \mathcal{S}$. A linear subspace $\mathcal{S}$ in $\Re^n$ is ***finite-dimensional*** iff there exists a finite set $\{\boldsymbol{v}_1, \ldots, \boldsymbol{v}_k\}$ of n-vectors such that $\mathcal{S} = \text{sp}(\boldsymbol{v}_1, \ldots, \boldsymbol{v}_k)$.

We shall be dealing only with subsets of $\Re^n$, where n is an arbitrary positive integer. Because $\Re^n$ is spanned by the columns $\boldsymbol{e}_1, \ldots, \boldsymbol{e}_n$ of I_n, $\Re^n$ is a finite-dimensional vector space. For a linear subspace $\mathcal{S}$ of $\Re^n$, while it is clear that $\text{sp}(\boldsymbol{e}_1, \ldots, \boldsymbol{e}_n)$ contains $\mathcal{S}$, some more argument is required to show that there exists a finite set of vectors such that $\text{sp}(\boldsymbol{v}_1, \ldots, \boldsymbol{v}_k) = \mathcal{S}$, that is, that every linear subspace of $\Re^n$ is finite-dimensional.

4.2 Exercises

Use the matrices and vectors defined in Section 3.3, p. 28, for Exercises 1 through 13. Use only the definitions and results covered so far in this book to address these questions.

1. With two tuples of vectors defined as $\mathcal{A} = \{\boldsymbol{a}_1, \boldsymbol{a}_2\}$ and $\mathcal{B} = \{\boldsymbol{b}_1, \boldsymbol{b}_2, \boldsymbol{b}_3\}$, find $\mathcal{A} + \mathcal{B}$.

2. Is $\boldsymbol{b}_1$ in $\text{sp}(B)$?

   Answer this "yes" by identifying scalars c_1, c_2, and c_3 such that $\boldsymbol{b}_1 = c_1\boldsymbol{b}_1 + c_2\boldsymbol{b}_2 + c_3\boldsymbol{b}_3$. Or answer "no" by showing that no such set of scalars exists: one way to do that is to show that their existence would imply something that is demonstrably untrue, like "$2 = 3$."

3. Is $\boldsymbol{x}$ in $\text{sp}(A)$?

4. Is $\boldsymbol{a}_1$ in $\text{sp}(AA')$?

   Since $\boldsymbol{a}_1 = A\boldsymbol{e}_1$, one approach is to find $\boldsymbol{v}$ such that $A'\boldsymbol{v} = \boldsymbol{e}_1$.

5. Is $\boldsymbol{x} - \boldsymbol{a}_2$ in $\text{sp}(A)$?

6. Is $\boldsymbol{a}_1 + \mathbf{1}_3$ in $\text{sp}(A)$?

7. Is $\boldsymbol{b}_1$ in $\text{sp}(A)$?

8. Does there exist a 2-vector $\boldsymbol{z} = (z_1, z_2)'$ such that

$$A\boldsymbol{z} = \begin{pmatrix} 3 \\ 7 \\ 11 \end{pmatrix}?$$

9. Does there exist a 2-vector $\boldsymbol{z}$ such that $A\boldsymbol{z} = \boldsymbol{b}_1$? Note that this is the same as asking whether $\boldsymbol{b}_1$ is in $\text{sp}(A)$.

10. Find a non-zero vector $\boldsymbol{c}$ that is in $\text{sp}(\mathbf{1}_3)^{\perp}$.

11. Find a non-zero vector $\boldsymbol{c}$ that is in $\text{sp}(\boldsymbol{x})^{\perp}$.

12. Let $\boldsymbol{z}$ be a 3-vector. Prove: In order that $\boldsymbol{z}$ be in $\text{sp}(A)^{\perp}$, it is necessary and sufficient that $A'\boldsymbol{z} = 0$. Note that $A'\boldsymbol{z} = 0$ iff $\boldsymbol{a}_1'\boldsymbol{z} = 0$ and $\boldsymbol{a}_2'\boldsymbol{z} = 0$.

13. (a) Find a vector $\boldsymbol{u}_1$ in $\text{sp}(\boldsymbol{a}_1)$ such that $\boldsymbol{u}_1'\boldsymbol{u}_1 = 1$.
    (b) Find $\hat{\boldsymbol{a}}_2 = t\boldsymbol{a}_1$ in $\text{sp}(\boldsymbol{a}_1)$ such that $\boldsymbol{a}_2 - \hat{\boldsymbol{a}}_2$ is in $\text{sp}(\boldsymbol{a}_1)^{\perp}$.
    (c) Find $\boldsymbol{u}_2$ in $\text{sp}(A)$ such that $\boldsymbol{u}_2'\boldsymbol{u}_2 = 1$ and $\boldsymbol{u}_2'\boldsymbol{u}_1 = 0$.
    (d) Find $\hat{\boldsymbol{x}} = \boldsymbol{u}_1\boldsymbol{u}_1'\boldsymbol{x} + \boldsymbol{u}_2\boldsymbol{u}_2'\boldsymbol{x}$. See 3 above.
    (e) Find $\hat{\boldsymbol{b}}_1 = \boldsymbol{u}_1\boldsymbol{u}_1'\boldsymbol{b}_1 + \boldsymbol{u}_2\boldsymbol{u}_2'\boldsymbol{b}_1$. See 7 above.

14. The *norm* or *length* of a vector $\boldsymbol{z}$ is defined to be $\|\boldsymbol{z}\| = \sqrt{\boldsymbol{z}'\boldsymbol{z}}$. Show that the Theorem of Pythagoras holds in n dimensions, that is, show that if $\boldsymbol{u}$ and $\boldsymbol{v}$ are orthogonal n-vectors, then

$$\|\boldsymbol{u} + \boldsymbol{v}\|^2 = \|\boldsymbol{u}\|^2 + \|\boldsymbol{v}\|^2.$$

15. Let $\mathcal{U} = \{\boldsymbol{u}_1, \ldots, \boldsymbol{u}_k\}$ be a tuple of $k \geqslant 1$ n-vectors. Show that $\text{sp}(\mathcal{U})$ is a linear subspace. Show that $\mathcal{U}^\perp$ is a linear subspace.

16. Let $\mathcal{A}$ and $\mathcal{B}$ be sets of n-vectors. Show that

    (a) $\mathcal{A}^\perp = [\text{sp}(\mathcal{A})]^\perp$.

    (b) $\text{sp}(\mathcal{A}+\mathcal{B}) \subset \text{sp}(\mathcal{A})+\text{sp}(\mathcal{B})$ and, if both $\mathcal{A}$ and $\mathcal{B}$ contain $\mathbf{0}$, $\text{sp}(\mathcal{A}+\mathcal{B}) \supset \text{sp}(\mathcal{A}) + \text{sp}(\mathcal{B})$.

    (c) If both $\mathcal{A}$ and $\mathcal{B}$ contain $\mathbf{0}$, then $(\mathcal{A} + \mathcal{B})^\perp = \mathcal{A}^\perp \cap \mathcal{B}^\perp$.

    (d) If $\text{sp}(\mathcal{A}) \supset \mathcal{B}$ then $\text{sp}(\mathcal{A}) \supset \text{sp}(\mathcal{B})$.

17. Let $\mathbf{1}_n$ denote an n-vector of ones, and let $\mathcal{S} = \{\mathbf{1}_n\}$ denote a set comprising only the vector $\mathbf{1}_n$. Describe $\mathcal{S}^\perp$.

18. Prove: If $\mathcal{U}$ and $\mathcal{V}$ are linear subspaces of $\Re^n$, then

    (a) $\mathcal{U} + \mathcal{V}$ and

    (b) $\mathcal{U} \cap \mathcal{V}$ are linear subspaces of $\Re^n$; and

    (c) $\mathbf{0} \in \mathcal{U}$.

19. Let $U = (\boldsymbol{u}_1, \ldots, \boldsymbol{u}_c)$ and $W = (\boldsymbol{w}_1, \ldots, \boldsymbol{w}_m)$ be matrices with r rows, and let $\mathcal{U} = \text{sp}(U)$ and $\mathcal{W} = \text{sp}(W)$. Let $B = (\boldsymbol{u}_1, \ldots, \boldsymbol{u}_c, \boldsymbol{w}_1, \ldots, \boldsymbol{w}_m)$, that is, $B = (U, W)$, the matrix concatenating the columns of U and W. Prove: $\mathcal{U} + \mathcal{W} = \text{sp}(B)$.

20. Let A be an $r \times n$ matrix. Let $\mathcal{N} = \{\boldsymbol{z} \in \Re^n : A\boldsymbol{z} = \mathbf{0}\}$. Show that $\mathcal{N}$ is a linear subspace. Show that $\mathcal{N} = \text{sp}(A')^\perp$.

21. Let $\boldsymbol{y}$ be an n-vector, and let $\mathcal{S}$ be a set of n-vectors.

    (a) Prove: $\boldsymbol{y}'\boldsymbol{y} = 0$ iff $\boldsymbol{y} = 0$.

    (b) Prove: If $\mathbf{0} \in \mathcal{S}$, then $\boldsymbol{y} \in \mathcal{S} \cap \mathcal{S}^\perp$ iff $\boldsymbol{y} = \mathbf{0}$.

22. Let A be an $r \times c$ matrix. Show that $\text{sp}(A) \cap \{\boldsymbol{x} \in \Re^r : A'\boldsymbol{x} = \mathbf{0}\} = \{\mathbf{0}\}$. That is, if $\boldsymbol{x} = A\boldsymbol{b}$ for some vector $\boldsymbol{b}$, then $A'\boldsymbol{x} = \mathbf{0}$ if and only if $\boldsymbol{x} = 0$. [Note that this can be formulated as $\text{sp}(A) \cap \text{sp}(A)^\perp = \{\mathbf{0}\}$.]

23. Let A be an $r \times c$ matrix, and let $\boldsymbol{x}$ be a c-vector.

    (a) Prove: $A\boldsymbol{x} = \mathbf{0}$ iff $A'A\boldsymbol{x} = \mathbf{0}$.

    (b) Prove: $\text{sp}(A')^\perp = \text{sp}(A'A)^\perp$.

24. With $\mathcal{S} = \text{sp}(\boldsymbol{u}_1, \ldots, \boldsymbol{u}_k)$, a linear subspace of $\Re^n$, show that the orthogonal complement of $\mathcal{S}$ can be described as

    $$\mathcal{S}^\perp = \{\boldsymbol{x} \in \Re^n : U'\boldsymbol{x} = 0\},$$

    where U is the $n \times k$ matrix $(\boldsymbol{u}_1, \ldots, \boldsymbol{u}_k)$.

To be a member of $\mathcal{S}^\perp$, a vector $\boldsymbol{x}$ must be orthogonal to *all* vectors in $\mathcal{S}$. This result shows that it is enough for $\boldsymbol{x}$ to be orthogonal to the k vectors $\boldsymbol{u}_1, \ldots, \boldsymbol{u}_k$ that span $\mathcal{S}$.

25. Let $\mathcal{S}$ be a set of n-vectors. Show that:

    (a) $\mathcal{S}^\perp$ is a linear subspace.
    (b) $\mathcal{S} \subset \text{sp}(\mathcal{S})$.
    (c) $(\mathcal{S}^\perp)^\perp \supset \text{sp}(\mathcal{S})$.
    (d) If $\mathcal{U}$ is a linear subspace that contains $\mathcal{S}$, then $\mathcal{U}$ contains $\text{sp}(\mathcal{S})$.
    (e) If $\mathcal{T}$ is a set of vectors and $\text{sp}(\mathcal{T}) \supset \text{sp}(\mathcal{S})$, then $\mathcal{T}^\perp \subset \mathcal{S}^\perp$.

26. Let A and C be matrices with r rows. Show that $\text{sp}(C) \subset \text{sp}(A)$ iff there exists a matrix B such that $AB = C$.

27. Let A be an $r \times c$ matrix, let $\boldsymbol{b}$ be an r-vector in $\text{sp}(A)$, and let $\boldsymbol{x}_0$ be a c-vector such that $A\boldsymbol{x}_0 = \boldsymbol{b}$.

    (a) Show that $\{\boldsymbol{x} \in \Re^c : A\boldsymbol{x} = \boldsymbol{0}\} = \text{sp}(A')^\perp$.
    (b) Show that $A\{\text{sp}(A')^\perp\} = \{\boldsymbol{0}\}$.
    (c) Show that $\{\boldsymbol{x} \in \Re^c : A\boldsymbol{x} = \boldsymbol{b}\} = \{\boldsymbol{x}_0\} + \text{sp}(A')^\perp$.

28. Let A be an $r \times c$ matrix, let B be an $m \times c$ matrix, and let $\boldsymbol{x}_0$ be a c-vector.

    (a) Show that $\mathcal{S}_0 = \{A\boldsymbol{x} : \boldsymbol{x} \in \Re^c \text{ and } B\boldsymbol{x} = \boldsymbol{0}\}$ is a linear subspace.
    (b) Show that $\mathcal{S}_0 = A\{\text{sp}(B')^\perp\} = A\{\text{sp}(A')^\perp + \text{sp}(B')^\perp\}$.
    (c) Show that $\{A\boldsymbol{x} : \boldsymbol{x} \in \Re^c \text{ and } B\boldsymbol{x} = B\boldsymbol{x}_0\} = \mathcal{S}_0 + \{A\boldsymbol{x}_0\}$.

29. Show that the sum of linear subspaces $\mathcal{S}$ and $\mathcal{T}$ is a direct sum iff $\mathcal{S} \cap \mathcal{T} = \{\boldsymbol{0}\}$.

30. An ***affine combination*** of n-vectors $\boldsymbol{x}_1, \ldots, \boldsymbol{x}_k$ is a linear combination $c_1\boldsymbol{x}_1 + \cdots + c_k\boldsymbol{x}_k$ such that $c_1 + \cdots + c_k = 1$. An ***affine set*** in $\Re^n$ is a set of n-vectors that is closed under all finite affine combinations (that is, for any number k of terms). It is clear that linear subspaces are affine sets, and that there are affine sets that are not linear subspaces.

    (a) If $\mathcal{S}$ is a linear subspace in $\Re^n$ and $\boldsymbol{x}_0$ is an n-vector, then $\mathcal{A} = \{\boldsymbol{x}_0\} + \mathcal{S}$ is an affine set.
    Proof: Let $\boldsymbol{z}_1 = \boldsymbol{x}_0 + \boldsymbol{s}_1, \ldots, \boldsymbol{z}_k = \boldsymbol{x}_0 + \boldsymbol{s}_k$ be vectors in $\mathcal{A}$, and let $c_1, \ldots, c_k$ be scalars such that $\sum_i c_i = 1$. Then

    $$\sum_i c_i \boldsymbol{z}_i = \boldsymbol{x}_0 + \sum_i c_i \boldsymbol{s}_i = \boldsymbol{x}_0 + \boldsymbol{s} \in \mathcal{A}$$

    because $\boldsymbol{s} = \sum_i c_i \boldsymbol{s}_i \in \mathcal{S}$.

(b) If $\mathcal{A}$ is an affine set and $\boldsymbol{x}_0 \in \mathcal{A}$, then

$$\mathcal{T} = \{\boldsymbol{x} - \boldsymbol{x}_0 : \boldsymbol{x} \in \mathcal{A}\}$$

is a linear subspace, and

$$\mathcal{A} = \{\boldsymbol{x}_0\} + \mathcal{T}.$$

It follows as a corollary that if $\mathbf{0} \in \mathcal{A}$, then $\mathcal{A}$ is a linear subspace.

Proof: That $\mathcal{A} = \{\boldsymbol{x}_0\} + \mathcal{T}$ is clear. Let $\boldsymbol{x}_1 = \boldsymbol{a}_1 - \boldsymbol{x}_0$ and $\boldsymbol{x}_2 = \boldsymbol{a}_2 - \boldsymbol{x}_0$, with $\boldsymbol{a}_1$ and $\boldsymbol{a}_2$ in $\mathcal{A}$, be vectors in $\mathcal{T}$, and let c_1 and c_2 be scalars. Then

$$c_1\boldsymbol{x}_1 + c_2\boldsymbol{x}_2 = [c_1\boldsymbol{a}_1 + c_2\boldsymbol{a}_2 - (c_1 + c_2)\boldsymbol{x}_0 + \boldsymbol{x}_0] - \boldsymbol{x}_0$$

is in $\mathcal{T}$ because $c_1\boldsymbol{a}_1 + c_2\boldsymbol{a}_2 - (c_1 + c_2)\boldsymbol{x}_0 + \boldsymbol{x}_0$ is an affine combination of members of $\mathcal{A}$, and hence it is in $\mathcal{A}$.

(c) Suppose $\mathcal{A} = \{\boldsymbol{x}_0\} + \mathcal{T}$, where $\boldsymbol{x}_0$ is a vector and $\mathcal{T}$ is a linear subspace in $\Re^n$. If $\boldsymbol{x}_*$ is a vector and $\mathcal{S}$ is a linear subspace such that $\mathcal{A} = \{\boldsymbol{x}_*\} + \mathcal{S}$, then $\boldsymbol{x}_* \in \mathcal{A}$ and $\mathcal{S} = \mathcal{T}$.

Proof. Because $\mathbf{0} \in \mathcal{S}$, it follows that $\boldsymbol{x}_* \in \mathcal{A}$. To prove that $\mathcal{S} \subset \mathcal{T}$, let $\boldsymbol{s} \in \mathcal{S}$. Then $\boldsymbol{x}_* + \boldsymbol{s} \in \mathcal{A} \Longrightarrow \exists\ \boldsymbol{z}_* \in \mathcal{A}$ such that $\boldsymbol{x}_* + \boldsymbol{s} = \boldsymbol{z}_* \Longrightarrow$

$$\boldsymbol{s} = \boldsymbol{z}_* - \boldsymbol{x}_* = (\boldsymbol{z}_* - \boldsymbol{x}_0 + \boldsymbol{x}_*) - \boldsymbol{x}_0 \in \mathcal{T}.$$

To prove that $\mathcal{T} \subset \mathcal{S}$, let $\boldsymbol{t} \in \mathcal{T}$. Then $\boldsymbol{x}_0 + \boldsymbol{t} \in \mathcal{A} = \{\boldsymbol{x}_*\} + \mathcal{S} \Longrightarrow \exists$ $\boldsymbol{s} \in \mathcal{S}$ such that $\boldsymbol{x}_0 + \boldsymbol{t} = \boldsymbol{x}_* + \boldsymbol{s} \Longrightarrow \boldsymbol{t} = (\boldsymbol{x}_* - \boldsymbol{x}_0) + \boldsymbol{s}$. Because both $\boldsymbol{x}_*$ and $\boldsymbol{x}_0$ are in $\mathcal{A} = \{\boldsymbol{x}_*\} + \mathcal{S}$, $\exists\ \boldsymbol{s}_0 \in \mathcal{S}$ such that $\boldsymbol{x}_0 = \boldsymbol{x}_* + \boldsymbol{s}_0$, and hence $\boldsymbol{x}_* - \boldsymbol{x}_0 = \boldsymbol{s}_0$, and therefore $\boldsymbol{t} = \boldsymbol{s}_0 + \boldsymbol{s}_* \in \mathcal{S}$.

(d) If $\mathcal{S}_1$ and $\mathcal{S}_2$ are linear subspaces and $\boldsymbol{x}_1$ and $\boldsymbol{x}_2$ are vectors, then $\{\boldsymbol{x}_1\} + \mathcal{S}_1 = \{\boldsymbol{x}_2\} + \mathcal{S}_2$ iff $\mathcal{S}_1 = \mathcal{S}_2$ and $\boldsymbol{x}_1 - \boldsymbol{x}_2 \in \mathcal{S}_1$.

5

Orthogonal Projection

This chapter establishes the existence and uniqueness of orthogonal projections in linear subspaces. It also shows that computing them and the other statistics associated with the method of least squares is straightforward with output from the Gram-Schmidt construction, the subject of Chapter 6.

5.1 Introduction

For a linear subspace $\mathcal{S}$ of $\Re^n$ and an n-vector $\boldsymbol{y}$, an ***orthogonal projection*** (OP) of $\boldsymbol{y}$ in $\mathcal{S}$, if it exists, is a vector $\hat{\boldsymbol{y}}$ in $\mathcal{S}$ such that $\boldsymbol{y} - \hat{\boldsymbol{y}}$ is in $\mathcal{S}^{\perp}$. It is straightforward to establish that, if it exists, it is unique (see Exercise 3a in Chapter 7).

Examples of OPs abound. If $\boldsymbol{z} \in \mathcal{S}$, then $\boldsymbol{z} - \boldsymbol{z} = \mathbf{0}$ is in $\mathcal{S}^{\perp}$, and therefore $\boldsymbol{z}$ is the OP of $\boldsymbol{z}$ in $\mathcal{S}$. If P is a symmetric, idempotent, $n \times n$ matrix, then, for any n-vector $\boldsymbol{y}$, $\boldsymbol{y} - P\boldsymbol{y}$ is in $\text{sp}(P)^{\perp}$, and therefore $P\boldsymbol{y}$ is the OP of $\boldsymbol{y}$ in $\text{sp}(P)$.

Chapter 6 establishes computational steps that, for any (non-zero) matrix X, produce matrices Q and T such that $\text{sp}(Q) = \text{sp}(X)$, $Q'Q = \text{I}$, and $Q = XT$. Then QQ' is symmetric and idempotent, and so, for any n-vector $\boldsymbol{y}$, $QQ'\boldsymbol{y}$ is the OP of $\boldsymbol{y}$ in $\text{sp}(QQ')$, which is shown later to be equal to $\text{sp}(X)$. That is, if the linear subspace $\mathcal{S}$ in $\Re^n$ has a finite spanning set (e.g., the columns of X), then every n-vector has an OP in $\mathcal{S}$.

All the models considered in this book can be expressed in terms of linear subspaces of the form $\mathcal{S} = \text{sp}(X)$. Then, as just outlined, every n-vector $\boldsymbol{y}$ has an OP in $\mathcal{S}$, namely $QQ'\boldsymbol{y}$.

The next section provides a sequence of propositions that establishes that every linear subspace of $\Re^n$ has a finite spanning set. That implies that, for any linear subspace $\mathcal{S}$ of $\Re^n$, each n-vector $\boldsymbol{y}$ has a unique orthogonal projection in $\mathcal{S}$. In the process, it establishes several other important properties and relations among linear subspaces and their orthogonal complements.

5.2 Basic Propositions

Propositions 5.1–5.3 develop relations between orthonormal sets of n-vectors and the linear subspaces that they span. Proofs are given in Appendix A, rather than in-line here, to encourage the reader to work through them independently. Throughout this section, Q and R denote matrices with n ($\geqslant 1$) rows and at least one column such that $Q'Q = \mathrm{I}_\nu$ and $R'R = \mathrm{I}_\eta$.

Proposition 5.1.

a. $\boldsymbol{z} \in \mathrm{sp}(Q)$ *iff* $QQ'\boldsymbol{z} = \boldsymbol{z}$

and

b. $\mathrm{sp}(Q) = \mathrm{sp}(QQ')$.

Recall that for a symmetric matrix A, $\mathrm{tr}(AA) = \sum_i \sum_j a_{ij}^2 \geqslant 0$, and $\mathrm{tr}(AA) = 0 \Longrightarrow A = 0$.

Proposition 5.2 establishes a fundamental correspondence between the linear subspaces $\mathrm{sp}(Q)$ and $\mathrm{sp}(R)$, on the one hand, and the matrices QQ' and RR' on the other.

Proposition 5.2. $\mathrm{sp}(Q) \subset \mathrm{sp}(R)$ *iff*

$$RR' - QQ' = (RR' - QQ')(RR' - QQ'),$$

and $\mathrm{sp}(Q) = \mathrm{sp}(R)$ *iff* $RR' = QQ'$.

Propositions 5.3 establishes relations between set inclusion and numbers of vectors in Q and R.

Proposition 5.3. *If* $\mathrm{sp}(Q) \subset \mathrm{sp}(R)$, *then (a)* $\eta \geqslant \nu$ *and (b)* $\mathrm{sp}(Q) = \mathrm{sp}(R)$ *iff* $\eta = \nu$.

With $\mathrm{sp}(\mathrm{I}_n) = \Re^n$, it follows that any $n \times \nu$ matrix Q with $Q'Q = \mathrm{I}_\nu$ has at most n columns. Our next objective is to use this simple fact to demonstrate that for any linear subspace $\mathcal{S} \neq \{\mathbf{0}\}$ of $\Re^n$, there exists a matrix Q with $\nu \leqslant n$ columns and $Q'Q = \mathrm{I}_\nu$ such that $\mathrm{sp}(Q) = \mathcal{S}$. Proposition 5.4 does that, and, together with Proposition 5.1, this establishes that each n-vector $\boldsymbol{y}$ has an orthogonal projection in $\mathcal{S}$.

Proposition 5.4. *If $\mathcal{S}$ is a linear subspace of $\Re^n$ and $\mathcal{S} \neq \{\mathbf{0}\}$, then there exists a matrix Q such that $Q'Q = \mathrm{I}_\nu$ and* $\mathrm{sp}(Q) = \mathcal{S}$.

Now we may conclude that, for any linear subspace $\mathcal{S}$ of $\Re^n$, there exists a matrix Q with $\nu \leqslant n$ columns such that $\mathrm{sp}(Q) = \mathcal{S}$ and $Q'Q = \mathrm{I}_\nu$. By the last part of Proposition 5.2, if Q_* is a matrix such that $Q_*'Q_* = \mathrm{I}$ and $\mathrm{sp}(Q_*) = \mathcal{S}$, then $Q_*Q_*' = QQ'$. By Proposition 5.3, Q_* also has ν columns. The ***dimension*** of a linear subspace $\mathcal{S}$ is defined to be ν, the number of vectors in an orthonormal set of vectors that spans $\mathcal{S}$. One consequence of these arguments is that every linear subspace of $\Re^n$ is finite-dimensional.

Illustrate by examples that, for a given linear subspace $\mathcal{S}$, generally there are many matrices Q such that $Q'Q = \mathrm{I}$ and $\mathrm{sp}(Q) = \mathcal{S}$ (e.g., by permuting columns of Q). All produce the same matrix QQ'.

Proposition 5.5. *Let $\mathcal{S}$ be a non-trivial linear subspace of $\Re^n$, and let $\boldsymbol{y}$ be an n-vector. There exists exactly one vector $\hat{\boldsymbol{y}}$ in $\mathcal{S}$ such that $\boldsymbol{y} - \hat{\boldsymbol{y}}$ is in $\mathcal{S}^\perp$.*

Proposition 5.5 establishes that, for any linear subspace $\mathcal{S}$ of $\Re^n$, each vector $\boldsymbol{y}$ in $\Re^n$ has a unique orthogonal projection in $\mathcal{S}$, and, for any matrix Q such that $\mathrm{sp}(Q) = \mathcal{S}$ and $Q'Q = \mathrm{I}$, it is $QQ'\boldsymbol{y}$. It was noted above that QQ' is unique. Suppose P is a matrix such that, for each $\boldsymbol{y} \in \Re^n$, $P\boldsymbol{y} \in \mathcal{S}$ and $\boldsymbol{y} - P\boldsymbol{y} \in \mathcal{S}^\perp$. Then $P\boldsymbol{y}$ is the orthogonal projection of $\boldsymbol{y}$ onto $\mathcal{S}$, and therefore $P\boldsymbol{y} = QQ'\boldsymbol{y}$. That this is true for all $\boldsymbol{y} \in \Re^n$ implies that $P = QQ'$. For a given linear subspace $\mathcal{S}$, there is exactly one matrix P such that, for each $\boldsymbol{y} \in \Re^n$, $P\boldsymbol{y}$ is the orthogonal projection of $\boldsymbol{y}$ onto $\mathcal{S}$. Denote this matrix by $\mathbf{P}_{\mathcal{S}}$. It follows that two linear subspaces $\mathcal{S}$ and $\mathcal{T}$ of $\Re^n$ are the same iff $\mathbf{P}_{\mathcal{S}} = \mathbf{P}_{\mathcal{T}}$.

For a matrix A, the notation $\mathbf{P}_A$ will be used as equivalent to $\mathbf{P}_{\mathrm{sp}(A)}$. However, keep in mind that the same linear subspace can be represented by different spanning sets. Thus $\mathbf{P}_A = \mathbf{P}_B$ implies that $\mathrm{sp}(A) = \mathrm{sp}(B)$, but it does not imply that $A = B$. As a simple example, let B have the same columns as A, but in different order.

For each linear subspace $\mathcal{S}$ of $\Re^n$ and vector $\boldsymbol{y} \in \Re^n$, there exist unique vectors $\boldsymbol{z}_1 \in \mathcal{S}$ (namely, $\boldsymbol{z}_1 = \hat{\boldsymbol{y}}$) and $\boldsymbol{z}_2 \in \mathcal{S}^\perp$ ($\boldsymbol{z}_2 = \boldsymbol{y} - \hat{\boldsymbol{y}}$) such that

$$\boldsymbol{y} = \boldsymbol{z}_1 + \boldsymbol{z}_2. \tag{5.1}$$

This breaks loose several relations involving orthogonal complements.

Proposition 5.6. *Let $\mathcal{S}$ and $\mathcal{T}$ denote linear subspaces of $\Re^n$.*

1. $(\mathcal{S}^\perp)^\perp = \mathcal{S}$.
2. $\mathcal{S} \subset \mathcal{T} \iff \mathcal{T}^\perp \subset \mathcal{S}^\perp$.
3. $(\mathcal{S} \cap \mathcal{T})^\perp = \mathcal{S}^\perp + \mathcal{T}^\perp$.

It was shown in Exercise 23b, p. 34, that, if A is a matrix, then $\mathrm{sp}(A')^\perp = \mathrm{sp}(A'A)^\perp$. Now, by Proposition 5.6.1, it follows that $\mathrm{sp}(A') = \mathrm{sp}(A'A)$.

In use, linear subspaces that we're interested in usually are described in terms of matrices, like $\mathrm{sp}(X) = \{X\boldsymbol{\beta} : \boldsymbol{\beta} \in \Re^{k+1}\}$, for example. The propositions in this section establish that there exists an orthonormal matrix Q such that $\mathrm{sp}(Q) = \mathrm{sp}(X)$, and the proof of Proposition 5.4 hints at how to construct Q. From the fact that $\mathrm{sp}(Q) = \mathrm{sp}(X)$, it is clear that there exists a matrix T such that $Q = XT$. Chapter 6 details the construction of Q and T from X, a sequence of steps known widely as the Gram-Schmidt (GS) algorithm. We may say that GS on X yields Q and T such that $\mathrm{sp}(Q) = \mathrm{sp}(X)$, $Q'Q = \mathrm{I}$, and $Q = XT$.

5.3 Bridges

The terms defined next are widely used in accounts of linear algebra and linear statistical models. They are not needed much in the developments here, but they are important in other settings.

Let $\mathcal{U} = \{\boldsymbol{u}_1, \ldots, \boldsymbol{u}_k\}$ be a k-tuple of n-vectors, and let $U = (\boldsymbol{u}_1, \ldots, \boldsymbol{u}_k)$ be the $n \times k$ matrix with these vectors as its columns.

The set $\mathcal{U}$ is said to be ***linearly dependent*** iff there exist scalars $c_1, \ldots, c_k$, not all 0, such that $c_1\boldsymbol{u}_1 + \cdots + c_k\boldsymbol{u}_k = 0$. Equivalently, the columns of U are linearly dependent iff the solution set $\{\boldsymbol{c} \in \Re^k : U\boldsymbol{c} = \boldsymbol{0}\}$ contains a non-zero vector.

$\mathcal{U}$ is said to be ***linearly independent*** iff it is not linearly dependent, that is, iff $c_1\boldsymbol{u}_1 + \cdots + c_k\boldsymbol{u}_k = 0$ implies that $c_1 = \cdots = c_k = 0$. Equivalently, the columns of U are linearly independent iff $U\boldsymbol{c} = \boldsymbol{0}$ implies that $\boldsymbol{c} = \boldsymbol{0}$ or, equivalently, the solution set $\{\boldsymbol{c} \in \Re^k : U\boldsymbol{c} = \boldsymbol{0}\}$ contains only $\boldsymbol{0}$, the zero vector.

If Q is an orthonormal matrix ($Q'Q = \mathrm{I}$), then the columns of Q are linearly independent, because $Q\boldsymbol{x} = \boldsymbol{0}$ implies that $Q'Q\boldsymbol{x} = \boldsymbol{x} = \boldsymbol{0}$.

If $\mathcal{U}$ is linearly dependent, then (you may show that) for any vector $\boldsymbol{y}$ in $\mathrm{sp}(\mathcal{U})$ there are multiple k-vectors $\boldsymbol{c}$ such that $U\boldsymbol{c} = \boldsymbol{y}$. If $\mathcal{U}$ is linearly independent, then (you may show that) for each vector $\boldsymbol{y}$ in $\mathrm{sp}(\mathcal{U})$, there is exactly one vector $\boldsymbol{c}$ such that $U\boldsymbol{c} = \boldsymbol{y}$. It is shown in Chapter 7 that the columns of U are linearly independent iff the number of columns in Q from GS on U is k.

If $\mathcal{S}$ is a linear subspace of $\Re^n$, and if $\{\boldsymbol{u}_1, \ldots, \boldsymbol{u}_k\}$ is a linearly independent spanning set for $\mathcal{S}$, then $\{\boldsymbol{u}_1, \ldots, \boldsymbol{u}_k\}$ is said to be a ***basis*** for $\mathcal{S}$. In this use, the plural of "basis" is *bases* (pronounced bā′sēz′). If Q is an orthonormal matrix and $\mathrm{sp}(Q) = \mathcal{S}$, then the columns of Q constitute an orthonormal basis for $\mathcal{S}$.

An $n \times n$ matrix A is said to have an ***inverse*** G iff there exists an $n \times n$ matrix G such that $GA = AG = \mathrm{I}_n$. If A has an inverse, its inverse is denoted A^{-1}. If A has an inverse, A is said to be ***non-singular***, and otherwise A is said to be ***singular***. Take the attitude here that matrices are singular (don't have inverses) until proven non-singular.

The following results are established in Chapter 7. A square matrix has an inverse iff its columns are linearly independent. The columns of a matrix U are linearly independent iff the columns of $U'U$ are linearly independent.

If A is an $n \times n$ non-singular matrix, with Q and T from GS on A, then $A^{-1} = TQ'$.

For an $n \times k$ matrix A, a ***generalized inverse*** of A is a matrix G such that $AGA = A$. In contrast to inverses, every matrix has a generalized inverse, and most have many.

A ***reflexive*** generalized inverse of A is a matrix G such that $AGA = A$ and $GAG = G$. Exercises in Chapter 7 establish that, with Q and T from GS on A, TQ' is a reflexive generalized inverse of A, and TT' is a symmetric, reflexive generalized inverse of $A'A$.

Some fundamental results in linear algebra have to do with dimensions of linear subspaces and ranks of matrices. Briefly, every basis of a linear subspace $\mathcal{S}$ has the same number (say ν) of member vectors, and any linearly independent set of ν vectors in $\mathcal{S}$ spans $\mathcal{S}$. The ***dimension*** of $\mathcal{S}$ is defined to be ν. The ***column rank*** of a matrix A is defined to be the dimension of $\text{sp}(A)$, and the ***row rank*** of A is the dimension of $\text{sp}(A')$. The column rank of A and the row rank of A are equal. These facts are established as exercises in Chapter 7.

A note on notation: Matrices can be used to define linear operators or linear transformations, and vice versa. Because of this association, $\text{sp}(U)$ is also denoted $\boldsymbol{R}(U)$, referring to the ***range*** of the linear mapping that defines U. In that setting, $\boldsymbol{N}(U)$ refers to the ***null space*** or ***kernel*** of the linear transformation; it is the set $\boldsymbol{N}(U) = \{\boldsymbol{z} \in \Re^c : U\boldsymbol{z} = 0\}$. To add further confusion to the notation, $\boldsymbol{C}(U)$ is used to mean the same as $\text{sp}(U)$ and $\boldsymbol{R}(U)$; it is called the ***column space*** of U. Some authors have used $\boldsymbol{C}(U)$ and $\boldsymbol{R}(U)$ for the column space and ***row space*** of U.

6

The Gram-Schmidt Construction

The Gram-Schmidt construction is a sequence of simple computational steps applied to the columns of a matrix A to produce a matrix Q such that $Q'Q = \mathrm{I}$ and $\mathrm{sp}(Q) = \mathrm{sp}(A)$. The version described here produces also a matrix T such that $Q = AT$. It fulfills practically all the computational needs for the material covered in this book.

6.1 The GS Construction

The earliest description of the Gram-Schmidt algorithm seems to be Laplace (1816). It was later described by Gram (1883) and Schmidt (1907).[1] The source of these citations is Miller (2018).

Given a non-zero $n \times m$ matrix A, or the n-vectors $\{\boldsymbol{a}_1, \ldots, \boldsymbol{a}_m\}$ constituting its columns, the Gram-Schmidt construction (GS) constructs a matrix Q and a matrix T such that $\mathrm{sp}(Q) = \mathrm{sp}(A)$, $Q'Q = \mathrm{I}$, and $Q = AT$. For its uses here, it is described in detail in LaMotte (2014).

The proof of Proposition 5.4, applied specifically to the columns of A (in place of the linear subspace $\mathcal{S}$), outlines the GS construction. Let $\boldsymbol{e}_\ell$ denote the ℓ-th column of the $m \times m$ identity matrix.

Let i_1 denote the column index of the left-most non-zero column of A. Choose c such that $\boldsymbol{q} = c\boldsymbol{a}_{i_1}$ has $\boldsymbol{q}'\boldsymbol{q} = 1$. Define Q to be the matrix with $\boldsymbol{q}$ as its single column. Then $\mathrm{sp}(Q) = \mathrm{sp}(\boldsymbol{a}_1, \ldots, \boldsymbol{a}_{i_1})$. Define $\boldsymbol{t} = c\boldsymbol{e}_{i_1}$ and T as the matrix with $\boldsymbol{t}$ as its single column, so that $Q = AT$. Define $\boldsymbol{j}$ to be i_1.

Proceed left to right through the remaining columns of A. After the i-th, Q and T have been accumulated such that $Q'Q = \mathrm{I}$, $\mathrm{sp}(Q) = \mathrm{sp}(\boldsymbol{a}_1, \ldots, \boldsymbol{a}_i)$, and $Q = AT$.

For the next step, if $\boldsymbol{a}_{i+1}$ is in $\mathrm{sp}(Q)$, then do nothing. Then $\mathrm{sp}(Q) = \mathrm{sp}(\boldsymbol{a}_1, \ldots, \boldsymbol{a}_{i+1})$. Otherwise choose c such that $\boldsymbol{q} = c(\boldsymbol{a}_{i+1} - QQ'\boldsymbol{a}_{i+1})$ has $\boldsymbol{q}'\boldsymbol{q} = 1$. Define $\boldsymbol{t}$ as $c(\boldsymbol{e}_{i+1} - TQ'\boldsymbol{a}_{i+1})$, so that $\boldsymbol{q} = A\boldsymbol{t}$. Then $\mathrm{sp}(Q, \boldsymbol{q}) =$

[1] As a personal aside, Erhard Schmidt co-directed the Ph.D. dissertation (1934) of H. O. Hirschfeld (later Hartley), who directed my M.S. thesis (1966) and served on my Ph.D. dissertation committee (1969).

$\text{sp}(\boldsymbol{a}_1, \ldots, \boldsymbol{a}_{i+1})$ and $A(T, \boldsymbol{t}) = (Q, \boldsymbol{q})$. In this case, we shall say that the column $\boldsymbol{a}_{i+1}$ of A ***contributes*** the column $\boldsymbol{q}$ to Q.

Update Q and T by appending $\boldsymbol{q}$ and $\boldsymbol{t}$ as their right-most columns, that is, $Q \leftarrow (Q, \boldsymbol{q})$ and $T \leftarrow (T, \boldsymbol{t})$. Update the list of column indexes of A as $\boldsymbol{j} \leftarrow (\boldsymbol{j}, i+1)$.

After all the columns of A have been scanned, Q has ν columns, $\text{sp}(Q) = \text{sp}(A)$, $Q'Q = \text{I}_\nu$, and $Q = AT$. The list $\boldsymbol{j}$ has ν entries, which are the column indexes $(i_1, \ldots, i_\nu)$ of the columns of A that contributed columns to Q and T.

The construction is easy to program. Because digital computations produce truncation error, to check whether a vector $\boldsymbol{v}$ is $\boldsymbol{0}$, check whether $\boldsymbol{v}'\boldsymbol{v} < \epsilon$ instead, where ϵ is small, perhaps $\epsilon = 10^{-12}$. It is known that the GS construction presented here does not have very good numerical properties, but it seems to work satisfactorily for most examples. According to Gentle (2007), its numerical properties can be improved with a simple modification. At each step, replace A by $(\text{I} - Q_i Q_i')A$ and remove the first i columns. Then each step becomes equivalent to the initialization step, to find the first non-zero column of the current A.

6.2 Consequences of GS

Following are some properties and consequences of the GS construction, as presented in LaMotte (2014). Let $A = (A_1, A_2)$ be a matrix with n rows, partitioned into two sets of columns. GS on A produces $Q = (Q_1, Q_2)$ and $T = (T_1, T_2)$ such that $Q'Q = \text{I}$ and $Q = AT$ and $\text{sp}(Q) = \text{sp}(A)$. The parts of Q and T are those contributed by the corresponding parts of A. They may be void, which would require rather obvious modifications of the discussion here.

The left-right progression of GS means that $\text{sp}(Q_1) = \text{sp}(A_1)$. This results in $\mathbf{P}_A = Q_1Q_1' + Q_2Q_2'$, $\text{sp}(Q_2) = \text{sp}(A) \cap \text{sp}(A_1)^\perp$, and $Q_2Q_2' = \mathbf{P}_A - \mathbf{P}_{A_1}$. Thus, for example, from GS on (A, I), $\text{sp}(Q_2) = \text{sp}(A)^\perp$. Given a suitable matrix N, GS on (XN, X) yields $Q_2Q_2' = \mathbf{P}_X - \mathbf{P}_{XN}$. Given matrices A and B with the same number of columns, one way to find a matrix C such that $\text{sp}(C) = \text{sp}(A') \cap \text{sp}(B')$ is as $C = A'H$ with H such that $\mathbf{P}_H = \mathbf{P}_A - \mathbf{P}_{AN}$, where $\text{sp}(N) = \text{sp}(B)^\perp$. See Exercise 27 in the next chapter.

Consider the system $A\boldsymbol{x} = \boldsymbol{b}$ of linear equations. There exists a solution $\boldsymbol{x}_0$ such that $A\boldsymbol{x}_0 = \boldsymbol{b}$ (the system is ***consistent***) iff $\boldsymbol{b} \in \text{sp}(A)$. From GS on A, then, the system is consistent iff $QQ'\boldsymbol{b} = \boldsymbol{b}$. If it is, then $A(TQ'\boldsymbol{b}) = QQ'\boldsymbol{b} = \boldsymbol{b}$, and so $\boldsymbol{x}_0 = TQ'\boldsymbol{b}$ is a solution. Further, with $N = \text{I} - TQ'A$, it can be shown that $\text{sp}(N) = \{\boldsymbol{x} : A\boldsymbol{x} = \boldsymbol{0}\} = \text{sp}(A')^\perp$, and hence that the solution set can be represented as $\{\boldsymbol{x} : A\boldsymbol{x} = \boldsymbol{b}\} = \{\boldsymbol{x}_0\} + \text{sp}(N)$. Any matrix M such that $\text{sp}(M) = \text{sp}(N) = \text{sp}(A')^\perp$ would work as well as N; the advantage of N is that it can be had directly from the results of GS on A.

6.3 Tools

As we cover the topics in this book, there will be many opportunities to undertake numerical examples. As you go through the material, many others will occur to you to try out. For such computations, I have found it convenient to write function subroutines that, given matrices A and B, return Q such that $Q'Q = \mathrm{I}$ and, respectively: $\mathrm{sp}(Q) = \mathrm{sp}(A)$; $\mathrm{sp}(Q) = \mathrm{sp}(A)^{\perp}$; $\mathrm{sp}(Q) = \mathrm{sp}(A) \cap \mathrm{sp}(B)$; and $\mathrm{sp}(Q) = \mathrm{sp}(A, B) \cap \mathrm{sp}(A)^{\perp}$. It is convenient also to have a function subroutine that returns $\mathbf{P}_A$ from A. These all can be based on a GS function that, given A, returns Q, T, and $\boldsymbol{j}$, as described above, such that $\mathrm{sp}(Q) = \mathrm{sp}(A)$, $Q'Q = \mathrm{I}$, $Q = AT$, and $\boldsymbol{j} = (j_i)$ listing the columns j_i of A that contributed column i of Q.

I urge you to create your own set of such subroutines in your favorite package (SAS IML, MATLAB, R), and to devise examples to illustrate each step described in this book. As a starter, use these tools to address exercises 1–13 in Section 4.2.

Exercises that establish further useful properties of the GS construction are included in the next chapter.

7

Further Results as Exercises

This chapter presents many useful results as exercises, with the expectation that the reader will work through all of their proofs and solutions. Some are mainly for drill and practice, while others establish results that are used and cited in the rest of the book. In particular, Exercises 31, 32, and 33 develop basic results and properties of linear least squares.

7.1 Exercises

Prove each assertion in the following exercises.

1. (a) If $\mathcal{S}$ and $\mathcal{T}$ are linear subspaces of $\Re^q$ and $\boldsymbol{s}_0$ and $\boldsymbol{t}_0$ are vectors in $\Re^q$, then $\{\boldsymbol{s}_0\} + \mathcal{S} = \{\boldsymbol{t}_0\} + \mathcal{T}$ iff $\mathcal{S} = \mathcal{T}$ and $\boldsymbol{s}_0 - \boldsymbol{t}_0 \in \mathcal{S}$.

   (b) If $\mathcal{S}_0$ and $\mathcal{S}$ are linear subspaces in $\Re^q$, then $\mathcal{S} = \mathcal{S}_0 + \mathcal{S} \cap \mathcal{S}_0^\perp$ iff $\mathcal{S}_0 \subset \mathcal{S}$.

2. Let G and G_* be matrices with r rows such that $\mathrm{sp}(G) = \mathrm{sp}(G_*)$, and let $\boldsymbol{b}_0$ be an r-vector. Then

$$\{\boldsymbol{b} \in \Re^r : G'\boldsymbol{b} = G'\boldsymbol{b}_0\} = \{\boldsymbol{b} \in \Re^r : G_*'\boldsymbol{b} = G_*'\boldsymbol{b}_0\}.$$

3. Most of the fundamental and consequential properties of orthogonal projection are inherent in the definition itself: If $\boldsymbol{y} \in \Re^n$ and $\mathcal{S}$ is a linear subspace of $\Re^n$, then $\hat{\boldsymbol{y}} \in \mathcal{S}$ is an orthogonal projection of $\boldsymbol{y}$ in $\mathcal{S}$ iff $\boldsymbol{y} - \hat{\boldsymbol{y}} \in \mathcal{S}^\perp$.

   Let $\boldsymbol{y} \in \Re^n$. Let $\mathcal{S}$ and $\mathcal{T}$ be linear subspaces of $\Re^n$.

   (a) If $\hat{\boldsymbol{y}}$ and $\tilde{\boldsymbol{y}}$ are both OPs of $\boldsymbol{y}$ in $\mathcal{S}$, then $\tilde{\boldsymbol{y}} = \hat{\boldsymbol{y}}$. That is, there is at most one orthogonal projection of $\boldsymbol{y}$ in $\mathcal{S}$.

   (b) The OP of $\boldsymbol{y}$ in $\mathcal{S}$ is $\boldsymbol{y}$ iff $\boldsymbol{y} \in \mathcal{S}$.

   (c) The OP of $\boldsymbol{y}$ in $\mathcal{S}$ is $\boldsymbol{0}$ iff $\boldsymbol{y} \in \mathcal{S}^\perp$.

   (d) $\hat{\boldsymbol{y}}$ is the OP of $\boldsymbol{y}$ in $\mathcal{S}$ iff $\boldsymbol{y} - \hat{\boldsymbol{y}}$ is the OP of $\boldsymbol{y}$ in $\mathcal{S}^\perp$.

(e) Let $\boldsymbol{y}_1$ and $\boldsymbol{y}_2$ be vectors in $\Re^n$, and let c_1 and c_2 be scalars. If $\hat{\boldsymbol{y}}_1$ and $\hat{\boldsymbol{y}}_2$ are the respective OPs of $\boldsymbol{y}_1$ and $\boldsymbol{y}_2$ in $\mathcal{S}$, then $c_1\hat{\boldsymbol{y}}_1 + c_2\hat{\boldsymbol{y}}_2$ is the orthogonal projection of $c_1\boldsymbol{y}_1 + c_2\boldsymbol{y}_2$ in $\mathcal{S}$.

(f) Suppose $\boldsymbol{y}$, $\hat{\boldsymbol{y}}_s$, $\hat{\boldsymbol{y}}_t$, $\hat{\boldsymbol{y}}_{st}$, $\hat{\boldsymbol{y}}_{ts}$, and $\hat{\boldsymbol{y}}_{s+t}$ are n-vectors such that

i. $\hat{\boldsymbol{y}}_s$ is the OP of $\boldsymbol{y}$ in $\mathcal{S}$,

ii. $\hat{\boldsymbol{y}}_t$ is the OP of $\boldsymbol{y}$ in $\mathcal{T}$,

iii. $\hat{\boldsymbol{y}}_{st}$ is the OP of $\hat{\boldsymbol{y}}_t$ in $\mathcal{S}$,

iv. $\hat{\boldsymbol{y}}_{ts}$ is the OP of $\hat{\boldsymbol{y}}_s$ in $\mathcal{T}$,

v. $\hat{\boldsymbol{y}}_{s+t}$ is the OP of $\boldsymbol{y}$ in $\mathcal{S} + \mathcal{T}$.

Then $\hat{\boldsymbol{y}}_{s+t} = \hat{\boldsymbol{y}}_s + \hat{\boldsymbol{y}}_t$ iff $\hat{\boldsymbol{y}}_{st} = \hat{\boldsymbol{y}}_{ts} = \mathbf{0}$.

4. Continuing Exercise 3: Let $\boldsymbol{e}_1, \ldots, \boldsymbol{e}_n$ denote the respective columns of I_n. For each $i = 1, \ldots, n$, let $\hat{\boldsymbol{e}}_i$ be the OP of $\boldsymbol{e}_i$ in $\mathcal{S}$. Let $C = (\hat{\boldsymbol{e}}_1, \ldots, \hat{\boldsymbol{e}}_n)$.

(a) For each $\boldsymbol{y} \in \Re^n$, $C\boldsymbol{y}$ is the orthogonal projection of $\boldsymbol{y}$ in $\mathcal{S}$.

(b) $\mathrm{sp}(C) = \mathcal{S}$.

(c) $C = C' = CC$.

(d) If M is a matrix such that, for each $\boldsymbol{y} \in \Re^n$, $M\boldsymbol{y}$ is the orthogonal projection of $\boldsymbol{y}$ in $\mathcal{S}$, then $M = C$.

(e) In order that each vector $\boldsymbol{y}$ in $\Re^n$ have an orthogonal projection in $\mathcal{S}$, it is necessary and sufficient that each $\boldsymbol{e}_i$, $i = 1, \ldots, n$, have an orthogonal projection in $\mathcal{S}$.

5. Suppose $\boldsymbol{w}_1, \ldots, \boldsymbol{w}_m$ is a spanning set for $\Re^n$. Let $\mathcal{S}$ be a linear subspace of $\Re^n$, and let $\hat{\boldsymbol{w}}_1, \ldots, \hat{\boldsymbol{w}}_m$ be the respective orthogonal projections of $\boldsymbol{w}_1, \ldots, \boldsymbol{w}_m$ in $\mathcal{S}$. Then $\mathrm{sp}(\hat{\boldsymbol{w}}_1, \ldots, \hat{\boldsymbol{w}}_m) = \mathcal{S}$.

6. Let $\mathcal{S}$ be a linear subspace of $\Re^n$. Suppose $\{\boldsymbol{q}_1, \ldots, \boldsymbol{q}_k\}$ is an orthonormal spanning set for $\mathcal{S}$. Let $Q = (\boldsymbol{q}_1, \ldots, \boldsymbol{q}_k)$, so that $\mathrm{sp}(Q) = \mathrm{sp}(\{\boldsymbol{q}_1, \ldots, \boldsymbol{q}_k\}) = \mathcal{S}$.

(a) $Q'Q = \mathrm{I}$.

(b) If $\boldsymbol{y} \in \Re^n$, then

$$\hat{\boldsymbol{y}} = (\boldsymbol{q}_1'\boldsymbol{y})\boldsymbol{q}_1 + \cdots + (\boldsymbol{q}_k'\boldsymbol{y})\boldsymbol{q}_k = QQ'\boldsymbol{y}$$

is the orthogonal projection of $\boldsymbol{y}$ in $\mathcal{S}$.

7. Another development of orthogonal projection is through generalized inverses. Let A be a non-zero $n \times c$ matrix. Suppose G is a generalized inverse of $A'A$, so that $(A'A)G(A'A) = A'A$.

(a) G' is also a generalized inverse of $A'A$. [Note that $(A'A)G(A'A)$ is symmetric.]

(b) $AGA'A = AG'A'A = A$.

(c) AGA' is symmetric. [Columns of $AGA' - AG'A'$ are in $\text{sp}(A)$ and they are in $\text{sp}(A)^{\perp}$.]

(d) AGA' is idempotent.

(e) $\text{sp}(AGA) = \text{sp}(A)$.

(f) If G_1 and G_2 are generalized inverses of $A'A$, then $AG_1A' = AG_2A'$.

(g) For each $\boldsymbol{y} \in \Re^n$, $(AGA')\boldsymbol{y}$ is the orthogonal projection of $\boldsymbol{y}$ in $\text{sp}(A)$.

8. Let $\mathcal{S}$ and $\mathcal{T}$ be linear subspaces of $\Re^n$, and let $\mathbf{P}_{\mathcal{S}}$ and $\mathbf{P}_{\mathcal{T}}$ denote their respective orthogonal projection matrices.

   (a) $\text{sp}(\mathbf{P}_{\mathcal{S}}) = \mathcal{S}$.

   (b) $\mathcal{S} = \mathcal{T}$ iff $\mathbf{P}_{\mathcal{S}} = \mathbf{P}_{\mathcal{T}}$.

9. Let $\mathcal{S}$ be a linear subspace of $\Re^n$, and let $A = (\boldsymbol{a}_1, \ldots, \boldsymbol{a}_m)$ be an $n \times m$ matrix such that $\mathcal{S} = \text{sp}(A)$. Suppose GS on A produces Q and T such that $Q'Q = \text{I}_\nu$, $\text{sp}(Q) = \mathcal{S}$, and $Q = AT$. Let $\boldsymbol{j} = (j_1, \ldots, j_\nu)$ be the list of indexes of columns of A that contribute columns to Q in the GS construction. Let $\boldsymbol{b}_i = \boldsymbol{a}_{j_i}$, $i = 1, \ldots, \nu$, and let $B = (\boldsymbol{b}_1, \ldots, \boldsymbol{b}_\nu)$.

   (a) $\mathbf{P}_A = QQ'$.

   (b) The columns of Q are linearly independent.

   (c) The columns of T are linearly independent.

   (d) The columns of B are linearly independent.

   (e) If A is symmetric and idempotent, then $A = QQ'$.

   (f) GS on B yields Q.

   (g) $\text{sp}(B) = \text{sp}(Q) = \mathcal{S}$: the columns of B comprise a basis for $\mathcal{S}$.

   (h) $\nu \leqslant m$.

   (i) $\nu \leqslant n$.

   (j) If $\boldsymbol{v}_1, \ldots, \boldsymbol{v}_\nu$ are linearly independent vectors in $\text{sp}(A)$, then $\text{sp}(\boldsymbol{v}_1, \ldots, \boldsymbol{v}_\nu) = \text{sp}(A)$.

10. Let $\mathcal{S}$ be a linear subspace of $\Re^n$.

    (a) Every basis of $\mathcal{S}$ has the same number of member vectors. Call this number ν.

    (b) Every linearly independent set of ν vectors in $\mathcal{S}$ spans $\mathcal{S}$ and hence is a basis for $\mathcal{S}$.

11. A square matrix has an inverse iff its columns are linearly independent.

12. The expressions that follow may be useful in the few instances where matrix inverses appear in this book.

    Suppose A, B, C, and D are matrices conformable to the operations in the following expressions. Show the following identities, assuming that the matrix inverses in them exist.

    (a)
    $$(A + BCD)^{-1} = A^{-1} - A^{-1}B(C^{-1} + DA^{-1}B)^{-1}DA^{-1}.$$

    (b)
    $$(\mathrm{I} + BD)^{-1} = \mathrm{I} - B(\mathrm{I} + DB)^{-1}D.$$

    (c)
    $$D(\mathrm{I} + BD)^{-1} = (\mathrm{I} + DB)^{-1}D.$$

    (d)
    $$(\mathrm{I} + BD)^{-1}B = B(\mathrm{I} + DB)^{-1}.$$

    (e) If $\mathrm{I}_r + BD$ is non-singular, then $|\mathrm{I}_r + BD| = |\mathrm{I}_c + DB|$, where r and c are the row and column dimensions of B.

    (f) If c is a scalar and $c > -1/n$, then
    $$(\mathrm{I}_n + c\mathbf{1}_n\mathbf{1}_n')^{-1} = \mathrm{I}_n - (1 + cn)^{-1}\mathbf{1}_n\mathbf{1}_n'.$$

13. Let A be an $n \times m$ matrix.

    (a) The columns of A are linearly independent iff the columns of $A'A$ are linearly independent.

    (b) If the columns of A are linearly independent, then $\mathbf{P}_A = A(A'A)^{-1}A'$.

14. For any matrix M, define $\nu_M = \mathrm{tr}(\mathbf{P}_M)$, that is, ν_M is the rank of M. Let A and B be matrices with b columns and b rows, respectively, so that the matrix product AB is defined.

    (a) $\mathbf{P}_A - \mathbf{P}_{AB}$ is the orthogonal projection matrix onto $\mathrm{sp}(A) \cap \mathrm{sp}(AB)^{\perp}$.

    (b) $\nu_A = \nu_{A'}$. That is, the column rank of A is the same as the row rank of A.

    (c) $\nu_{AB} \leqslant \nu_A$ and $\nu_{AB} \leqslant \nu_B$.

15. Let $A \neq 0_{n\times c}$ be an $n \times c$ matrix. Let Q and T be matrices such that $Q'Q = \mathrm{I}_\nu$, $\mathrm{sp}(Q) = \mathrm{sp}(A)$, and $Q = AT$. Let $\boldsymbol{b}$ be a vector in $\mathrm{sp}(A)$, let $N = \mathrm{I} - TQ'A$, and let $\boldsymbol{x}_* = TQ'\boldsymbol{b}$.

(a) $QQ'A = A$.

(b) $\mathbf{P}_A = QQ' = A(TT')A'$.

(c) TQ' is a reflexive generalized inverse of A, that is, $A(TQ')A = A$ and $(TQ')A(TQ') = TQ'$.

(d) TT' is a symmetric, reflexive generalized inverse of $A'A$.

(e) If columns of A are linearly independent, then $TQ' = (A'A)^{-1}A'$.

(f) If A is square and non-singular, then $A^{-1} = TQ'$.

(g) $\text{sp}(N) = \text{sp}(A')^{\perp}$.

(h) $A\boldsymbol{x}_* = \boldsymbol{b}$.

(i) $\{\boldsymbol{x} \in \Re^m : A\boldsymbol{x} = \boldsymbol{b}\} = \{\boldsymbol{x}_*\} + \text{sp}(N)$.

16. Suppose P is an $n \times n$ symmetric idempotent matrix. Show that P is the orthogonal projection matrix onto $\text{sp}(P)$ and that the dimension of $\text{sp}(P)$ is $\nu = \text{tr}(P)$.

17. Suppose $A = (A_1, A_2)$ is a matrix, and its columns are partitioned so that A_1 is non-zero and $\text{sp}(A_2)$ is not contained in $\text{sp}(A_1)$.

    Let $Q = (Q_1, Q_2)$ be a matrix such that $Q'Q = \text{I}$, $\text{sp}(Q) = \text{sp}(A)$, and $\text{sp}(Q_1) = \text{sp}(A_1)$.

    Show that $\text{sp}(Q_2) = \text{sp}(A) \cap \text{sp}(A_1)^{\perp}$.

18. Let A be a matrix with n rows. Then

    (1) M is $n \times n$, $\text{sp}(M) \subset \text{sp}(A)$, and $\text{sp}(\text{I} - M) \subset \text{sp}(A)^{\perp}$

    is equivalent to

    (2) $M = M' = MM$, $\text{sp}(M) = \text{sp}(A)$, and $\text{sp}(\text{I} - M) = \text{sp}(A)^{\perp}$.

19. Let $\mathcal{S}$ be a linear subspace of $\Re^n$. Then the following three statements are equivalent.

    (1) For each $\boldsymbol{y} \in \Re^n$ there exists a vector $\hat{\boldsymbol{y}} \in \mathcal{S}$ such that $\boldsymbol{y} - \hat{\boldsymbol{y}} \in \mathcal{S}^{\perp}$.

    (2) There exists a matrix M such that $\text{sp}(M) \subset \mathcal{S}$ and $\text{sp}(\text{I} - M) \subset \mathcal{S}^{\perp}$.

    (3) There exists a matrix Q such that $\text{sp}(Q) = \mathcal{S}$ and $Q'Q = \text{I}$.

20. (a) If $\mathcal{A}$ and $\mathcal{B}$ are linear subspaces of $\Re^n$, then $\mathbf{P}_{\mathcal{A}+\mathcal{B}} = \mathbf{P}_{\mathcal{A}} + \mathbf{P}_{\mathcal{B}}$ iff $\mathbf{P}_{\mathcal{A}}\mathbf{P}_{\mathcal{B}} = 0$, that is, iff $\mathcal{A} \subset \mathcal{B}^{\perp}$.

    (b) If A and B are matrices with n rows, $\mathbf{P}_{(A,B)} = \mathbf{P}_A + \mathbf{P}_B$ iff $A'B = 0$.

21. Let $\mathcal{S}$ and $\mathcal{S}_0$ be linear subspaces of $\Re^n$ such that $\mathcal{S}_0 \subset \mathcal{S}$. Prove the following statements.

    (a) $\mathbf{P}_{\mathcal{S}}\mathbf{P}_{\mathcal{S}_0} = \mathbf{P}_{\mathcal{S}_0} = \mathbf{P}_{\mathcal{S}_0}\mathbf{P}_{\mathcal{S}}$.

    (b) $\mathbf{P}_{\mathcal{S}} - \mathbf{P}_{\mathcal{S}_0}$ is symmetric and idempotent.

    (c) $\mathbf{P}_{\mathcal{S} \cap \mathcal{S}_0^{\perp}} = \mathbf{P}_{\mathcal{S}} - \mathbf{P}_{\mathcal{S}_0}$.

(d) If $\boldsymbol{z} \in \mathcal{S}$, then $\boldsymbol{z} \in \mathcal{S}_0$ iff $(\mathbf{P}_{\mathcal{S}} - \mathbf{P}_{\mathcal{S}_0})\boldsymbol{z} = \mathbf{0}$.

(e) $\mathbf{P}_{\mathcal{S}_0^\perp} = \mathbf{P}_{\mathcal{S}^\perp} + \mathbf{P}_{\mathcal{S}_0^\perp \cap \mathcal{S}}$.

(f) For $\boldsymbol{y} \in \Re^n$, let $\hat{\boldsymbol{y}}_{\mathcal{S}_0} = \mathbf{P}_{\mathcal{S}_0}\boldsymbol{y}$ and $\hat{\boldsymbol{y}}_{\mathcal{S}} = \mathbf{P}_{\mathcal{S}}\boldsymbol{y}$. Show that

$$(\boldsymbol{y} - \hat{\boldsymbol{y}}_{\mathcal{S}_0})'(\hat{\boldsymbol{y}} - \hat{\boldsymbol{y}}_{\mathcal{S}_0}) = (\boldsymbol{y} - \hat{\boldsymbol{y}}_{\mathcal{S}})'(\boldsymbol{y} - \hat{\boldsymbol{y}}_{\mathcal{S}}) + (\hat{\boldsymbol{y}}_{\mathcal{S}} - \hat{\boldsymbol{y}}_{\mathcal{S}_0})'(\hat{\boldsymbol{y}}_{\mathcal{S}} - \hat{\boldsymbol{y}}_{\mathcal{S}_0}).$$

22. Let A and B be matrices with q columns. Prove the following propositions.

(a) If $\boldsymbol{x}$ is a vector such that $A\boldsymbol{x}$ is in $\{A\boldsymbol{x} : B\boldsymbol{x} = \mathbf{0}\}$, then $\boldsymbol{x} \in \text{sp}(A')^\perp + \text{sp}(B')^\perp$.

(b) If N is a matrix such that $[\text{sp}(A') \cap \text{sp}(B')]^\perp \supset \text{sp}(N) \supset \text{sp}(B')^\perp$, then $\text{sp}(AN) = \{A\boldsymbol{x} : B\boldsymbol{x} = \mathbf{0}\}$.

(c) If $\mathcal{S}$ is a linear subspace of $\Re^q$, then $\{A\boldsymbol{x} : B\boldsymbol{x} = \mathbf{0}\} = \{A\boldsymbol{x} : \boldsymbol{x} \in \mathcal{S}\}$ iff $\text{sp}(B')^\perp \subset \mathcal{S} \subset \text{sp}(A')^\perp + \text{sp}(B')^\perp$.

23. If A, B, and C are matrices with q columns, then
(I) $\text{sp}(A') \cap \text{sp}(B') = \text{sp}(A') \cap \text{sp}(C')$
iff
(II) $\{A\boldsymbol{x} : B\boldsymbol{x} = \mathbf{0}\} = \{A\boldsymbol{x} : C\boldsymbol{x} = \mathbf{0}\}$

24. Let A and B be matrices, both with c columns. Prove:

$$\mathbf{P}_{\binom{A}{B}} = \begin{pmatrix} \mathbf{P}_A & 0 \\ 0 & \mathbf{P}_B \end{pmatrix}$$

if and only if $\text{sp}(A') \cap \text{sp}(B') = \{\mathbf{0}\}$.

25. Let A and B be matrices with c columns. Each assertion below holds also with A and B switched.

(a)

$$\text{sp}(A') \cap \text{sp}(B') = \text{sp}(A') \cap \{\text{sp}[(\mathrm{I} - \mathbf{P}_{B'})A']\}^\perp. \tag{7.1}$$

It follows that

$$\text{sp}(A')^\perp + \text{sp}(B')^\perp = \text{sp}(A')^\perp + \text{sp}[(\mathrm{I} - \mathbf{P}_{B'})A']. \tag{7.2}$$

(b) The sum on the right is direct, that is,

$$\text{sp}(A')^\perp \cap \text{sp}[(\mathrm{I} - \mathbf{P}_{B'})A'] = \{\mathbf{0}\}.$$

(c)

$$\{A\boldsymbol{x} : \boldsymbol{x} \in \Re^c \text{ and } B\boldsymbol{x} = \mathbf{0}\} = \{A\boldsymbol{x} : \boldsymbol{x} \in \text{sp}[(\mathrm{I} - \mathbf{P}_{B'})A']\}.$$

Equivalently,

$$A\{\text{sp}(B')^\perp\} = A\{\text{sp}[(\mathrm{I} - \mathbf{P}_{B'})A']\}.$$

26. Let A and B be matrices with c columns.

    (a) $\text{sp}(A')^{\perp} \subset \text{sp}(B')^{\perp}$ iff $\text{sp}[(\mathrm{I} - \mathbf{P}_{A'})B'] = \{\mathbf{0}\}$.

    (b) There exists $\boldsymbol{z}$ such that $A\boldsymbol{z} = \mathbf{0}$ and $B\boldsymbol{z} \neq \mathbf{0}$ iff $\text{sp}(A')^{\perp} \not\subset \text{sp}(B')^{\perp}$.

    (c) Let $\boldsymbol{z} \in \text{sp}[(\mathrm{I} - \mathbf{P}_{A'})B']$, so that $A\boldsymbol{z} = \mathbf{0}$. Then $B\boldsymbol{z} = \mathbf{0}$ iff $\boldsymbol{z} = \mathbf{0}$.

    (d) Prove: $AB^{-}B = A$ iff $\text{sp}(A') \subset \text{sp}(B')$.

    (e) Prove: $\text{sp}(A) + \text{sp}(B) = \text{sp}(A) + \text{sp}[(\mathrm{I} - \mathbf{P}_{A'})B]$.

27. Let A and B be matrices with c columns. Let N be a matrix such that $\text{sp}(B')^{\perp} \subset \text{sp}(N) \subset [\text{sp}(A') \cap \text{sp}(B')]^{\perp}$. Let H be a matrix such that $\mathbf{P}_H = \mathbf{P}_A - \mathbf{P}_{AN}$. Let $A|H = (\mathrm{I} - \mathbf{P}_H)A$. Prove the following propositions.

    (a) $\text{sp}(H) \subset \text{sp}(A)$ and $\text{sp}(H, A|H) = \text{sp}(A)$.

    (b)
$$\begin{aligned}\{A\boldsymbol{x} : B\boldsymbol{x} = \mathbf{0}\} &= \text{sp}(A) \cap \text{sp}(H)^{\perp} \\ &= \{A\boldsymbol{x} : H'A\boldsymbol{x} = \mathbf{0}\} \\ &= \text{sp}(A|H).\end{aligned}$$

    (c) $\text{sp}(A'H) = \text{sp}(A') \cap \text{sp}(B')$.

    (d) $\{A\boldsymbol{x} : B\boldsymbol{x} = \mathbf{0}\} = \text{sp}(A)$ iff $\text{sp}(A') \cap \text{sp}(B') = \{\mathbf{0}\}$.

    (e) $\{A\boldsymbol{x} : B\boldsymbol{x} = \mathbf{0}\} = \{\mathbf{0}\}$ iff $\text{sp}(A') \subset \text{sp}(B')$.

    (f) $\mathbf{P}_{A|H} = \mathbf{P}_A - \mathbf{P}_H = \mathbf{P}_{AN}$.

    (g) $\text{sp}[(A|H)'] \cap \text{sp}(A'H) = \{\mathbf{0}\}$.

    (h) $\text{sp}[(A|H)'] \cap \text{sp}(B') = \{\mathbf{0}\}$.

    (i) $\text{sp}[(A|H)'] \cap \text{sp}(A')^{\perp} = \{\mathbf{0}\}$.

    (j) From these results, show that
$$\mathbf{P}_{\binom{A|H}{B}}\begin{pmatrix}\boldsymbol{z}\\ \mathbf{0}\end{pmatrix} = \begin{pmatrix}\mathbf{P}_{A|H}\boldsymbol{z}\\ \mathbf{0}\end{pmatrix} = \begin{pmatrix}(A|H)\boldsymbol{x}\\ B\boldsymbol{x}\end{pmatrix} = \begin{pmatrix}A\boldsymbol{x}\\ B\boldsymbol{x} = \mathbf{0}\end{pmatrix}.$$
    This specifies a vector $\boldsymbol{x}$ such that $\mathbf{P}_{A|H}\boldsymbol{z} = (A|H)\boldsymbol{x}$ with $B\boldsymbol{x} = \mathbf{0}$. In addition, $B\boldsymbol{x} = \mathbf{0} \Longrightarrow \mathbf{P}_H A\boldsymbol{x} = \mathbf{0}$ and hence that $\mathbf{P}_{A|H}\boldsymbol{z} = A\boldsymbol{x}$ with $B\boldsymbol{x} = \mathbf{0}$.

28. Let A be a matrix, and let L and H be matrices such that $\text{sp}(H) \subset \text{sp}(A)$, $\text{sp}(L) \subset \text{sp}(A)$, and $\text{sp}(A'H) = \text{sp}(A'L)$. Then $\text{sp}(L) = \text{sp}(H)$.

29. Suppose G and G_* have r rows, $\text{sp}(G) = \text{sp}(G_*)$, and $\boldsymbol{b}_0 \in \Re^r$. Then $\{\boldsymbol{b} : G'\boldsymbol{b} = G'\boldsymbol{b}_0\} = \{\boldsymbol{b} : G_*'\boldsymbol{b} = G_*'\boldsymbol{b}_0\}$.

30. **Kronecker Product.** The Kronecker product of matrices A and B, denoted $A \otimes B$, is the matrix formed by replacing each entry a_{ij} of A by the matrix $a_{ij}B$. For the propositions that follow, assume that the matrices have dimensions such that the operations are defined. Prove the following propositions.

    (a) $(A \otimes B)(C \otimes D) = (AC) \otimes (BD)$.
    (b) $A \otimes (B + C) = A \otimes B + A \otimes C$.
    (c) $(A + B) \otimes C = A \otimes C + B \otimes C$.
    (d) $(A \otimes B)' = A' \otimes B'$.
    (e) $\mathbf{P}_{A \otimes B} = \mathbf{P}_A \otimes \mathbf{P}_B$.
    (f) $\text{tr}(A \otimes B) = \text{tr}(A)\text{tr}(B)$.
    (g) $(A, B) \otimes C = (A \otimes C, B \otimes C)$.
    (h) $\text{sp}[C \otimes (A, B)] = \text{sp}(C \otimes A, C \otimes B)$.

31. **Linear Least Squares.** Let X be an $n \times (k+1)$ matrix, and let $\mathcal{S} = \text{sp}(X)$ be the linear subspace of $\Re^n$ spanned by the columns of X. Let $\boldsymbol{y}$ be an n-vector. For each $\boldsymbol{y} \in \Re^n$, let $\hat{\boldsymbol{y}}$ be the orthogonal projection of $\boldsymbol{y}$ in $\mathcal{S}$, that is, $\hat{\boldsymbol{y}} = \mathbf{P}_X \boldsymbol{y}$.

    (a) Show that $\hat{\boldsymbol{y}}$ is the unique vector in $\mathcal{S}$ that, among vectors in $\mathcal{S}$, is closest to $\boldsymbol{y}$. That is, show that for any $\boldsymbol{s} \in \mathcal{S}$,

    $$(\boldsymbol{y} - \boldsymbol{s})'(\boldsymbol{y} - \boldsymbol{s}) \geqslant (\boldsymbol{y} - \hat{\boldsymbol{y}})'(\boldsymbol{y} - \hat{\boldsymbol{y}}),$$

    with equality iff $\boldsymbol{s} = \hat{\boldsymbol{y}}$.
    [Prove this using only the defining properties of orthogonal projection, that $\hat{\boldsymbol{y}} \in \mathcal{S}$ and $\boldsymbol{y} - \hat{\boldsymbol{y}} \in \mathcal{S}^{\perp}$. Note that, for any $\boldsymbol{s} \in \mathcal{S}$, $\boldsymbol{s} - \hat{\boldsymbol{y}}$ is in $\mathcal{S}$: Why?]

    (b) The squared distance from $\boldsymbol{y}$ to the closest vector in $\mathcal{S}$ is $(\boldsymbol{y} - \hat{\boldsymbol{y}})'(\boldsymbol{y} - \hat{\boldsymbol{y}})$. Show that

    $$(\boldsymbol{y} - \hat{\boldsymbol{y}})'(\boldsymbol{y} - \hat{\boldsymbol{y}}) = \boldsymbol{y}'(\mathrm{I} - \mathbf{P}_X)\boldsymbol{y}. \tag{7.3}$$

    This sum of squares is called ***Error Sum of Squares*** and denoted SSE. That is, for a response $\boldsymbol{y}$ and the model $\mathrm{E}(\boldsymbol{Y}) = X\boldsymbol{\beta}$, $SSE = \boldsymbol{y}'(\mathrm{I} - \mathbf{P}_X)\boldsymbol{y}$.

    (c) Let $\check{\boldsymbol{b}}(\boldsymbol{y})$ denote a $(k + 1)$-vector-valued function of $\boldsymbol{y}$. (Read $\check{\boldsymbol{b}}$ as "b-check.") $\check{\boldsymbol{b}}(\boldsymbol{y})$ is said to be a ***least-squares solution*** iff, for each $\boldsymbol{y} \in \Re^n$, it minimizes (with respect to $\boldsymbol{b}$)

    $$\sum_{i=1}^{n}[y_i - (b_0 + b_1 x_{1i} + \cdots + b_k x_{ki})]^2 = (\boldsymbol{y} - X\boldsymbol{b})'(\boldsymbol{y} - X\boldsymbol{b}).$$

    For brevity, denote $\check{\boldsymbol{b}}(\boldsymbol{y})$ by $\check{\boldsymbol{b}}$ with the understanding that it is a function of $\boldsymbol{y}$.

i. Prove that $\check{\boldsymbol{b}}$ is a least-squares solution iff, for each $\boldsymbol{y} \in \Re^n$, $X\check{\boldsymbol{b}} = \hat{\boldsymbol{y}}$.

ii. Prove that $\check{\boldsymbol{b}}$ is a least-squares solution iff, for each $\boldsymbol{y} \in \Re^n$,

$$X'(\boldsymbol{y} - X\check{\boldsymbol{b}}) = \mathbf{0}.$$

Consequently, $\check{\boldsymbol{b}}$ is a least-squares solution iff, for each $\boldsymbol{y} \in \Re^n$,

$$X'X\check{\boldsymbol{b}} = X'\boldsymbol{y}. \tag{7.4}$$

Equation (7.4) is called the ***normal equation***. As a system of linear equations in $\boldsymbol{b}$, i.e., $X'X\boldsymbol{b} = X'\boldsymbol{y}$, existence of a solution $\check{\boldsymbol{b}}$ is equivalent to the existence of the orthogonal projection $\hat{\boldsymbol{y}}$ of $\boldsymbol{y}$ in $\text{sp}(X)$.

Note further that existence of a solution to the normal equation (for $\boldsymbol{b}$ given any $\boldsymbol{y}$, and for $\boldsymbol{y}$ given any $\boldsymbol{b}$) is equivalent to the relation $\text{sp}(X') = \text{sp}(X'X)$, which was noted as a consequence of Proposition 5.6.

iii. Let Q and T be matrices such that $Q'Q = \text{I}_\nu$, $\text{sp}(Q) = \text{sp}(X)$, and $Q = XT$, as from GS on X. Show that $\boldsymbol{b} = TQ'\boldsymbol{y}$ is a least-squares solution.

iv. Let $(X'X)^-$ be a generalized inverse of $X'X$. Show that $\boldsymbol{b} = (X'X)^-X'\boldsymbol{y}$ is a least-squares solution.

v. Let A be an $n \times (k+1)$ matrix. Prove that $A'\boldsymbol{y}$ is a least-squares solution iff $XA' = \mathbf{P}_X$. That is, all least-squares solutions that are linear in $\boldsymbol{y}$ take the form $A'\boldsymbol{y}$ with A such that $XA' = \mathbf{P}_X$.

vi. Prove: $\boldsymbol{g}'\check{\boldsymbol{b}}(\boldsymbol{y})$ is the same for all least-squares solutions $\check{\boldsymbol{b}}(\boldsymbol{y})$ iff $\boldsymbol{g} \in \text{sp}(X')$. ["If" is clear. For "only if," show that $\boldsymbol{g} \notin \text{sp}(X')$ implies that there exists a $\boldsymbol{z}$ such that $\boldsymbol{g}'\boldsymbol{z} \neq 0$ and, if $\check{\boldsymbol{b}}(\boldsymbol{y})$ is a least-squares solution, then $\check{\boldsymbol{b}}(\boldsymbol{y}) + \boldsymbol{z}$ is too.]

32. **Affine Least Squares.** Let X be an $n \times (k+1)$ matrix, and let $\boldsymbol{y}$ and $\boldsymbol{m}_0$ be n-vectors.

Prove: For any $\boldsymbol{m} \in \{\boldsymbol{m}_0\} + \text{sp}(X)$,

$$(\boldsymbol{y} - \boldsymbol{m})'(\boldsymbol{y} - \boldsymbol{m}) \;\geqslant\; (\boldsymbol{y} - \boldsymbol{m}_0)'(\text{I} - \mathbf{P}_X)(\boldsymbol{y} - \boldsymbol{m}_0),$$

and equality holds iff $\boldsymbol{m} = \boldsymbol{m}_0 + \mathbf{P}_X(\boldsymbol{y} - \boldsymbol{m}_0)$.

33. **Restricted Least Squares.** Let X be an $n \times (k+1)$ matrix. Let R be a $(k+1) \times r$ matrix, and let $\boldsymbol{r}_0$ be a vector in $\text{sp}(R')$. Let

$$\mathcal{M} = \{X\boldsymbol{b} : \boldsymbol{b} \in \Re^{(k+1)} \text{ and } R'\boldsymbol{b} = \boldsymbol{r}_0\}.$$

Let N be a matrix such that $[\text{sp}(R) \cap \text{sp}(X')]^\perp \supset \text{sp}(N) \supset \text{sp}(R)^\perp$. Let $\boldsymbol{b}_0$ be a vector such that $R'\boldsymbol{b}_0 = \boldsymbol{r}_0$.

Let $\boldsymbol{y}$ be an n-vector. A ***Restricted Least Squares*** (RLS) solution is a vector $\tilde{\boldsymbol{b}}$ such that $X\tilde{\boldsymbol{b}}$ minimizes $(\boldsymbol{y} - X\boldsymbol{b})'(\boldsymbol{y} - X\boldsymbol{b})$ over $\mathcal{M}$.

(a) Show that $\mathcal{M} = \{X\boldsymbol{b}_0\} + \mathrm{sp}(XN)$.

(b) Prove that $\tilde{\boldsymbol{b}}$ is an RLS solution iff

$$X\tilde{\boldsymbol{b}}(\boldsymbol{y}) = X\boldsymbol{b}_0 + \mathbf{P}_{XN}(\boldsymbol{y} - X\boldsymbol{b}_0)$$

for all $\boldsymbol{y} \in \Re^n$.

(c) Describe computational steps to find an RLS solution in this setting.

(d) Show that

$$(\boldsymbol{y} - X\tilde{\boldsymbol{b}})'(\boldsymbol{y} - X\tilde{\boldsymbol{b}}) = (\boldsymbol{y} - X\boldsymbol{b}_0)'(\mathrm{I} - \mathbf{P}_{XN})(\boldsymbol{y} - X\boldsymbol{b}_0).$$

Part II

Inference

8

Linear Models, Least Squares, and the Gauss-Markov Theorem

This chapter formulates the linear models of mean vectors that are the main topic of this book. It shows that they can be transformed into multiple regression models, which in turn are linear subspaces. They are widely known as Gauss-Markov models.

Least-squares estimators (LSEs) of the mean vectors in these models are orthogonal projections of the response vector in the model space. The Gauss-Markov Theorem establishes that LSEs of linear functions of the mean vector have minimum variance (are Best) among linear unbiased estimators (LUEs) of their expected values: *LSEs are BLUEs.*

8.1 Linear Models

We shall be dealing with methods of statistical inference in the following setting. Real-valued responses $y_1, \ldots, y_n$ are observed from n subjects (or sampling units or experimental units). Denote the corresponding random variables by $Y_1, \ldots, Y_n$, that is, $Y_1, \ldots, Y_n$ are random variables and $y_1, \ldots, y_n$ are their realized values.

It will be assumed that the n subjects are sampled in such a way that the covariances between pairs of random variables Y_i and Y_j are all zero. For example, this is accomplished if subjects are sampled independently from some population. It is assumed further that the variances of all the responses are the same, that is, that $\text{Var}(Y_i) = \sigma^2$, $i = 1, \ldots, n$, where σ^2 is some positive real number. Variances are then said to be ***homoscedastic*** or ***homogeneous***. Let $\boldsymbol{Y} = (Y_1, \ldots, Y_n)'$ and $\boldsymbol{y} = (y_1, \ldots, y_n)'$. Then, with these assumptions, the variance-covariance matrix of the n-variate random variable $\boldsymbol{Y}$ is $\text{Var}(\boldsymbol{Y}) = \sigma^2 \text{I}_n$.

Denote the expected value of Y_i by $\mu_i = \text{E}(Y_i)$, and let $\boldsymbol{\mu} = \text{E}(\boldsymbol{Y}) = (\text{E}(Y_1), \ldots, \text{E}(Y_n))' = (\mu_1, \ldots, \mu_n)'$ denote the mean vector of $\boldsymbol{Y}$. We shall be concerned here with statistical inference about linear functions of the mean vector in models for $\boldsymbol{\mu}$.

A ***model*** for the mean vector, as the word is used here, is the set of potential mean vectors under consideration. While this set could be practically any subset of vectors in $\Re^n$, it can be argued that such sets and their affine closures are essentially equivalent in terms of methods of linear statistical inference, and so there is little loss of generality by restricting attention to models that are affine sets.

All the models that we shall deal with here are affine sets in $\Re^n$, sets that take the general form $\mathcal{A} = \boldsymbol{a}_0 + \mathcal{S}$. In such formulations, $\boldsymbol{a}_0$ is an n-vector, sometimes called an ***offset***, and $\mathcal{S}$ is a linear subspace of $\Re^n$. In any specific setting, both $\boldsymbol{a}_0$ and $\mathcal{S}$ are known. Transforming $\mathcal{A}$ to $\mathcal{S} = \mathcal{A} - \boldsymbol{a}_0$, there is very little loss of generality in considering only models that are linear subspaces.

For ***multiple regression models***, the model is specified as the set of all linear combinations of columns of a given, known matrix X, so that it takes the form

$$\{\boldsymbol{m} = X\boldsymbol{b} : \boldsymbol{b} \in \Re^{k+1}\} = \text{sp}(X), \tag{8.1}$$

which is a linear subspace. Any model $\mathcal{S}$ that is a linear subspace can be considered to be a multiple regression model with columns of X comprising a set of vectors that span $\mathcal{S}$.

Keep in mind the general framework here. The model for $\boldsymbol{\mu}$ is formulated in the context of the problem at hand, as in the examples in Chapter 2, producing the model matrix X, which is then fixed and known.

The general object of inference is to assess whether, and how, responses are affected by the different conditions that are quantified in the different rows of X. Under the models assumed here, effects can show up only as differences among the population means. In that case, the relevant questions are about the mean vector $\boldsymbol{\mu}$ and relations among its components. The assumption that $\boldsymbol{\mu}$ is a member of the model sp(X) posits that there exists a vector $\boldsymbol{\beta}$ such that $\boldsymbol{\mu} = X\boldsymbol{\beta}$. In this model, $\boldsymbol{\beta}$ is an unknown parameter, and, given X, inference about $\boldsymbol{\mu}$ is equivalent to inference about $\boldsymbol{\beta}$.

The functions of $\boldsymbol{\mu}$ addressed here are linear functions, taking the form $G'\boldsymbol{\beta}$. The questions take the form of propositions, like $G'\boldsymbol{\beta} = \boldsymbol{c}_0$, with G and $\boldsymbol{c}_0$ fixed and known. The objective is to assess such propositions in light of the responses in $\boldsymbol{y}$. The only real fact here is $\boldsymbol{y}$.

In light of the facts, $\boldsymbol{y}$, the objective of inference in the model $X\boldsymbol{\beta}$ is to assess whether it is unreasonable to think that $G'\boldsymbol{\beta}$ might be equal to $\boldsymbol{c}_0$. Here, that assessment entails choosing a ***test statistic*** $F_P(\boldsymbol{y})$ and computing a ***p-value*** as the probability of a greater value if the proposition were true.

The models for $\boldsymbol{\mu}$ that we shall deal with can take other forms. As we shall see, though, they are essentially the same as (8.1). Most generally, they may incorporate linear restrictions $R'\boldsymbol{b} = \boldsymbol{r}_0$ and an offset $\boldsymbol{m}_0$, so that they take the form

$$\mathcal{M} = \{\boldsymbol{m}_0 + X\boldsymbol{b} : \boldsymbol{b} \in \Re^{k+1} \text{ and } R'\boldsymbol{b} = \boldsymbol{r}_0\}. \tag{8.2}$$

X $(n \times (k+1))$, $\boldsymbol{m}_0$ (an n-vector), R $((k+1) \times r)$, and $\boldsymbol{r}_0$ (an r-vector) are fixed and known, specific to the particular setting at hand. Let M be a matrix such that $\text{sp}(M) = \text{sp}(R)^{\perp}$ and let $\boldsymbol{b}_0$ be a vector such that $R'\boldsymbol{b}_0 = \boldsymbol{r}_0$. Then

$$\begin{aligned} \mathcal{M} - \{\boldsymbol{m}_0 + X\boldsymbol{b}_0\} &= \{X\boldsymbol{b} : \boldsymbol{b} \in \Re^{k+1} \text{ and } R'\boldsymbol{b} = \boldsymbol{0}\} \\ &= \text{sp}(XM), \end{aligned}$$

which is a multiple regression model. Translating between $\mathcal{M}$ and $\text{sp}(XM)$ requires only adding or subtracting the known vector $\boldsymbol{m}_0 + X\boldsymbol{b}_0$. Thus, if the model for the mean vector of $\boldsymbol{Y}$ is $\mathcal{M}$, then the model for the mean vector of $\boldsymbol{Y} - (\boldsymbol{m}_0 + X\boldsymbol{b}_0)$ is a multiple regression model.

There is a considerable range of possible models in the form of (8.2). For example, (8.1) can be viewed as (8.2) with $\boldsymbol{m}_0 = \boldsymbol{0}$, $R = 0$, and $\boldsymbol{r}_0 = \boldsymbol{0}$. At the other extreme, the model may be specified entirely by linear restrictions as $\{\boldsymbol{m} \in \Re^n : R'\boldsymbol{m} = \boldsymbol{r}_0\} = \boldsymbol{m}_0 + \text{sp}(M)$, with $\boldsymbol{m}_0$ such that $R'\boldsymbol{m}_0 = \boldsymbol{r}_0$. In this case, the model for the mean vector of $\boldsymbol{Y} - \boldsymbol{m}_0$ is $\text{sp}(M)$, which is a multiple regression model.

The multiple regression model is the model for the mean vector in what is widely termed the ***Gauss-Markov*** (GM) model. The model is denoted as $\boldsymbol{Y} \sim (X\boldsymbol{\beta}, \sigma^2\text{I})$, $\boldsymbol{\beta} \in \Re^{k+1}$, $\sigma^2 > 0$. This signifies that the random n-vector $\boldsymbol{Y}$ has a mean vector $\boldsymbol{\mu}$ in $\text{sp}(X)$, and its variance-covariance matrix is $\sigma^2\text{I}$ for some positive number σ^2. Under this model, procedures most widely used for inference about the mean vector have known distributional properties if $\boldsymbol{Y}$ follows a multivariate normal distribution, that is, if $\boldsymbol{Y} \sim \mathbf{N}(X\boldsymbol{\beta}, \sigma^2\text{I})$.

It would not be reasonable to assert, in practically any real application, that the response actually follows a multivariate normal distribution. However, due to well-established asymptotic results, like several versions of the Central Limit Theorem, it is not unreasonable to expect that normal-based inferential procedures have the probabilistic properties that they would have under normality, to a reasonable approximation. See Casella and Berger (2002), for example.

8.2 Least Squares and Least-Squares Solutions

The ***method of least squares*** (***LS***) identifies the vector in the model $\text{sp}(X)$ that fits the response $\boldsymbol{y}$ best in the sense of LS, that it minimizes the ***LS criterion*** $(\boldsymbol{y} - \boldsymbol{m})'(\boldsymbol{y} - \boldsymbol{m})$ over all vectors $\boldsymbol{m}$ in $\text{sp}(X)$. Properties of LS in multiple regression models are established in Exercises 31, 32, and 33 of Chapter 7.

The ***LS fit*** to $\boldsymbol{y}$ in the model $X\boldsymbol{\beta}$ is the orthogonal projection $\hat{\boldsymbol{y}} = \mathbf{P}_X\boldsymbol{y}$ of $\boldsymbol{y}$ on $\text{sp}(X)$. For it, the LS criterion has the value

$$SSE = (\boldsymbol{y} - \hat{\boldsymbol{y}})'(\boldsymbol{y} - \hat{\boldsymbol{y}}) = \boldsymbol{y}'(\text{I} - \mathbf{P}_X)\boldsymbol{y}. \tag{8.3}$$

See Equation (7.3), p. 54.

A function $\check{\boldsymbol{b}}(\boldsymbol{y})$ such that $X\check{\boldsymbol{b}}(\boldsymbol{y}) = \mathbf{P}_X \boldsymbol{y}$ for all $\boldsymbol{y} \in \Re^n$ is called a ***least-squares solution.*** Each satisfies the ***normal equation*** $X'X\check{\boldsymbol{b}}(\boldsymbol{y}) = X'\boldsymbol{y}$ for each $\boldsymbol{y} \in \Re^n$. The normal equation is equivalent to the definition of the orthogonal projection onto sp(X): $X\check{\boldsymbol{b}}(\boldsymbol{y}) \in \text{sp}(X)$ and $\boldsymbol{y} - X\check{\boldsymbol{b}}(\boldsymbol{y}) \in \text{sp}(X)^{\perp}$.

8.3 Linear Statistics and Unbiased Linear Estimators

The theory and practice of linear statistical models are built mainly on properties of the GM model, either $\boldsymbol{Y} \sim (X\boldsymbol{\beta}, \sigma^2\mathrm{I})$ or $\boldsymbol{Y} \sim \mathbf{N}(X\boldsymbol{\beta}, \sigma^2\mathrm{I})$. In this setting, the targets of inference are most often linear functions of $\boldsymbol{\beta}$, and methods for that purpose are built mostly on linear statistics and sums of squares related to them.

A univariate (real-valued) linear statistic takes the form $\boldsymbol{\ell}'\boldsymbol{Y}$, where $\boldsymbol{\ell}$ is a fixed, given vector. Multivariate linear statistics take the form $L'\boldsymbol{Y}$, where L is a fixed, given matrix.

In GM models the expected value and variance of a univariate linear statistic $\boldsymbol{\ell}'\boldsymbol{Y}$ are

$$\begin{aligned} \mathrm{E}(\boldsymbol{\ell}'\boldsymbol{Y}) &= \boldsymbol{\ell}'X\boldsymbol{\beta} \text{ and} \\ \mathrm{Var}(\boldsymbol{\ell}'\boldsymbol{Y}) &= \sigma^2\boldsymbol{\ell}'\boldsymbol{\ell}. \end{aligned} \tag{8.4}$$

The expected value and variance-covariance matrix of a multivariate linear statistic are

$$\begin{aligned} \mathrm{E}(L'\boldsymbol{Y}) &= L'X\boldsymbol{\beta} \text{ and} \\ \mathrm{Var}(L'\boldsymbol{Y}) &= \sigma^2 L'L. \end{aligned} \tag{8.5}$$

Linear statistics are used for statistical inference about linear functions of the coefficient vector $\boldsymbol{\beta}$. Such functions can be real-valued, like $\boldsymbol{g}'\boldsymbol{\beta}$, or vector-valued, like $G'\boldsymbol{\beta}$.

Consider using a linear statistic $\boldsymbol{\ell}'\boldsymbol{y}$ to estimate the value of $\boldsymbol{g}'\boldsymbol{\beta}$, where both $\boldsymbol{g}$ and $\boldsymbol{\ell}$ are given. How close $\boldsymbol{\ell}'\boldsymbol{y}$ is on average to its target $\boldsymbol{g}'\boldsymbol{\beta}$ is measured by its ***Mean Squared Error*** (***MSE***) ,

$$\begin{aligned} MSE(\boldsymbol{\ell}'\boldsymbol{Y}|\boldsymbol{g}, \boldsymbol{\beta}, \sigma^2) &= \mathrm{E}(\boldsymbol{\ell}'\boldsymbol{Y} - \boldsymbol{g}'\boldsymbol{\beta})^2 \\ &= \mathrm{Var}(\boldsymbol{\ell}'\boldsymbol{Y}) + [\mathrm{E}(\boldsymbol{\ell}'\boldsymbol{Y}) - \boldsymbol{g}'\boldsymbol{\beta}]^2 \\ &= \sigma^2\boldsymbol{\ell}'\boldsymbol{\ell} + (\boldsymbol{\ell}'X\boldsymbol{\beta} - \boldsymbol{g}'\boldsymbol{\beta})^2. \end{aligned} \tag{8.6}$$

Generally, less MSE is better. If both $\boldsymbol{\ell}_1'\boldsymbol{y}$ and $\boldsymbol{\ell}_2'\boldsymbol{y}$ are used to estimate $\boldsymbol{g}'\boldsymbol{\beta}$, then $\boldsymbol{\ell}_1'\boldsymbol{y}$ is ***as good as*** $\boldsymbol{\ell}_2'\boldsymbol{y}$ iff

$$MSE(\boldsymbol{\ell}_1'\boldsymbol{Y}|\boldsymbol{g}, \boldsymbol{\beta}, \sigma^2) \leqslant MSE(\boldsymbol{\ell}_2'\boldsymbol{Y}|\boldsymbol{g}, \boldsymbol{\beta}, \sigma^2) \text{ for all } \boldsymbol{\beta} \in \Re^{k+1} \text{ and } \sigma^2 > 0;$$

and it is ***better than*** $\boldsymbol{\ell}_2'\boldsymbol{y}$ iff in addition strict inequality holds for some combination of parameter values. It should be clear that this is a partial ordering, that not all pairs $\boldsymbol{\ell}_1$ and $\boldsymbol{\ell}_2$ are such that one is as good as the other.

The expression $\mathrm{E}(\boldsymbol{\ell}'\boldsymbol{Y}) - \boldsymbol{g}'\boldsymbol{\beta}$ is the ***bias*** of $\boldsymbol{\ell}'\boldsymbol{Y}$ as an estimator of $\boldsymbol{g}'\boldsymbol{\beta}$. The result in the second line of (8.6) is that MSE (of $\boldsymbol{\ell}'\boldsymbol{Y}$ used as an estimator of $\boldsymbol{g}'\boldsymbol{\beta}$) is equal to (its) variance plus (its) squared bias (from $\boldsymbol{g}'\boldsymbol{\beta}$).

The estimator $\boldsymbol{\ell}'\boldsymbol{Y}$ is a linear ***unbiased*** estimator of $\boldsymbol{g}'\boldsymbol{\beta}$ iff the bias term is zero for all $(k+1)$-vectors $\boldsymbol{\beta}$, which is equivalent to the relation $X'\boldsymbol{\ell} = \boldsymbol{g}$. While "unbiased" sounds good and wholesome, it can happen in some settings that allowing a little bias can permit estimators with lesser variance, enough so that MSE is less. Such settings will be touched on later in this book. For now, and in the conventional development of inferential methods in linear models, if an estimator of a linear function of the mean vector is needed, it will be an unbiased linear estimator, unless indicated otherwise.

Let $\boldsymbol{g}$ be a vector in $\mathrm{sp}(X')$. Then $X'\boldsymbol{\ell} = \boldsymbol{g}$ is a linear equation in $\boldsymbol{\ell}$ that has a solution: it is a consistent linear equation. The next proposition is an old and well-known result on solutions to consistent linear equations.

Proposition 8.1. *Let* $\boldsymbol{g} \in \mathrm{sp}(X')$, *and let* $\mathcal{L} = \{\boldsymbol{\ell} : X'\boldsymbol{\ell} = \boldsymbol{g}\}$. *Then:*

1. *There exists exactly one vector* $\boldsymbol{\ell}_*$ *in* $\mathrm{sp}(X)$ *such that* $X'\boldsymbol{\ell}_* = \boldsymbol{g}$.
2. $\boldsymbol{\ell} \in \mathcal{L}$ *iff* $\mathbf{P}_X\boldsymbol{\ell} = \boldsymbol{\ell}_*$.
3. *If* $\boldsymbol{\ell} \in \mathcal{L}$, *then* $\boldsymbol{\ell}'\boldsymbol{\ell} \geqslant \boldsymbol{\ell}_*'\boldsymbol{\ell}_*$, *with equality iff* $\boldsymbol{\ell} = \boldsymbol{\ell}_*$.

Suppose that $\boldsymbol{Y} \sim (X\boldsymbol{\beta}, \sigma^2\mathrm{I})$. Let $\boldsymbol{g}$ be a $(k+1)$-vector. Then (prove that) there exists a linear unbiased estimator of $\boldsymbol{g}'\boldsymbol{\beta}$ iff $\boldsymbol{g} \in \mathrm{sp}(X')$.

Let $\boldsymbol{g}$ be a vector in $\mathrm{sp}(X')$. Proposition 8.1 establishes that, among all linear unbiased estimators of $\boldsymbol{g}'\boldsymbol{\beta}$, $\boldsymbol{\ell}_*'\boldsymbol{Y}$, and only $\boldsymbol{\ell}_*'\boldsymbol{Y}$, has least variance. In that case, $\boldsymbol{\ell}_*'\boldsymbol{Y}$ is said to be the ***B****est* ***L****inear* ***U****nbiased* ***E****stimator of* $\boldsymbol{g}'\boldsymbol{\beta}$, the ***BLUE*** of $\boldsymbol{g}'\boldsymbol{\beta}$.

A ***least-squares estimator*** of $\boldsymbol{g}'\boldsymbol{\beta}$ is defined to be $\boldsymbol{g}'\check{\boldsymbol{b}}(\boldsymbol{y})$, where $\check{\boldsymbol{b}}(\boldsymbol{y})$ is a LS solution. If $\boldsymbol{g} \in \mathrm{sp}(X')$, then $\boldsymbol{g}'\check{\boldsymbol{b}}(\boldsymbol{y})$ is linear in $\boldsymbol{y}$, it is the same for all LS solutions, and its expected value is $\boldsymbol{g}'\boldsymbol{\beta}$ for all $\boldsymbol{\beta} \in \Re^{k+1}$. Proposition 8.2 establishes that if $\boldsymbol{g} \in \mathrm{sp}(X')$, then the LS estimator of $\boldsymbol{g}'\boldsymbol{\beta}$ is identical to the BLUE of $\boldsymbol{g}'\boldsymbol{\beta}$. Originally established by Gauss (1823), it was described later by Markov (1912).

Proposition 8.2. The Gauss-Markov Theorem. *If* $\boldsymbol{Y} \sim (X\boldsymbol{\beta}, \sigma^2\mathrm{I})$, *and* $\check{\boldsymbol{b}}(\boldsymbol{y})$ *satisfies* $X\check{\boldsymbol{b}}(\boldsymbol{y}) = \mathbf{P}_X\boldsymbol{y}$ *for every* $\boldsymbol{y} \in \Re^n$, *and* $\boldsymbol{g} \in \mathrm{sp}(X')$, *then the least-squares estimator* $\boldsymbol{g}'\check{\boldsymbol{b}}(\boldsymbol{Y})$ *is an unbiased linear estimator of* $\boldsymbol{g}'\boldsymbol{\beta}$ *and, among all unbiased linear estimators of* $\boldsymbol{g}'\boldsymbol{\beta}$, *it has minimum variance.*

Correspondingly, consider estimating $G'\boldsymbol{\beta}$, where G is a given $(k+1) \times c$ matrix. There exists a linear unbiased estimator $L'\boldsymbol{Y}$ of $G'\boldsymbol{\beta}$ iff $\mathrm{sp}(G) \subset \mathrm{sp}(X')$, that is, iff there exists a matrix L such that $X'L = G$. Proposition 8.3 extends Proposition 8.1 to solutions to $X'L = G$.

Proposition 8.3. *Let G be a matrix such that* $\mathrm{sp}(G) \subset \mathrm{sp}(X')$, *and let* $\mathcal{L}_G = \{L : X'L = G\}$.

1. *There exists exactly one matrix L_* such that* $\mathrm{sp}(L_*) \subset \mathrm{sp}(X)$ *and* $X'L_* = G$.
2. *If $L \in \mathcal{L}_G$, then $\mathbf{P}_X L = L_*$.*
3. *If $L \in \mathcal{L}_G$, then $L'L - L_*'L_*$ is nnd, and $L'L - L_*'L_* = 0$ iff $L = L_*$.*

In the last part of the proposition, that $L'L - L_*'L_*$ is nnd, and it is 0 iff $L = L_*$, means that, for any c-vector $\boldsymbol{z}$, $\boldsymbol{z}'(L'L)\boldsymbol{z} \geqslant \boldsymbol{z}'(L_*'L_*)\boldsymbol{z}$, and there exists a vector $\boldsymbol{z}_*$ such that strict inequality holds unless $L = L_*$. In this case $L'L$ is said to succeed $L_*'L_*$ *in Löwner ordering.*

Proposition 8.3 parallels Proposition 8.1. It follows that $L_*'\boldsymbol{Y}$, and only $L_*'\boldsymbol{Y}$, minimizes $\mathrm{Var}(L'\boldsymbol{Y}) = \sigma^2 L'L$ (in Löwner ordering) among unbiased linear estimators of $G'\boldsymbol{\beta}$, that is, it is the BLUE of $G'\boldsymbol{\beta}$.

Suppose X, G, and L are such that $\mathrm{sp}(G) \subset \mathrm{sp}(X')$ and $X'L = G$. Given a least-squares solution $\check{\boldsymbol{b}}(\boldsymbol{y})$, a least-squares estimator of $G'\boldsymbol{\beta}$ is $G'\check{\boldsymbol{b}}(\boldsymbol{Y})$. The BLUE of $G'\boldsymbol{\beta}$ is $L_*'\boldsymbol{Y}$, where $L_* = \mathbf{P}_X L$. It follows that

$$G'\check{\boldsymbol{b}}(\boldsymbol{y}) = (X'L)'\check{\boldsymbol{b}}(\boldsymbol{y}) = L'X\check{\boldsymbol{b}}(\boldsymbol{y}) = L'\mathbf{P}_X\boldsymbol{y} = L_*'\boldsymbol{y}$$

for all $\boldsymbol{y} \in \Re^n$, that is, the least-squares estimator is BLUE.

8.4 Conclusion

The Gauss-Markov Theorem, Proposition 8.2, establishes a connection between least-squares, which is aimed at finding the best fit to $\boldsymbol{y}$ in the model, and estimation of linear functions of $\boldsymbol{\beta}$.

As a mathematical result, the GM Theorem is simple, and it is easy to prove. It is the gateway and validation for the ubiquitous use of LS methods in practice. Sums of squares are central to those methods, which then lead to chi-squared random variables and to t- and F-statistics for inference, the subject of Chapter 10.

8.5 Exercises

In these exercises, X is a fixed, known $n \times (k+1)$ matrix and $\boldsymbol{y}$ is an n-vector. For two functions $f_1(\boldsymbol{y})$ and $f_2(\boldsymbol{y})$ on $\Re^n$, that $f_1(\boldsymbol{y}) \equiv f_2(\boldsymbol{y})$ means that $f_1(\boldsymbol{y}) = f_2(\boldsymbol{y})$ for all $\boldsymbol{y} \in \Re^n$; and $f_1(\boldsymbol{y}) \not\equiv f_2(\boldsymbol{y})$ means that there exists a $\boldsymbol{y}_* \in \Re^n$ such that $f_1(\boldsymbol{y}_*) \neq f_2(\boldsymbol{y}_*)$.

1. Let $X = \mathbf{1}_n$, and let $\boldsymbol{y} = (y_i)$ be the observed response of the n-variate random variable $\boldsymbol{Y}$, which has mean vector $\mathrm{E}(\boldsymbol{Y}) = X\beta_0$ for some $\beta_0 \in \Re$. Assume in addition that $\mathrm{Var}(\boldsymbol{Y}) = \sigma^2 \mathrm{I}_n$ for some $\sigma^2 > 0$. Let $\bar{y} = (1/n)\sum_i y_i$. Recall that $U_n = (1/n)\mathbf{1}_n\mathbf{1}_n'$ and $S_n = \mathrm{I}_n - U_n$.

   (a) Show that $\mathbf{P}_X = U_n$ and that hence $\mathbf{P}_X\boldsymbol{y} = \bar{y}\mathbf{1}_n$ for all $\boldsymbol{y} \in \Re^n$.

   (b) Find a matrix A such that $XA' = \mathbf{P}_X$.

   (c) Show that $\hat{\beta}_0(\boldsymbol{y}) = A'\boldsymbol{y} = \bar{y} = (1/n)\sum_i y_i$ is a least-squares solution, that is, that $X\hat{\beta}_0(\boldsymbol{y}) = \mathbf{P}_X\boldsymbol{y}$ for all $\boldsymbol{y} \in \Re^n$. Is there any other function $\tilde{\beta}_0(\boldsymbol{y})$ such that $X\tilde{\beta}_0(\boldsymbol{y}) = \mathbf{P}_X\boldsymbol{y}$ for all $\boldsymbol{y} \in \Re^n$?

   (d) Show that $\mathrm{E}[\hat{\beta}_0(\boldsymbol{Y})] = \beta_0$ for each $\beta_0 \in \Re$.

   (e) Show that $[\boldsymbol{y} - X\hat{\beta}_0(\boldsymbol{y})]'[\boldsymbol{y} - X\hat{\beta}_0(\boldsymbol{y})] = \boldsymbol{y}'(\mathrm{I} - \mathbf{P}_X)\boldsymbol{y} = \sum_{i=1}^n (y_i - \bar{y})^2$.

   (f) Show that $\mathrm{Cov}[\hat{\beta}_0(\boldsymbol{Y}), (\mathrm{I} - \mathbf{P}_X)\boldsymbol{Y}] = \mathbf{0}$.

   (g) Find $\mathrm{Var}[\hat{\beta}_0(\boldsymbol{Y})]$ and show that, if $\tilde{\beta}_0(\boldsymbol{Y}) = \boldsymbol{\ell}'\boldsymbol{Y}$ is an unbiased linear estimator of β_0, then $\mathrm{Var}[\tilde{\beta}_0(\boldsymbol{Y})] \geqslant \mathrm{Var}[\hat{\beta}_0(\boldsymbol{Y})]$, with equality iff $\tilde{\beta}_0(\boldsymbol{y}) \equiv \hat{\beta}(\boldsymbol{y})$.

2. Let $X = (\mathbf{1}_n, \mathrm{Diag}(\mathbf{1}_{n_1}, \mathbf{1}_{n_2}))$, where n_1 and n_2 are positive integers and $n = n_1 + n_2$. Assume that $\boldsymbol{Y} \sim (X\boldsymbol{\beta}, \sigma^2\mathrm{I}_n)$, and denote its realized value by $\boldsymbol{y} = (y_{ij})$, with $i = 1, 2$, $j = 1, \ldots, n_i$. Let $\boldsymbol{\beta} = (\beta_0, \beta_1, \beta_2)'$, $\boldsymbol{\beta} \in \Re^3$. Let $\bar{y}_{i\cdot} = (1/n_i)\sum_j y_{ij}$.

   (a) Show that $\mathrm{sp}(X) = \mathrm{sp}[\mathrm{Diag}(\mathbf{1}_{n_1}, \mathbf{1}_{n_2})]$.

   (b) Show that $\mathbf{P}_X = \mathrm{Diag}(U_{n_1}, U_{n_2})$.

   (c) Find an expression for a least-squares solution, $\hat{\boldsymbol{b}}(\boldsymbol{y})$. Identify two different least-squares solutions.

   (d) Find a least-squares solution that is not linear in $\boldsymbol{y}$.

   (e) Show that $\mathrm{tr}(\mathrm{I} - \mathbf{P}_X) = n_1 + n_2 - 2$ and

   $$\boldsymbol{y}'(\mathrm{I} - \mathbf{P}_X)\boldsymbol{y} = \sum_{i=1}^{2}\sum_{j=1}^{n_i}(y_{ij} - \bar{y}_{i\cdot})^2.$$

   (f) Express $\bar{y}_{1\cdot} - \bar{y}_{2\cdot}$ as a linear function of $\boldsymbol{y}$. That is, identify $\boldsymbol{\ell} = (\boldsymbol{\ell}_1', \boldsymbol{\ell}_2')'$ such that $\boldsymbol{\ell}'\boldsymbol{y} = \bar{y}_{1\cdot} - \bar{y}_{2\cdot}$ for all $\boldsymbol{y} = (\boldsymbol{y}_1', \boldsymbol{y}_2')'$ in $\Re^n$.

   (g) What are the expected value and variance of $\boldsymbol{\ell}'\boldsymbol{Y} = \bar{Y}_{1\cdot} - \bar{Y}_{2\cdot}$?

   (h) Show that $\mathrm{Cov}[\boldsymbol{\ell}'\boldsymbol{Y}, (\mathrm{I} - \mathbf{P}_X)\boldsymbol{Y}] = \mathbf{0}$.

   (i) Define $\hat{\sigma}^2 = \boldsymbol{y}'(\mathrm{I} - \mathbf{P}_X)\boldsymbol{y}/(n_1 + n_2 - 2)$.
   Justify the assertion that, if $\boldsymbol{Y} \sim \mathbf{N}(X\boldsymbol{\beta}, \sigma^2\mathrm{I}_n)$, then $T^2(\boldsymbol{Y})$, defined by

   $$T^2(\boldsymbol{y}) = \frac{(\bar{y}_{1\cdot} - \bar{y}_{2\cdot})^2}{\hat{\sigma}^2(1/n_1 + 1/n_2)},$$

is distributed as an F random variable with 1 and n_1+n_2-2 degrees of freedom and non-centrality parameter $(\beta_1-\beta_2)^2/[\sigma^2(1/n_1+1/n_2)]$.

3. Let: R be a given $(k+1)\times r$ matrix, $\boldsymbol{r}_0$ be a given r-vector $\in \text{sp}(R')$, and $\boldsymbol{m}_0$ be a given n-vector, respectively. Let $\mathcal{M} = \{\boldsymbol{m} = \boldsymbol{m}_0 + X\boldsymbol{b} : \boldsymbol{b} \in \Re^{k+1}$ and $R'\boldsymbol{b} = \boldsymbol{r}_0\}$.

   Given $\boldsymbol{y}$, develop computable expressions that lead to the LSE of $\boldsymbol{m}$ in $\mathcal{M}$.

4. Define a ***linear least-squares solution*** to be a linear function of $\boldsymbol{y}$ that is a least-squares solution.

   (a) Show that there exists a matrix A such that $XA' = \mathbf{P}_X$.

   (b) Describe steps by which A could be computed from X.

   (c) Given a matrix A such that $XA' = \mathbf{P}_X$, verify that $A'\boldsymbol{y}$ is a linear LS solution in the model $X\boldsymbol{\beta}$.

   (d) Prove: $\hat{\boldsymbol{b}}(\boldsymbol{y})$ is a linear least-squares solution iff there exists a matrix A such that $XA' = \mathbf{P}_X$ and $\hat{\boldsymbol{b}}(\boldsymbol{y}) \equiv A'\boldsymbol{y}$.

   (e) Prove: There exists a constant n-vector $\boldsymbol{c}$ such that $\boldsymbol{g}'\check{\boldsymbol{b}}(\boldsymbol{y}) \equiv \boldsymbol{c}'\boldsymbol{y}$ for every linear least-squares solution $\check{\boldsymbol{b}}(\boldsymbol{y})$ iff $\boldsymbol{g} \in \text{sp}(X')$. That is, $\boldsymbol{g}'A'\boldsymbol{y} \equiv \boldsymbol{c}'\boldsymbol{y}$ for all matrices A such that $XA' = \mathbf{P}_X$ iff $\boldsymbol{g} \in \text{sp}(X')$.

5. Suppose $\boldsymbol{g}$ is in $\text{sp}(X')$. Let $\boldsymbol{\ell}$ be a vector such that $X'\boldsymbol{\ell} = \boldsymbol{g}$. Then, for any least-squares solution $\hat{\boldsymbol{b}}(\boldsymbol{y})$, $\boldsymbol{g}'\hat{\boldsymbol{b}}(\boldsymbol{y}) \equiv \boldsymbol{\ell}'X\hat{\boldsymbol{b}}(\boldsymbol{y}) \equiv \boldsymbol{\ell}'\mathbf{P}_X\boldsymbol{y} \equiv \boldsymbol{\ell}_*'\boldsymbol{y}$, where $\boldsymbol{\ell}_*$ is the unique vector in $\text{sp}(X)$ such that $X'\boldsymbol{\ell}_* = \boldsymbol{g}$. That is, if $\boldsymbol{g} \in \text{sp}(X')$, then $\boldsymbol{g}'\hat{\boldsymbol{b}}(\boldsymbol{y})$ is invariant to the choice of least-squares solution.

   Prove the converse: If $\boldsymbol{g} \notin \text{sp}(X')$, then there exist least-squares solutions $\hat{\boldsymbol{b}}(\boldsymbol{y})$ and $\check{\boldsymbol{b}}(\boldsymbol{y})$ such that $\boldsymbol{g}'\hat{\boldsymbol{b}}(\boldsymbol{y}) \not\equiv \boldsymbol{g}'\check{\boldsymbol{b}}(\boldsymbol{y})$.

   Together these establish that $\boldsymbol{g}'\check{\boldsymbol{b}}(\boldsymbol{y})$ is invariant to the choice of least-squares solution iff $\boldsymbol{g} \in \text{sp}(X')$.

9

Estimability

This chapter is an extensive discussion of estimability, the property that determines what linear functions of regression coefficients can cause differences in models for the mean vector. Inference is possible about estimable functions, and it is impossible about functions that are not estimable. This chapter defines that property simply and explicitly in the context of models for the mean vector, mentions some of the heuristics provided in extant texts, and presents a list of equivalent formulations. The first section is essential for its definitions. Much of the subsequent discussion is background that is not essential for understanding the material in later chapters.

9.1 Introduction and Definition

A model for additive effects of combinations of two factors on cell means η_{ij} is traditionally formulated as $\boldsymbol{\eta} = (\eta_{ij}) = (\eta_0+\alpha_i+\beta_j)$, over the ab combinations i, j of a levels of factor A and b levels of factor B. The factor effects α_i and β_j are the main targets of interest, but they can be seen only through the cell means. Even if we knew $\boldsymbol{\eta}$, it would not be possible to deduce all the values of the $1 + a + b$ parameters in the model. However, it is clear that we would know some linear combinations of them: $\eta_0 + \alpha_3 + \beta_1$, for example, must equal η_{31}, and $\alpha_1 - \alpha_2$ can be deduced from $\eta_{13} - \eta_{23}$. But other linear functions of the parameters, like η_0 or α_2 or $\beta_1 + \beta_2$, cannot be determined from $\boldsymbol{\eta}$. Given $\boldsymbol{\eta}$, the numerical value of $\alpha_1 - \alpha_2$ is determined uniquely, but α_2 could be any real number whatsoever. If we knew $\boldsymbol{\eta}$ we would know $\alpha_1 - \alpha_2$ exactly, but we would know nothing about α_2.

If a function $\boldsymbol{g}'\boldsymbol{\beta}$ of $\boldsymbol{\beta}$ is determined by the mean vector $X\boldsymbol{\beta}$, then inference on it from the observed response $\boldsymbol{y}$ is possible through the connection that $\mathrm{E}(\boldsymbol{Y}) = X\boldsymbol{\beta}$. On the other hand, if even the true value of $X\boldsymbol{\beta}$ does not determine $\boldsymbol{g}'\boldsymbol{\beta}$, then any attempt at inference about $\boldsymbol{g}'\boldsymbol{\beta}$ from $\boldsymbol{y}$ is misguided and futile, because there is no functional or probabilistic relation between $\boldsymbol{Y}$ and $\boldsymbol{g}'\boldsymbol{\beta}$. If we are to enjoy the natural appeal of formulations like $\eta_{ij} = \eta_0 + \alpha_i + \beta_j$, it is essential to be able to tell which linear functions of its parameters can be addressed through $X\boldsymbol{\beta}$ and which cannot.

In the linear model $\boldsymbol{\mu} = X\boldsymbol{\beta}$, the linear function $\boldsymbol{g}'\boldsymbol{\beta}$ is said to be ***estimable*** iff it is a linear function of the mean vector. That is, $\boldsymbol{g}'\boldsymbol{\beta}$ is estimable in the model $X\boldsymbol{\beta}$ iff $\boldsymbol{g} \in \text{sp}(X')$. A vector $\boldsymbol{g}$ is either estimable or it is not estimable.

This definition extends to multiple linear functions of $\boldsymbol{\beta}$ as follows. Let $G = (\boldsymbol{g}_1, \ldots, \boldsymbol{g}_c)$ be a $(k+1) \times c$ matrix. The linear function $G'\boldsymbol{\beta}$ is said to be estimable iff each $\boldsymbol{g}_i'\boldsymbol{\beta}$ is estimable, which is equivalent to $\text{sp}(G) \subset \text{sp}(X')$.

Clearly there are matrices G such that part of sp(G) is estimable and part is not, that is, such that sp(G) includes some vectors in sp(X') and some not in sp(X'). The ***estimable part*** of sp(G) is $\{\boldsymbol{g} \in \text{sp}(G) : \boldsymbol{g} \in \text{sp}(X')\}$, that is, it is $\text{sp}(G) \cap \text{sp}(X')$.

It should be clear that questions of estimability arise only if the columns of X are linearly dependent. If they are linearly independent, then $X\boldsymbol{\beta} = \boldsymbol{\mu}$ determines $\boldsymbol{\beta}$ uniquely, and hence every linear function $\boldsymbol{g}'\boldsymbol{\beta}$ is estimable.

9.2 Background and Commentaries

The definition of "estimable" given in the Apple Inc. Dictionary (2019) is "worthy of great respect." That is clearly not the meaning intended in the context of linear models. As originally coined by Bose (1944) (see Seely 1977), the term was intended to mean that the function possesses an unbiased linear estimator, and so "estimatable" might have been the better appellation. Now, though, "estimable" is so deeply entrenched in the practice and theory of linear statistical models that it is futile to think that it could be supplanted by a name that more clearly suggests its core meaning.

Otherwise, if it were possible to start all over again, "identifiable" might be better. It is already used with a broader meaning to describe meaningful functions of parameters that index families of probability distributions. Specialized to the model $X\boldsymbol{\beta}$ and linear functions of $\boldsymbol{\beta}$, it would be equivalent to "$X\boldsymbol{\beta} = \boldsymbol{0}$ implies that $\boldsymbol{g}'\boldsymbol{\beta} = \boldsymbol{0}$," which, as you can see, is equivalent to $\boldsymbol{g} \subset \text{sp}(X')$. Then a linear function $\boldsymbol{g}'\boldsymbol{\beta}$ would be said to be identifiable if its value could be deduced from $X\boldsymbol{\beta}$. In the examples above, $\eta_0 + \alpha_3 + \beta_1$ and $\alpha_1 - \alpha_2$ are identifiable, while η_0 and $\alpha_1 + \alpha_2$ are not.

Notice that, in this definition, there is no mention, not even a hint, of any necessary connection to random variables or probability distributions or estimation. It is a linear-algebraic property of X and $\boldsymbol{g}$. For that reason, the term "estimable" leads to mis- and over-interpretation because it suggests a connection with estimation and inference. Nevertheless, I shall continue to use it here in order to conform to precedent and its ubiquitous usage.

While "estimable" sounds wholesome and good, the reasons given in the literature and textbooks for eschewing non-estimable functions are not always compelling or convincing. One reason that is widely cited is that their estimators or test statistics differ with the choice of least-squares solution. In an early, important, thorough, and influential discussion of estimability, Searle (1966) states this justification emphatically.

> Elston and Bush (1964) define a hypothesis $\boldsymbol{q}'\boldsymbol{b} = \boldsymbol{m}$ as being testable if and only if $\boldsymbol{q}'\boldsymbol{b}$ is estimable. This, of course, is correct. But it gives no answer to the question "why can we not test a hypothesis about $\boldsymbol{q}'\boldsymbol{b}$ if $\boldsymbol{q}'\boldsymbol{b}$ is not estimable"? To some this question may appear trite, but to experimenters steeped in data it is not a question asked lightly, because it is only a special case of the more general question "why can't I test any hypothesis I want to"? There is one overriding reason; only if $\boldsymbol{q}'\boldsymbol{b}$ is estimable will $\boldsymbol{q}'\hat{\boldsymbol{b}}$ be invariant to the choice of $\hat{\boldsymbol{b}}$; and by the manner in which $\boldsymbol{q}'\boldsymbol{b}$ is involved in the customary F-test for testing the hypothesis $\boldsymbol{q}'\boldsymbol{b} = \boldsymbol{m}$ it is necessary to have $\boldsymbol{q}'\hat{\boldsymbol{b}}$ invariant for F to be invariant also. Clearly a test using F would be of no value unless it was so invariant. For this reason, then, as we will show, the hypothesis $\boldsymbol{q}'\boldsymbol{b}$ is testable only when $\boldsymbol{q}'\boldsymbol{b}$ is estimable.

More recently, Stroup (2013, p. 150) asks, "Why does estimability matter?" Again, it is non-invariance to the choice of least-squares solution that causes problems.

> Simply put, in models that are not of full rank – for example, all ANOVA-type effects models – estimating equation solutions for the effects themselves have no intrinsic meaning. Their solution depends entirely on the generalized inverse used, and theory tells us that there are infinitely many ways to construct a generalized inverse. On the other hand, estimable functions are invariant to choice of generalized inverse, and therefore have an assignable meaning. The effect estimates *per se* do *not* have *any* legitimate interpretation; estimable functions *do*.

While "why can't I test any hypothesis I want to?" and "Why does estimability matter?" are reasonable questions, the "overriding reason" given is not compelling. In several contexts in statistics, the availability of multiple choices is viewed as a good thing, an opportunity to optimize, not a fatal flaw. Why, then, is non-invariance so bad here?

From a slightly different angle, Rencher and Schaalje (2008, p. 303) point out that, when columns of X are linearly dependent, $\hat{\boldsymbol{\beta}} = (X'X)^{-}X'\boldsymbol{y}$ is not an unbiased estimator of $\boldsymbol{\beta}$ "[s]ince $(X'X)^{-}X'X \neq \mathrm{I}$" and that "$\mathrm{E}(\hat{\boldsymbol{\beta}})$ is different for each choice of $(X'X)^{-}$," and "[t]hus, $\hat{\boldsymbol{\beta}}$ ... does not estimate $\boldsymbol{\beta}$."

Among Searle's (1966) examples are illustrations that the expected value of $\boldsymbol{g}'\hat{\boldsymbol{\beta}}$ is not uniformly equal to $\boldsymbol{g}'\boldsymbol{\beta}$ if it is not estimable, with the insinuation that attempts at inference about $\boldsymbol{g}'\boldsymbol{\beta}$ based on $\boldsymbol{g}'\hat{\boldsymbol{\beta}}$ are therefore spurious. Rencher and Schaalje (2008) make the same point.

Searle (1966) makes another telling point, that when $\boldsymbol{q}'\boldsymbol{\beta}$ is not estimable, the Restricted Model – Full Model difference in SSE (RMFM SS) is zero, and so "[w]e therefore conclude that the hypothesis $\boldsymbol{q}'\boldsymbol{\beta} = m$ is not testable" and hence "the only hypotheses that can be tested are those involving estimable functions."

As you can see, these authors state their cases emphatically, but in the end their conclusions are based on vague assertions and non sequiturs: "would be of no value," "do *not* have *any* legitimate interpretation," "does not estimate $\boldsymbol{\beta}$," "is not testable."

Seely (1977), in a cogent discussion of estimability, acknowledges the "not well defined" numerator SS problem with non-estimable functions and offers an "alternative justification" for restricting inference to estimable functions "as the requirement that the class of distributions under the null hypothesis be disjoint from the class of distributions under the alternative hypothesis." In his main proposition, he established four conditions equivalent to estimability, one of which links estimability to an identifiability condition due to Reiersol (1963). These equivalencies, plus several more, are established in Section 9.3.

9.3 Estimability Equivalencies

The context of this discussion is inference on linear functions $G'\boldsymbol{\beta}$ of the mean vector in the model $\mathcal{M} = \{\boldsymbol{\mu} = X\boldsymbol{\beta} : \boldsymbol{\beta} \in \Re^{k+1}\} = \mathrm{sp}(X)$. X is a given $n \times (k+1)$ matrix and G is a matrix (or vector) with $(k+1)$ rows. To avoid dealing with trivialities, assume that both X and G are non-zero. In order to emphasize that estimability is a linear-algebraic property, without any necessary connection to probability distributions or statistical inference, replace the parameter vector $\boldsymbol{\beta}$ throughout with $\boldsymbol{b}$, representing just an arbitrary vector in $\Re^{k+1}$.

With $\mathcal{M} = \mathrm{sp}(X)$, let $\mathcal{M}_0 = \{X\boldsymbol{b} : \boldsymbol{b} \in \Re^{k+1} \text{ and } G'\boldsymbol{b} = \boldsymbol{0}\} = X\{\mathrm{sp}(G)^\perp\}$. Let N be a matrix such that $\mathrm{sp}(N) = \mathrm{sp}(G)^\perp$, so that $\mathcal{M}_0 = \mathrm{sp}(XN)$, and let H be a matrix such that $\mathbf{P}_H = \mathbf{P}_X - \mathbf{P}_{XN}$. See Exercise 11, p. 74. Then $\mathrm{sp}(X'H) = \mathrm{sp}(X') \cap \mathrm{sp}(G)$, $\mathcal{M}_0 = X\{\mathrm{sp}(X'H)^\perp\}$, and $\{\boldsymbol{b} \in \Re^{k+1} : X\boldsymbol{b} \in \mathcal{M}_0\} = [\mathrm{sp}(X') \cap \mathrm{sp}(G)]^\perp = \mathrm{sp}(X'H)^\perp$.

If $\mathrm{sp}(G) \subset \mathrm{sp}(X')$, there exists a matrix L such that $X'L = G$, and then $X\boldsymbol{b}$ determines $G'\boldsymbol{b}$ $(= L'X\boldsymbol{b})$ uniquely. The purpose of this section is to describe this and other equivalent properties of estimable functions. Extensions to linear models that are affine sets and incorporate linear conditions, and to non-homogeneous conditions $G'\boldsymbol{b} = \boldsymbol{c}_0$, are shown as exercises at the end of this chapter.

Proposition 9.1. Estimability equivalencies. *The following statements are equivalent.*

- ⋆ *There exists a matrix L such that, for each $\boldsymbol{b}$ in $\Re^{k+1}$, $G'\boldsymbol{b} = L'X\boldsymbol{b}$.*

1. $\mathrm{sp}(G) \subset \mathrm{sp}(X')$.

2. *For any two vectors* $\boldsymbol{b}$ *and* $\boldsymbol{b}_*$ *in* $\Re^{k+1}$*, if* $X\boldsymbol{b} = X\boldsymbol{b}_*$*, then* $G'\boldsymbol{b} = G'\boldsymbol{b}_*$*. That is,* $X\boldsymbol{b}$ *determines* $G'\boldsymbol{b}$ *uniquely.*

3. *If* $\boldsymbol{b}_0$ *is a vector such that* $X\boldsymbol{b}_0 \in \mathcal{M}_0$*, then* $G'\boldsymbol{b}_0 = \mathbf{0}$*.*
   [Note that this is equivalent to $[\mathrm{sp}(X') \cap \mathrm{sp}(G)]^\perp \subset \mathrm{sp}(G)^\perp$*.]*

4. *For any matrix* A *such that* $XA' = \mathbf{P}_X$*,* $G = X'AG$*.*

5. *For each non-zero vector* $\boldsymbol{g}$ *in* $\mathrm{sp}(G)$*,*
   $\{X\boldsymbol{b} : \boldsymbol{b} \in \Re^{k+1}$ *and* $\boldsymbol{g}'\boldsymbol{b} = 0\} \neq \mathcal{M}$*.*

6. $\{\boldsymbol{b} : G'\boldsymbol{b} = \mathbf{0}\} = \{\boldsymbol{b} : H'X\boldsymbol{b} = \mathbf{0}\}$*. Equivalently,* $\mathrm{sp}(G)^\perp = \mathrm{sp}(X'H)^\perp$*, or* $\mathrm{sp}(G) = \mathrm{sp}(X'H)$*.*

7. *For any matrices* A_1 *and* A_2 *such that* $XA_1' = XA_2' = \mathbf{P}_X$ *and any vector* $\boldsymbol{g} \in \mathrm{sp}(G)$*,* $\boldsymbol{g}'A_1'\boldsymbol{y} \equiv \boldsymbol{g}'A_2'\boldsymbol{y}$*.*

Property 1, equivalent to $(\mathrm{I} - \mathbf{P}_{X'})G = 0$, is useful for identifying algebraically whether $G'\boldsymbol{b}$ is estimable. Property 2 is that estimable functions are uniquely determined by the mean vector.

Property 3 is that there are no vectors $\boldsymbol{b}$ with $G'\boldsymbol{b} \neq \mathbf{0}$ such that $X\boldsymbol{b}$ is in $\mathcal{M}_0$; knowing that $\boldsymbol{m}$ is in $\mathcal{M}_0$ is sufficient to conclude that, for any $\boldsymbol{b}$ such that $\boldsymbol{m} = X\boldsymbol{b}$, $G'\boldsymbol{b} = \mathbf{0}$. This is equivalent to condition (d) in the proposition in Seely (1977).

Property 4 says that $G'A'X\boldsymbol{b} = G'\boldsymbol{b}$ for all $\boldsymbol{b} \in \Re^{k+1}$. When $\mathrm{E}(\boldsymbol{Y}) = X\boldsymbol{\beta}$, this property guarantees that $G'A'\boldsymbol{Y}$ is an unbiased linear estimator of $G'\boldsymbol{\beta}$.

Property 5 is that, for each non-zero $\boldsymbol{g} \in \mathrm{sp}(G)$, $\boldsymbol{g}'\boldsymbol{b}$ carries information about the mean vector; given $\boldsymbol{g}'\boldsymbol{b}$, the set of possibilities for $\boldsymbol{m}$ is reduced. With estimability defined by $\mathrm{sp}(G) \subset \mathrm{sp}(X')$, that $G'\boldsymbol{b}$ be estimable requires that $\boldsymbol{g}_j'\boldsymbol{b}$ be estimable for every column $\boldsymbol{g}_j$ of G. The negation of this is that some column of G is not estimable, in which case $\{X\boldsymbol{b} : \boldsymbol{b} \in \Re^{k+1}$ and $\boldsymbol{g}_j'\boldsymbol{b} = 0\} = \mathrm{sp}(X)$. The set of possible values of $X\boldsymbol{b}$ is not changed by the condition that $\boldsymbol{g}_j'\boldsymbol{b} = 0$. Correspondingly, knowing $X\boldsymbol{b}$ carries no information about $\boldsymbol{g}_j'\boldsymbol{b}$.

Property 7 is that, for each $\boldsymbol{g} \in \mathrm{sp}(G)$, $\boldsymbol{g}'\check{\boldsymbol{b}}$ is the same for every linear LS solution $\check{\boldsymbol{b}}$. This equivalence was shown in Exercise 5, p. 66.

With $\mathrm{sp}(N) = \mathrm{sp}(G)^\perp$ and H such that $\mathbf{P}_H = \mathbf{P}_X - \mathbf{P}_{XN}$, $\mathrm{sp}(X'H) = \mathrm{sp}(X') \cap \mathrm{sp}(G)$. The estimable conditions $H'X\boldsymbol{b} = \mathbf{0}$ reduce $\mathcal{M}$ to $\mathcal{M}_0$ in the same way as $G'\boldsymbol{b} = \mathbf{0}$ does. These are called ***equivalent estimable conditions***, and $\mathrm{sp}(X'H) = \mathrm{sp}(G) \cap \mathrm{sp}(X')$ is the estimable part of $G'\boldsymbol{b}$. Identifying such conditions was the subject of a series of papers spanning some 22 years: Peixoto (1986), Del Río (1989), von Rosen (1990), Chan and Li (1995), LaMotte (1997), Chan and Keung (1997), Kshirshagar (1998), and Hu and Shi (2008).

A matrix H such that $\mathrm{sp}(H) \subset \mathrm{sp}(X)$ and $\mathrm{sp}(X'H) = \mathrm{sp}(X') \cap \mathrm{sp}(G)$ can be constructed as follows. Find N such that $\mathrm{sp}(N) = \mathrm{sp}(G)^\perp$, then find H as Q_2 from $\mathrm{GS}(XN, X) \rightarrow (Q_1, Q_2)$.

9.4 Discussion

In the model $\mathcal{M} = \{X\boldsymbol{b} : \boldsymbol{b} \in \Re^{k+1}\}$, there seems to be no impediment to constructing a conventional test statistic, either based on the RMFM SS (10.3) or in the GLH form (10.7), for any non-void hypothesis $\mathrm{H}_0 : G'\boldsymbol{\beta} = \mathbf{0}$. Both processes can be followed through to a test statistic, although they might give different results. It does not seem to be necessary to consider questions of estimability at all. However, there is in any case the question of what exactly is then tested.

Note that $G'\boldsymbol{b} = \mathbf{0}$ implies that $X\boldsymbol{b} \in \mathcal{M}_0$. However, if $G'\boldsymbol{b}$ is not estimable, then the implication does not go in the other direction: there exist vectors $\boldsymbol{b}_*$ such that $G'\boldsymbol{b}_* \neq \mathbf{0}$ but $X\boldsymbol{b}_* \in \mathcal{M}_0$. For this reason, Seely (1977) asserted that inference about $G'\boldsymbol{\beta}$ should not be addressed at all unless $G'\boldsymbol{\beta}$ is estimable. Some statistical computing packages follow that rule.

In my opinion, simply shutting down if $G'\boldsymbol{\beta}$ is not estimable is less informative than it could be. Common forms of the F-statistic are developed fully in the next chapter. They test $G'\boldsymbol{\beta}$ when it is estimable, and otherwise they test the estimable part plus some other estimable functions that are not functions of $G'\boldsymbol{\beta}$. The user could be given the option to test only the estimable part of $G'\boldsymbol{\beta}$. In any case the statistical package could identify the functions actually tested (they are all estimable), distinguishing those that are functions of the target effect $G'\boldsymbol{\beta}$ and any additional ones that are not.

9.5 Examples and Exercises

For (1)–(9), use the following definitions.

$$
\begin{aligned}
X &= \begin{pmatrix} 1 & 1 & 0 & 0 \\ 1 & 0 & 1 & 0 \\ 1 & 0 & 0 & 1 \end{pmatrix}, \\
\boldsymbol{g}_0 &= \begin{pmatrix} 0 \\ 1 \\ 0 \\ -1 \end{pmatrix}, \quad G = (\boldsymbol{g}_1, \boldsymbol{g}_2) = \begin{pmatrix} 0 & 0 \\ 1 & 0 \\ 1 & 1 \\ 1 & 0 \end{pmatrix}, \\
\boldsymbol{b}_* &= \begin{pmatrix} 100 \\ 25 \\ 0 \\ 75 \end{pmatrix}, \quad \boldsymbol{m}_* = X\boldsymbol{b}_* = \begin{pmatrix} 125 \\ 100 \\ 175 \end{pmatrix}, \\
\mathcal{B}_* &= \{\boldsymbol{b} : X\boldsymbol{b} = X\boldsymbol{b}_*\}, \text{ and } \boldsymbol{c}_0 = \begin{pmatrix} 100 \\ 0 \end{pmatrix} = G'\boldsymbol{b}_*.
\end{aligned}
$$

1. Show that

$$\mathcal{B}_* = \{\boldsymbol{b}_*\} + \mathrm{sp}(N),$$

where $N = (-1, 1, 1, 1)'$.

[Note that $\mathrm{sp}(N) = \mathrm{sp}(X')^\perp = \{\boldsymbol{z} : X\boldsymbol{z} = \boldsymbol{0}\}$. It follows that $\boldsymbol{z} \in \mathrm{sp}(X')$ iff $N'\boldsymbol{z} = 0$.]

2. Find a vector $\boldsymbol{b}_1$, different from $\boldsymbol{b}_*$, such that $X\boldsymbol{b}_1 = X\boldsymbol{b}_*$.

3. For each of $\boldsymbol{g}_0$, $\boldsymbol{g}_1$, and $\boldsymbol{g}_2$, is $\boldsymbol{g}_i'\boldsymbol{b}_1 = \boldsymbol{g}_i'\boldsymbol{b}_*$? Is $G'\boldsymbol{b}_1 = G'\boldsymbol{b}_*$?

[If not, then $X\boldsymbol{b}_1 = X\boldsymbol{b}_*$, and $G'\boldsymbol{b}_* = \boldsymbol{c}_0$ and $G'\boldsymbol{b}_1 \neq \boldsymbol{c}_0$.]

4. Is $\{G'\boldsymbol{b} : X\boldsymbol{b} = X\boldsymbol{b}_*\} = \{G'\boldsymbol{b}_*\}$? That is, does $X\boldsymbol{b} = X\boldsymbol{b}_*$ uniquely determine $G'\boldsymbol{b}$?

5. For each $i = 0, 1, 2$, describe the set $\{\boldsymbol{g}_i'\boldsymbol{b} : X\boldsymbol{b} = X\boldsymbol{b}_*\}$.

[In each case, $\mathrm{sp}(\boldsymbol{g}_i') = \Re$. So the main distinction is whether the restrictions $X\boldsymbol{b} = X\boldsymbol{b}_*$ reduce the set of possible values of $\boldsymbol{g}_i'\boldsymbol{b}$.]

6. For each $i = 0, 1, 2$, determine whether $\boldsymbol{g}_i \in \mathrm{sp}(X')$, and determine $\mathrm{sp}(\boldsymbol{g}_i) \cap \mathrm{sp}(X')$.

7. Show that $\mathrm{sp}(G) \cap \mathrm{sp}(X') = \mathrm{sp}[(0, 1, -2, 1)']$.

[That $\boldsymbol{z} \in \mathrm{sp}(G)$ requires that

$$\boldsymbol{z} = \begin{pmatrix} 0 \\ c_1 \\ c_1 + c_2 \\ c_1 \end{pmatrix},$$

and that $\boldsymbol{z} \in \mathrm{sp}(X')$ requires that $N'\boldsymbol{z} = 0$.]

8. Show that, in general, if $G'\boldsymbol{b}_* = \boldsymbol{c}_0$, then

$$\{X\boldsymbol{b} : G'\boldsymbol{b} = \boldsymbol{c}_0\} = X\boldsymbol{b}_* + \{X\boldsymbol{b} : \boldsymbol{b} \in [\mathrm{sp}(G) \cap \mathrm{sp}(X')]^\perp\}.$$

9. Show that, in general,

$$[\mathrm{sp}(G) \cap \mathrm{sp}(X')]^\perp = \mathrm{sp}(G)^\perp + \mathrm{sp}[(\mathrm{I} - \mathbf{P}_{X'})G];$$

and that in this case,

$$\mathrm{sp}(G)^\perp = \mathrm{sp}\begin{pmatrix} 1 & 0 \\ 0 & 1 \\ 0 & 0 \\ 0 & -1 \end{pmatrix} \text{ and } \mathrm{sp}[(\mathrm{I} - \mathbf{P}_{X'})G] = \mathrm{sp}\begin{pmatrix} -1 \\ 1 \\ 1 \\ 1 \end{pmatrix}.$$

Then $z \in [\text{sp}(G) \cap \text{sp}(X')]^\perp$ can be represented as $z = z_1 + z_2$ with $G'z_1 = \mathbf{0}$ and $Xz_2 = \mathbf{0}$; and $G'z = \mathbf{0}$ iff $z_2 = \mathbf{0}$. Thus z_1 can change $X\boldsymbol{b}$ but not $G'\boldsymbol{b}$; and z_2 can change $G'\boldsymbol{b}$ but not $X\boldsymbol{b}$. Thus there exist vectors $\boldsymbol{b} = \boldsymbol{b}_* + z_2$ such that $X\boldsymbol{b} = X\boldsymbol{b}_*$ and $G'\boldsymbol{b} \neq G'\boldsymbol{b}_*$ for any non-zero z_2 iff $\text{sp}[(\text{I} - \mathbf{P}_{X^\perp})G] \neq \{\mathbf{0}\}$, that is, iff $\text{sp}(G) \not\subset \text{sp}(X')$. Equivalently, $X\boldsymbol{b}$ uniquely determines $G'\boldsymbol{b}$ iff $\text{sp}(G) \subset \text{sp}(X')$. This is one of the four equivalences in Seely's (1977, p. 122) Proposition. This also shows that $\{X\boldsymbol{b} : G'\boldsymbol{b} = \boldsymbol{c}_0\}$ has no vectors in common with $\{X\boldsymbol{b} : G'\boldsymbol{b} \neq \boldsymbol{c}_0\}$ iff $\text{sp}(G) \subset \text{sp}(X')$, another of the equivalences in Seely's (1977) Proposition.

10. Recall that X has n rows and $k + 1$ columns. Let N and G be matrices with $k + 1$ rows.

    Prove: $\text{sp}(XN) = \{X\boldsymbol{b} : \boldsymbol{b} \in \Re^{k+1} \text{ and } G'\boldsymbol{b} = \mathbf{0}\}$ iff $\text{sp}(G)^\perp \subset \text{sp}(N) \subset \text{sp}(X')^\perp + \text{sp}(G)^\perp$.

11. Let N be a matrix such that $\text{sp}(G)^\perp \subset \text{sp}(N) \subset \text{sp}(X')^\perp + \text{sp}(G)^\perp$, and let $\mathbf{P}_H = \mathbf{P}_X - \mathbf{P}_{XN}$.

    (a) Prove that $\text{sp}(X'H) = \text{sp}(X'\mathbf{P}_H) = \text{sp}(G) \cap \text{sp}(X')$.

    (b) Prove that $\text{sp}(G) = \text{sp}(X'H)$ iff $\text{sp}(G) \subset \text{sp}(X')$.

12. Let $\boldsymbol{m}_0$ and X be fixed and given. Show that $\boldsymbol{g}'\boldsymbol{b}$ is estimable in the model $\{\boldsymbol{m}_0 + X\boldsymbol{b} : \boldsymbol{b} \in \Re^{k+1}\}$ iff $\boldsymbol{g} \in \text{sp}(X')$. That is, the offset $\boldsymbol{m}_0$ does not affect estimability.

13. Let: $\boldsymbol{b}_0$ be a $(k+1)$-vector, G be a $(k+1) \times c$ matrix, X be an $n \times (k+1)$ matrix, N be a matrix such that $\text{sp}(N) = \text{sp}(G)^\perp$, all constant and known. Let $\mathbf{P}_H = \mathbf{P}_X - \mathbf{P}_{XN}$. Let

    $$\Delta_0 = \{\boldsymbol{b} \in \Re^{k+1} : G'\boldsymbol{b} = G'\boldsymbol{b}_0\} \text{ and}$$

    $$\Delta_{0*} = \{\boldsymbol{b} \in \Re^{k+1} : \mathbf{P}_H X\boldsymbol{b} = \mathbf{P}_H X\boldsymbol{b}_0\}.$$

    (a) Prove: $\Delta_{0*} = \Delta_0 + \text{sp}[(\text{I} - \mathbf{P}_{X'})G]$, and this is a direct sum.

    (b) Prove: $\Delta_{0*} = \Delta_0$ iff $(\text{I} - \mathbf{P}_{X'})G = 0$.

    (c) Prove: $(\text{I} - \mathbf{P}_{X'})G = 0$ iff $\text{sp}(G) \subset \text{sp}(X')$.

14. Let $X = (X_1, X_2)$. That $\boldsymbol{g}_1'\boldsymbol{\beta}_1$ is *estimable in* X_1 means that the linear function $\boldsymbol{g}_1'\boldsymbol{\beta}_1$ is estimable in the model $\text{sp}(X_1)$.

    (a) Prove: Every linear function of $\boldsymbol{\beta}_1$ that is estimable in X_1 is a linear function of $X_1\boldsymbol{\beta}_1$, and every linear function of $X_1\boldsymbol{\beta}_1$ is estimable in X_1. That is, $X_1\boldsymbol{\beta}_1$ is estimable in X_1, every linear function of $X_1\boldsymbol{\beta}_1$ is estimable in X_1, and every linear function of $\boldsymbol{\beta}_1$ that is estimable in X_1 is a linear function of $X_1\boldsymbol{\beta}_1$.

    (b) Let N_2 be a matrix such that $\text{sp}(N_2) = \text{sp}(X_2)^\perp$. Prove: $X_1\boldsymbol{\beta}_1$ is estimable in X iff $\text{sp}(X_1') = \text{sp}(X_1'N_2)$.

(c) Prove: $\boldsymbol{g}_1'\boldsymbol{\beta}_1$ is estimable in X iff $\boldsymbol{g}_1 \in \text{sp}(X_1'N_2)$, where $\text{sp}(N_2) = \text{sp}(X_2)^\perp$.

(d) Prove: $X_1\boldsymbol{\beta}_1$ is estimable in X iff $\text{sp}(X_1) \cap \text{sp}(X_2) = \{\mathbf{0}\}$.

(e) Prove: The linear functions of $\boldsymbol{\beta}_1$ that are estimable in X_1 are the same as the linear functions of $\boldsymbol{\beta}_1$ that are estimable in X iff $X_1\boldsymbol{\beta}_1$ is estimable in X.

This means that expanding the model from X_1 to (X_1, X_2) (specifically such that $\text{sp}(X_1, X_2) \neq \text{sp}(X_1)$) either maintains the space of estimable functions of $\boldsymbol{\beta}_1$ or reduces it, but it cannot expand it.

Construct an example where expanding the model reduces the space of estimable functions of $\boldsymbol{\beta}_1$. (One example is straightforward: $X_1 = \boldsymbol{x}_1$ and $X_2 = \boldsymbol{x}_2$ with $\boldsymbol{x}_2$ proportional to $\boldsymbol{x}_1$.)

15. (a) Prove: $M'\boldsymbol{g} \in \text{sp}(M'\mathbb{K}')$ iff $\mathbf{P}_M\boldsymbol{g} \in \text{sp}(\mathbf{P}_M\mathbb{K}')$.

(b) Let $X_1 = \mathbb{K}M_1$ and $X_2 = \mathbb{K}M_2$. Suppose that $\text{sp}(M_2) \supset \text{sp}(M_1)$. Prove: If $\boldsymbol{g}'M_2\boldsymbol{\beta}_2$ is estimable in X_2, then $\boldsymbol{g}'M_1\boldsymbol{\beta}_1$ is estimable in X_1.

10

Inference on the Mean

In multiple regression models, effects of conditions on the response are defined by linear functions of the regression coefficients. The conventional F-statistic is the main tool for statistical inference. Its numerator sum of squares determines what effects it tests. This chapter establishes important properties of the F-statistic and describes and compares the two main methods for formulating its numerator sum of squares. It also surveys the rationales that have been used to justify its use.

10.1 Introduction

The setting of the developments in this chapter is the GM model for the random n-vector $\boldsymbol{Y}$, namely $\boldsymbol{Y} \sim (X\boldsymbol{\beta}, \sigma^2\mathrm{I})$. X is a fixed, known, non-zero $n \times (k+1)$ matrix. While it is customary to consider its first column to be $\mathbf{1}_n$, we shall not necessarily follow that convention.

The model for the mean vector $\boldsymbol{\mu} = \mathrm{E}(\boldsymbol{Y})$ is $\mathcal{M} = \{X\boldsymbol{b} : \boldsymbol{b} \in \Re^{k+1}\}$. It is posited that there is a particular, but unknown, vector $\boldsymbol{\beta}$ that generates the true mean vector as $X\boldsymbol{\beta}$, and that the observed response $\boldsymbol{y}$ came from a population with that mean vector. Based on $\boldsymbol{y}$, along with the known model matrix X, the objective of inference is to address properties of $\boldsymbol{\mu}$ or $\boldsymbol{\beta}$.

Rows of X quantify the conditions under which each response was observed. Each entry β_j of the coefficient vector $\boldsymbol{\beta}$ is the rate of change of the mean with respect to the j-th condition variable (the j-th column of X). ***Effects*** are changes in the response mean due to changes in the condition variables, and so effects of conditions on the response are seen through the coefficients in $\boldsymbol{\beta}$.

Questions about effects are questions about the unknown parameter vector $\boldsymbol{\beta}$. The class of questions considered here is limited to linear functions of $\boldsymbol{\beta}$, $G'\boldsymbol{\beta}$, where G is a fixed, known (in any particular context) $(k+1) \times r$ matrix. Questions or propositions are formulated as ***linear hypotheses*** in the form $\mathrm{H}_0 : G'\boldsymbol{\beta} = \boldsymbol{c}_0$, where $\boldsymbol{c}_0$ is a fixed, known r-vector.

Under the assumptions of such models, information from the response vector $\boldsymbol{y}$ about $\boldsymbol{\beta}$ comes indirectly, and only, through the mean vector. The probability distribution of $\boldsymbol{Y}$ depends on $\boldsymbol{\beta}$ only through $X\boldsymbol{\beta}$. However, it is usually easier, and more natural in the setting at hand, to formulate the questions of interest in terms of the coefficient vector $\boldsymbol{\beta}$ than in terms of the mean vector. For example, whether $\beta_1 = \beta_2 = \beta_3$ in $\boldsymbol{\beta}' = (\beta_0, \beta_1, \ldots, \beta_k)$ could be formulated with

$$G = \begin{pmatrix} 0 & 0 \\ 1 & 0 \\ -1 & 1 \\ 0 & -1 \\ \vdots & \vdots \end{pmatrix} \text{ and } \boldsymbol{c}_0 = \begin{pmatrix} 0 \\ 0 \end{pmatrix}.$$

The framework of tests of hypotheses can be viewed as the basic element of inference, whether or not one intends to follow it formally to a "reject" or "do not reject" conclusion. It encapsulates the formalism by which a proposition, like $\mathrm{H}_0 : G'\boldsymbol{\beta} = \boldsymbol{c}_0$, is assessed in light of the response $\boldsymbol{y}$. It is neatly expressed in a p-value p, a statistic that can be interpreted as: "If H_0 were true, the probability of results as extreme as those observed would not be greater than p." We may choose to designate a proposed value $\boldsymbol{c}_0$ of $G'\boldsymbol{\beta}$ as *untenable in light of the data* if p is too small (say, less than a cutoff value α, usually called the *level of significance*), and otherwise as tenable. See Kempthorne and Folks (1971) for a thorough discussion. They refer to this process as a test of "consonance" rather than a test of significance.

Ideally, to develop a statistical test of a hypothesis like $\mathrm{H}_0 : G'\boldsymbol{\beta} = \boldsymbol{c}_0$, we would describe criteria for comparing performances of tests, and then we would try to find a test statistic that optimizes those criteria. In classical, formal treatment of tests of hypotheses, tests are comparable only if they have the same size, and then one is better than another if it has greater power. The classic book by Lehmann (1959) on testing hypotheses gives a thorough account of this theory.

In the setting we are examining here, under the assumptions that $\boldsymbol{Y} \sim \mathrm{N}(\boldsymbol{\mu}, \sigma^2\mathrm{I})$, there is no test that is uniformly most powerful among size-α tests. Even restricting the class of tests to unbiased tests, following Lehmann (1986, p. 370), it can be demonstrated that there does not exist a uniformly most powerful unbiased test unless $\mathrm{sp}(G)$ has dimension 1. Restricting the class of tests further, though, Wichura (2006, Section 6.4) demonstrates that the conventional F-test is uniformly most powerful among tests that are invariant under a certain group of transformations.

No simple, direct logical path in terms of powers of size-α tests leads to exactly one, best statistic on which to base inference. However, it is established here that, if the field is restricted to F-statistics that all test exclusively the same linear functions of $\boldsymbol{\beta}$ in the same model $X\boldsymbol{\beta}$, then there is exactly one numerator sum of squares that performs best.

The next section discusses the formulation of the F-statistic and its properties. Textbooks on linear models have rationalized this statistic in three main ways, in terms of the likelihood-ratio test, comparison of SSEs for the full model and the restricted model, and by mimicking a chi-squared statistic. Those and other approaches are reviewed in Section 10.6 and exercises. The discussion here is for the homogeneous form $\mathrm{H}_0 : G'\boldsymbol{\beta} = \mathbf{0}$, with $\boldsymbol{c}_0 = \mathbf{0}$. Extending it to any $\boldsymbol{c}_0$ is fairly straightforward, as described in Section 10.7.

It is worth emphasizing again that information about $\boldsymbol{\beta}$ from the response $\boldsymbol{y}$ comes only through the mean vector $\boldsymbol{\mu} = X\boldsymbol{\beta}$. That $\boldsymbol{\mu}$ is fixed and given does not uniquely determine $\boldsymbol{\beta}$ unless the columns of X are linearly independent. More generally, a given value of $\boldsymbol{\mu}$ does not determine $G'\boldsymbol{\beta}$ uniquely unless $G'\boldsymbol{\beta}$ is a function of $\boldsymbol{\mu}$, that is, unless $G'\boldsymbol{\beta}$ is estimable. This can be seen explicitly in Exercise 4, p. 95.

10.2 The Conventional F-Statistic

Sums of squares (SSs) are prominent in methods of inference in linear models. As used here, a sum of squares based on a matrix M is defined as $SS_M \equiv SS_M(\boldsymbol{y}) = \boldsymbol{y}'\mathbf{P}_M\boldsymbol{y}$. Its degrees of freedom are $\nu_M = \mathrm{tr}(\mathbf{P}_M)$. Note that if P is a symmetric idempotent matrix, then $SS_P(\boldsymbol{y}) = \boldsymbol{y}'P\boldsymbol{y}$.

We start in terms of the multiple regression model for the mean vector $\boldsymbol{\mu} = X\boldsymbol{\beta}$ of the n-variate random variable $\boldsymbol{Y}$: $\mathcal{M} = \mathrm{sp}(X) = \{\boldsymbol{\mu} = X\boldsymbol{b} : \boldsymbol{b} \in \Re^{k+1}\}$. Let $\boldsymbol{y}$ denote a realized value of $\boldsymbol{Y}$. In this model, the conventional statistic on which inference is based takes the general form

$$F_P(\boldsymbol{y}) = \frac{SS_P/\nu_P}{SS_Q/\nu_Q}, \tag{10.1}$$

where P and Q are symmetric idempotent matrices, $PQ = 0$, $QX = 0$, $\nu_P = \mathrm{tr}(P)$, and $\nu_Q = \mathrm{tr}(Q)$. The ***numerator sum of squares*** (SS) and degrees of freedom (df) are SS_P and ν_P. The ***denominator*** SS and df are SS_Q and ν_Q. If $\boldsymbol{Y} \sim \mathbf{N}(\boldsymbol{\mu}, \sigma^2\mathrm{I})$, then $F_P(\boldsymbol{Y})$ follows an F distribution with ν_P and ν_Q degrees of freedom, respectively. Chapter 2 shows examples of the process of constructing an F-statistic .

See Section B.2, p. 240. The conventional choice for Q is $\mathrm{I} - \mathbf{P}_X$, so that $SS_Q = SSE$, $\nu_Q = \nu_E$, and $SS_Q/\nu_Q = MSE = \hat{\sigma}^2$. Other quadratic forms can replace SSE, so long as Q is symmetric and idempotent and $QX = 0$, to maintain centrality for all $\boldsymbol{\mu} \in \mathrm{sp}(X)$. For example, $Q = \mathrm{I} - \mathbf{P}_{X_*}$ with $\mathrm{sp}(X_*) \supset \mathrm{sp}(X)$ would work, too, but with fewer denominator degrees of freedom than ν_E.

The focus here is mainly on the numerator SS. For notational convenience, for a matrix M with n rows, let F_M denote the same as F_P with $P = \mathbf{P}_M$. It can happen that $P = 0_{n\times n}$ and hence $\nu_P = 0$: in that case define $F_P(\boldsymbol{y})$ to be identically 0.

The non-centrality parameter (ncp) of the distribution of $F_P(\boldsymbol{Y})$ is the ncp of the numerator SS, which is

$$\begin{aligned} \delta_P^2(\boldsymbol{\mu}/\sigma) &= \boldsymbol{\mu}'P\boldsymbol{\mu}/\sigma^2 \\ &= SS_P(X\boldsymbol{\beta})/\sigma^2 \\ &= \boldsymbol{\beta}'X'PX\boldsymbol{\beta}/\sigma^2. \end{aligned} \tag{10.2}$$

That the ncp is 0 is equivalent to $\boldsymbol{\beta}$ satisfying $PX\boldsymbol{\beta} = \mathbf{0}$, that is, that $\boldsymbol{\beta} \in \text{sp}(X'P)^{\perp}$. The distribution of F_P is ***central*** iff $\delta_P^2 = 0$.

10.2.1 Powers of F-tests, CDFs of p-values

Given the response $\boldsymbol{y}$ and P and Q, a formal test based on $F_P(\boldsymbol{y})$ rejects $\text{H}_0 : PX\boldsymbol{\beta} = \mathbf{0}$ at the α level of significance if $F_P(\boldsymbol{y}) > F_{1-\alpha;\nu_P,\nu_Q}$, where the *critical value* $F_{1-\alpha;\nu_P,\nu_Q}$ is the $1-\alpha$ quantile of the central F distribution with ν_P and ν_Q degrees of freedom, respectively. The p-value for H_0 based on the realized value $F_P(\boldsymbol{y})$ is the right-tail probability

$$p(\boldsymbol{y}) = \Pr[F > F_P(\boldsymbol{y})|\nu_P, \nu_Q] \tag{10.3}$$

from the central F-distribution.

If H_0 is true, if the response is normally distributed, then the p-value is uniformly distributed, and hence $\Pr(p \leqslant \alpha) = \alpha$ for each $\alpha \in [0, 1]$. The heuristic behind this process can be thought of as follows: If we conduct an experiment to assess H_0, and the outcome is something that would be impossible (probability 0) if H_0 were true, then the only logical conclusion is that H_0 is not true. In statistical inference, outcomes are never so clear, but the rationale is that occurrence of an outcome that would be very improbable if H_0 were true, and less improbable if H_0 were not true, is evidence against H_0. The strength of that evidence depends on the difference between the probabilities of the outcome if H_0 is true or not true. This translates into the CDF of the p-value. It should be uniform under H_0, hence known, and ideally the CDF should be greater than the uniform CDF if H_0 is not true, and the greater the better.

Given the matrices P and Q such that $F_P(\boldsymbol{Y})$ follows an F-distribution, define

$$\Phi(\alpha, \delta_P^2|\nu_P, \nu_Q) = \Pr[F_P(\boldsymbol{Y}) > F_{1-\alpha;\nu_P,\nu_Q}|\nu_P, \nu_Q, \delta_P^2]. \tag{10.4}$$

Given α (and ν_P and ν_Q), as a function of δ_P^2 it is the power function of the size-α test of H_0. Given δ_P^2, as a function of α it is the CDF of the p-value for H_0 based on F_P. Ghosh (1970, 1973) established that right-tail probabilities like (10.4) are monotone increasing in the ncp and, for fixed positive ncp, monotone increasing in denominator degrees of freedom and monotone decreasing in numerator degrees of freedom.

The ncp is 0, the distribution is central, and the power is α, for any $\boldsymbol{\beta}$ such that $PX\boldsymbol{\beta} = \mathbf{0}$. Furthermore, the ncp is strictly increasing in any direction (along any ray $c\boldsymbol{\beta}$, $c > 0$) such that $PX\boldsymbol{\beta}$ is not $\mathbf{0}$.

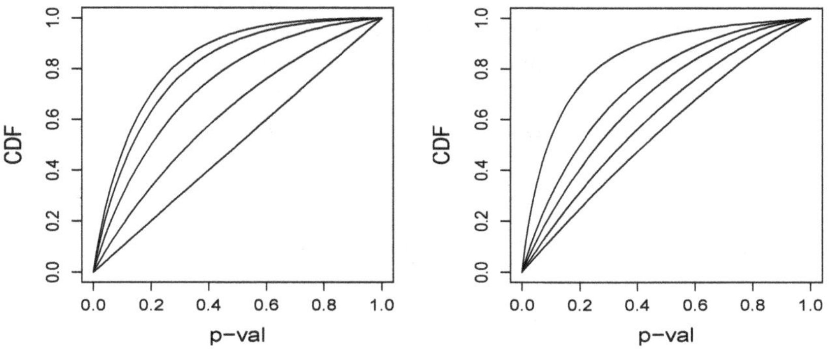

FIGURE 10.1: CDFs of p-values from $F_{\nu_P,\,\nu_Q;\,\delta^2}$ for $\nu_Q = 3$. Left: $\nu_P = 3$ and $\delta^2 = 0, 2, 5, 8, 10$, bottom up. Right: $\delta^2 = 5$ and $\nu_P = 1, 3, 5, 10, 25$, top down.

These facts also mean that, for fixed ν_P and ν_Q, the CDF of the p-value is, at each value α, increasing in δ_P^2: the probability of lesser p-values increases with δ_P^2. For fixed $\delta_P^2 > 0$, the CDF at each α is increasing in ν_Q and decreasing in ν_P. See Figure 10.1.

Because of these monotonicity properties, we may say that the probability of results more extreme than those observed is greater than p or not greater than p, depending on whether H_0 is false or true, respectively.

10.2.2 Inference Based on F

Consider now that we seek inference on a particular linear function $G'\boldsymbol{\beta}$ of $\boldsymbol{\beta}$, in the form of a p-value or a test of $H_0 : G'\boldsymbol{\beta} = \mathbf{0}$, based on the statistic $F_P(\boldsymbol{y})$. Considering Q given, the distribution of F_P depends only on the choice of P, and only as P affects the ncp and numerator degrees of freedom. Of these two, only the ncp is affected by $\boldsymbol{\beta}$.

It is desirable that δ_P^2 be *sensitive* to departures of $G'\boldsymbol{\beta}$ from $\mathbf{0}$, and that its value be *specific* to such departures. The first desideratum corresponds to the condition that $G'\boldsymbol{\beta} \neq \mathbf{0} \implies \delta_P^2(\boldsymbol{\beta}) > 0$, which is equivalent to $\mathrm{sp}(G) \subset \mathrm{sp}(X'P)$. The second, $\delta_P^2(\boldsymbol{\beta}) > 0 \implies G'\boldsymbol{\beta} \neq \mathbf{0}$, is equivalent to $\mathrm{sp}(X'P) \subset \mathrm{sp}(G)$. If the first holds, we might say that F_P ***tests*** $G'\boldsymbol{\beta}$; if the second holds, that F_P ***tests only*** $G'\boldsymbol{\beta}$. If both hold, we will say that F_P ***tests exclusively*** $G'\boldsymbol{\beta}$.

The ncp of F_P is 0 iff $PX\boldsymbol{\beta} = \mathbf{0}$, and thus F_P tests exclusively $PX\boldsymbol{\beta} = \mathbf{0}$. Note that if G is a matrix such that $\mathrm{sp}(G) = \mathrm{sp}(X'P)$, then F_P tests exclusively $G'\boldsymbol{\beta}$, because $PX\boldsymbol{\beta} = \mathbf{0}$ iff $G'\boldsymbol{\beta} = \mathbf{0}$. The matrix G used to specify the proposition is not unique, and any matrix that spans the same linear subspace specifies the same proposition. What is fixed and unique is the linear subspace $\mathcal{S}_0 = \text{sp}(X'P)^\perp = \text{sp}(G)^\perp$ to which $\boldsymbol{\beta}$ is hypothesized to belong.

For power and the distribution of p-values, greater ncp and denominator degrees of freedom and lesser numerator degrees of freedom are better. Stated more completely, suppose P_1 and P_2 are $n \times n$ symmetric idempotent matrices such that $\text{sp}(X'P_1) = \text{sp}(X'P_2)$. Let SS_j and ν_j denote SS_{P_j} and $\text{tr}(P_j)$, $j = 1, 2$.

Both SS_1 and SS_2 test exclusively $\text{H}_0 : G'\boldsymbol{\beta} = \mathbf{0}$, where $\text{sp}(G) = \text{sp}(X'P_1) = \text{sp}(X'P_2)$. By the properties of the F-distribution discussed above, the power function and the CDF of p-values for SS_1 are uniformly $\geqslant$ those functions for SS_2 if

$$X'P_1X - X'P_2X \text{ is nnd and } \nu_1 \leqslant \nu_2.$$

We may say, in that case, that SS_1 is ***as good as*** SS_2 for testing H_0, and that it is ***better than*** SS_2 if in addition either $X'P_1X - X'P_2X$ is non-zero or $\nu_1 < \nu_2$.

Suppose that F_P tests exclusively $G'\boldsymbol{\beta}$. Suppose further that $G = (G_1, G_2)$ with $\text{sp}(G_1) \neq \text{sp}(G)$. Then $G_1'\boldsymbol{\beta} \neq \mathbf{0}$ implies that $G'\boldsymbol{\beta} \neq \mathbf{0}$ and hence that $\delta_P^2 > 0$, and so F_P tests $G_1'\boldsymbol{\beta}$. But $\delta_P^2 > 0$ does not imply that $G_1'\boldsymbol{\beta} \neq \mathbf{0}$, and so F_P does not test only $G_1'\boldsymbol{\beta}$. That F_P tests $G'\boldsymbol{\beta}$ exclusively implies that it tests both, but it does not imply that it tests either component exclusively. That $\delta_P^2 > 0$ implies that either $G_1'\boldsymbol{\beta} \neq \mathbf{0}$ or $G_2'\boldsymbol{\beta} \neq \mathbf{0}$, but not that both are non-zero.

This definition of "tests" depends only on the ncp of the numerator SS, and the ncp is zero iff $PX\boldsymbol{\beta} = \mathbf{0}$. That means that the linear functions tested in the form (10.1) are all estimable even if the target $G'\boldsymbol{\beta}$ is not.

Consider a function $\boldsymbol{g}'\boldsymbol{\beta}$ that is not estimable, so that $\boldsymbol{g} \notin \text{sp}(X')$. Let $\boldsymbol{b}_* = (\text{I} - \mathbf{P}_{X'})\boldsymbol{g}$. Then $\boldsymbol{g}'\boldsymbol{b}_* \neq 0$, but $X\boldsymbol{b}_* = \mathbf{0}$ and hence $\delta_P^2(X\boldsymbol{b}_*) = 0$. This means that, if $\boldsymbol{g}'\boldsymbol{\beta}$ is not estimable, then no F_P tests $\boldsymbol{g}'\boldsymbol{\beta}$ in the meaning just defined.

10.2.3 Propositions with Non-zero RHSs, Confidence Sets

Hypotheses with non-zero right-hand sides (RHSs), like $\text{H}_0 : PX\boldsymbol{\beta} = \boldsymbol{c}_0$, with $\boldsymbol{c}_0 \in \text{sp}(PX)$, can be tested in the form (10.1) by replacing $\boldsymbol{y}$ by $\boldsymbol{y} - X\boldsymbol{b}_0$ where $X\boldsymbol{b}_0$ satisfies $PX\boldsymbol{b}_0 = \boldsymbol{c}_0$. The resulting ncp is

$$\delta^2 = (\boldsymbol{\beta} - \boldsymbol{b}_0)'X'PX(\boldsymbol{\beta} - \boldsymbol{b}_0)/\sigma^2,$$

and it is 0 iff $PX\boldsymbol{\beta} = \boldsymbol{c}_0$. That is, $F_P(\boldsymbol{y} - X\boldsymbol{b}_0)$ tests exclusively that $PX\boldsymbol{\beta} = \boldsymbol{c}_0$.

Given $\boldsymbol{y}$, a p-value $p(\boldsymbol{y}, \boldsymbol{c}_0)$ can be had for each possible value $\boldsymbol{c}_0$ of $PX\boldsymbol{\beta}$. It can be shown that a $100(1-\alpha)\%$ confidence set on $PX\boldsymbol{\beta}$ comprises all those values of $\boldsymbol{c}_0$ such that $p(\boldsymbol{y}, \boldsymbol{c}_0) \geqslant \alpha$. This in turn translates into the ellipsoidal set (in $\boldsymbol{c}_0$)

$$\{\boldsymbol{c}_0 : (P\boldsymbol{y} - \boldsymbol{c}_0)'(P\boldsymbol{y} - \boldsymbol{c}_0) \leqslant \nu_P\hat{\sigma}^2 F_{1-\alpha;\nu_P,\nu_E}\}. \tag{10.5}$$

10.3 The Unique Best Numerator SS

F_P tests exclusively $PX\boldsymbol{\beta} = (X'P)'\boldsymbol{\beta}$. Proposition 10.1 establishes that $G'\boldsymbol{\beta}$ can be tested exclusively iff it is estimable. Proposition 10.2 establishes that, given X and G such that $G'\boldsymbol{\beta}$ is estimable, there is a unique best numerator SS for F-statistics that test exclusively $G'\boldsymbol{\beta}$.

Proposition 10.1. *In order that there exist a symmetric idempotent matrix P such that* $\mathrm{sp}(X'P) = \mathrm{sp}(G)$ *it is necessary and sufficient that* $\mathrm{sp}(G) \subset \mathrm{sp}(X')$.

Given G such that $\mathrm{sp}(G) \subset \mathrm{sp}(X')$, consider the set of all matrices L such that $\mathrm{sp}(X'L) = \mathrm{sp}(G)$. Clearly, then there is among them a matrix H such that $\mathrm{sp}(H) \subset \mathrm{sp}(X)$ and $\mathrm{sp}(X'H) = \mathrm{sp}(G)$ ($H = \mathbf{P}_X L$, for example). Proposition 10.2 establishes that, among matrices L such that SS_L tests exclusively $G'\boldsymbol{\beta}$, SS_H is uniquely best.

For a matrix C, let $\nu_C = \dim \mathrm{sp}(C) = \mathrm{tr}(\mathbf{P}_C) = \mathrm{rank}(C)$.

Proposition 10.2. *Let X, H, and L be non-zero matrices with n rows such that* $\mathrm{sp}(H) \subset \mathrm{sp}(X)$ *and* $\mathrm{sp}(X'L) = \mathrm{sp}(X'H)$. *Then*

1. $\mathrm{sp}(\mathbf{P}_X L) = \mathrm{sp}(H)$,
2. $\mathrm{sp}(L) \subset \mathrm{sp}(H) + \mathrm{sp}(X)^{\perp}$,
3. $X'\mathbf{P}_H X - X'\mathbf{P}_L X$ *is nnd,*
4. $\nu_H = \nu_{\mathbf{P}_X L} \leqslant \nu_L \leqslant \nu_H + n - \nu_X$, *and*
5. *if* $\mathbf{P}_L \neq \mathbf{P}_H$ *and* $X'\mathbf{P}_H X - X'\mathbf{P}_L X = 0$, *then* $\nu_H < \nu_L$.

If L is a matrix such that $\mathrm{sp}(X'L) = \mathrm{sp}(X'H)$, then both SS_L and SS_H test exclusively $H'X\boldsymbol{\beta}$. If $\mathrm{sp}(L) \subset \mathrm{sp}(X)$, then $\mathbf{P}_L = \mathbf{P}_H$, by Proposition 10.2(1). The ncp of SS_H is everywhere $\geqslant$ the ncp of SS_L, and its df is $\leqslant$ the df of SS_L. By the properties of F_P noted above, no such SS_L is better than SS_H for inference about $H'X\boldsymbol{\beta}$. If $X'\mathbf{P}_H X - X'\mathbf{P}_L X \neq 0$, then there exist $\boldsymbol{\beta}$s for which $\delta_H^2(\boldsymbol{\beta}) > \delta_L^2(\boldsymbol{\beta})$, and so SS_H is better than SS_L. Proposition 10.2(5) establishes the same when $X'\mathbf{P}_H X - X'\mathbf{P}_L X = 0$ and $\mathbf{P}_L \neq \mathbf{P}_H$, because then $\nu_H < \nu_L$, and hence SS_H is better than SS_L. The proof given in Appendix A for Proposition 10.2(5) works, but it seems to be more complicated than necessary.

One of the interesting consequences of these results is that if $\nu_L = \nu_H$, then either $\mathbf{P}_L = \mathbf{P}_H$ or $X'\mathbf{P}_H X - X'\mathbf{P}_L X$ is nnd and non-zero, which would imply that SS_H is better than SS_L. These conditions occur, for example, when $G'\boldsymbol{\beta}_1$ is estimable both in the model $X_1\boldsymbol{\beta}_1$ (call this model 1) and in the model $X_1\boldsymbol{\beta}_1 + X_2\boldsymbol{\beta}_2$ (model 12). Denote its best numerator SSs in the two models by SS_1 and SS_{12} and their respective dfs correspondingly by ν_1 and

ν_{12}. Then both SSs test exclusively $G'\boldsymbol{\beta}_1$ in model 1, $\nu_1 = \nu_{12}$, and this implies that SS_1 is better than SS_{12} in model 1. This is one argument for parsimony in formulating multiple regression models. (However, that advantage may be confounded with the choice of denominator SS.)

Computing H, given X and an estimable G, is straightforward. Compute a solution L to $X'L = G$ and then compute $H = \mathbf{P}_X L$.

If G is not estimable, then only the estimable part of G can be tested. One way to identify the estimable part is to find H such that $\text{sp}(H) \subset \text{sp}(X)$ and $\text{sp}(X'H) = \text{sp}(X') \cap \text{sp}(G)$. Then SS_H is the best numerator SS for testing $H'X\boldsymbol{\beta}$. The next section establishes that, for any G, estimable or not, SS_H that tests exclusively the estimable part of G results naturally and directly as the extra SSE due to imposing the conditions $G'\boldsymbol{\beta} = \mathbf{0}$ on the model.

Note that, in this section, there is no fitting models by least squares and no estimation going on. The focus is instead on F-statistics and functions $G'\boldsymbol{\beta}$ that can be tested exclusively. That property is equivalent to estimability despite the absence of any pretense at estimation, and so I have used the same word rather than inventing a new one, like "testability." Note further that there is no mention of "unbiased" here, nor even of linear estimation. If we begin with F-statistics in the form (10.1), and if G is such that $G'\boldsymbol{\beta}$ can be tested exclusively, then SS_H is the uniquely best numerator SS. On the other hand, it can be argued that the F-statistic evolved from least squares, linear estimation, and the GM Theorem.

10.4 The RMFM SS

There are two general-purpose schemes that have long been used in practice to construct numerator sums of squares for F-statistics. This and the next section describe them and establish some of their properties.

The increase in SSE due to imposing the condition $G'\boldsymbol{\beta} = \mathbf{0}$ on the model $X\boldsymbol{\beta}$ is called here the ***Restricted Model–Full Model*** SS (***RMFM*** SS). Let N be a matrix such that the restricted model under the condition that $G'\boldsymbol{\beta} = \mathbf{0}$ is $\text{sp}(XN)$. That is, N is a matrix such that

$$\{X\boldsymbol{\beta} : \boldsymbol{\beta} \in \Re^{k+1} \text{ and } G'\boldsymbol{\beta} = \mathbf{0}\} = \text{sp}(XN). \tag{10.6}$$

It can be shown that any N such that $\text{sp}(G)^{\perp} \subset \text{sp}(N) \subset [\text{sp}(X') \cap \text{sp}(G)]^{\perp}$ will do. Then SSE for the full model is $\boldsymbol{y}'(\text{I} - \mathbf{P}_X)\boldsymbol{y}$, SSE for the restricted model is $\boldsymbol{y}'(\text{I} - \mathbf{P}_{XN})\boldsymbol{y}$, and the increase in SSE due to the restriction is $\boldsymbol{y}'(\mathbf{P}_X - \mathbf{P}_{XN})\boldsymbol{y}$. With H such that $\mathbf{P}_H = \mathbf{P}_X - \mathbf{P}_{XN}$, the RMFM SS is SS_H. Note particularly that $\text{sp}(H) \subset \text{sp}(X)$.

Proposition 10.3 establishes that the RMFM SS SS_H tests exclusively the estimable part of $G'\boldsymbol{\beta}$. By Proposition 10.2, SS_H is the unique best numerator SS for an F-statistic that tests exclusively the estimable part of $G'\boldsymbol{\beta}$.

Proposition 10.3. *Let X and G be $n \times (k+1)$ and $(k+1) \times r$ matrices, respectively, both non-zero. Let N be a matrix such that* $\text{sp}(XN) = \{X\boldsymbol{\beta} : \boldsymbol{\beta} \in \Re^{k+1}$ *and* $G'\boldsymbol{\beta} = \mathbf{0}\}$.

1. *If H is a matrix such that* $\text{sp}(H) \subset \text{sp}(X)$ *and* $\text{sp}(X'H) = \text{sp}(X') \cap \text{sp}(G)$, *then* $\mathbf{P}_H = \mathbf{P}_X - \mathbf{P}_{XN}$.

2. *If H is a matrix such that* $\mathbf{P}_H = \mathbf{P}_X - \mathbf{P}_{XN}$, *then* $\text{sp}(X'H) = \text{sp}(X') \cap \text{sp}(G)$.

This point is worth emphasizing. Given X and G, finding the unique best numerator SS for an F-statistic that tests exclusively the estimable part of $G'\boldsymbol{\beta}$ requires only SSEs for the full model and the restricted model. It does not require checking whether $G'\boldsymbol{\beta}$ is estimable, and it does not require identifying the estimable part.

SS_H also can be had directly as $\boldsymbol{y}'\mathbf{P}_H\boldsymbol{y}$, where $\text{sp}(H) = \text{sp}(X) \cap \text{sp}(XN)^{\perp}$. One sequence of steps to get H is: Find N with GS on (G, I), yielding (Q_1, Q_2), and $N = Q_2$. Find H as (a different) Q_2 from GS on (XN, X), so that $H = Q_2$ and $\mathbf{P}_H = Q_2Q_2'$.

If $G'\boldsymbol{\beta}$ is estimable, then its estimable part is $\text{sp}(G)$. In that case "$G'\boldsymbol{\beta}$" replaces "the estimable part of $G'\boldsymbol{\beta}$" in all the results established above.

It can happen that $\text{sp}(X') \cap \text{sp}(G) = \{\mathbf{0}\}$. In that case $\text{sp}(X) = \text{sp}(XN)$, $\mathbf{P}_H = 0_{n\times n}$, and hence $SS_H \equiv 0$.

If $G'\boldsymbol{\beta}$ is not estimable, then, as shown in Chapter 9, there are vectors $\boldsymbol{b}$ such that $H'X\boldsymbol{b} = \mathbf{0}$, and hence the ncp is 0, but $G'\boldsymbol{b} \neq \mathbf{0}$. In that case, the RMFM numerator SS does not test $\text{H}_0 : G'\boldsymbol{\beta} = \mathbf{0}$. It does test the estimable part of $G'\boldsymbol{\beta}$, that is, that $H'X\boldsymbol{\beta} = \mathbf{0}$.

In practical applications of multiple regression analysis, the simplest effects to test are those due to deleting a set of predictor variables from the full model. That is, with $X = (X_1, X_2)$, the extra SSE due to removing $X_2\boldsymbol{\beta}_2$ from the model is the easiest extra SSE to compute, because it requires only regressing $\boldsymbol{y}$ on X and regressing $\boldsymbol{y}$ on X_1. Now we can see that the extra SSE due to the conditions $H'X\boldsymbol{\beta} = \mathbf{0}$ can be computed as a deleted-variables extra SSE: the full model is (XN, H), and the restricted model is XN.

In the next proposition, the particular formulation of the models, $\mathbb{K}\boldsymbol{\eta}$, $\boldsymbol{\eta} \in \text{sp}(M_j)$, while entirely general, is related particularly to models that include effects of categorical factors, the subject of Part III in this book. The matrix $\mathbb{K}$ is special to that setting. See p. 138. As used here, though, it represents an arbitrary $n \times c$ matrix. Proposition 10.4 establishes that, if SS_L tests $G'\boldsymbol{\eta}$, $\boldsymbol{\eta} \in \text{sp}(M_2)$, in the larger model, then it tests $(\mathbf{P}_{M_1}G)'\boldsymbol{\eta}$, $\boldsymbol{\eta} \in \text{sp}(M_1)$, in the smaller model. In the larger model, SS_{H_2} also tests $G'\boldsymbol{\eta}$, and it has everywhere $\geqslant$ ncp and $\leqslant$ numerator degrees of freedom than any SS_L that tests the same hypothesis. SS_{H_1} has the same properties in model 1. In particular, in model 1, SS_{H_1} has $\geqslant$ ncp and $\leqslant$ numerator df than does SS_{H_2}.

Proposition 10.4. (Corollary to Proposition 10.2.) *Let* $\mathbb{K}$ *and* L *be matrices with* n *rows,* $\mathbb{K}$ *with* c *columns, and let* M_1 *and* M_2 *be matrices with* c *rows such that* $\text{sp}(M_1) \subset \text{sp}(M_2)$*. Let* $X_1 = \mathbb{K}M_1$*,* $X_2 = \mathbb{K}M_2$*,* $G = \mathbf{P}_{M_2}\mathbb{K}'L$*,* $H_2 = \mathbf{P}_{X_2}L$*, and* $H_1 = \mathbf{P}_{X_1}L = \mathbf{P}_{X_1}H_2$*.*

1. SS_L *tests* $G'M_2\boldsymbol{\beta}_2$ *in the model* $X_2\boldsymbol{\beta}_2$*.*

2. SS_L *tests* $(\mathbf{P}_{M_1}G)'(M_1\boldsymbol{\beta}_1)$ *in the model* $X_1\boldsymbol{\beta}_1$*.*

3. $\text{sp}(L) \subset \text{sp}(H_2) + \text{sp}(X_2)^{\perp}$.

4. $\text{sp}(L) \subset \text{sp}(H_1) + \text{sp}(X_1)^{\perp}$.

5. $\text{sp}(H_2) \subset \text{sp}(H_1) + \text{sp}(X_1)^{\perp}$.

10.5 The GLH Form

In the examples in his seminal paper, K. Pearson (1900) constructed test statistics by identifying a function such that its expected value would be zero if the proposition in question were true and not zero if it were not. He then considered the distribution of that function to be multivariate normal, and the statistic was the quadratic form in the exponent of that density function.

Let A be a matrix such that $XA' = \mathbf{P}_X$, so that $\hat{\boldsymbol{\beta}} = A'\boldsymbol{y}$ is a linear least-squares solution. A statistic like Pearson's that is intended to test $\text{H}_0 : G'\boldsymbol{\beta} = \mathbf{0}$ takes the form

$$\begin{aligned} \chi^2_G &= (G'\hat{\boldsymbol{\beta}})'[\hat{\text{V}}\text{ar}(G'\hat{\boldsymbol{\beta}})]^-(G'\hat{\boldsymbol{\beta}}) \\ &= (G'A'\boldsymbol{y})'[G'A'AG\hat{\sigma}^2]^-(G'A'\boldsymbol{y}) \\ &= \boldsymbol{y}'\mathbf{P}_{AG}\boldsymbol{y}/\hat{\sigma}^2 \\ &= \nu_{AG}F_{AG}(\boldsymbol{y}),\ \nu_{AG} = \text{tr}(\mathbf{P}_{AG}). \end{aligned} \tag{10.7}$$

The form $(G'\hat{\boldsymbol{\beta}})'[\text{Var}(G'\hat{\boldsymbol{\beta}})/\sigma^2]^-(G'\hat{\boldsymbol{\beta}}) = SS_{AG}(\boldsymbol{y}) = \boldsymbol{y}'\mathbf{P}_{AG}\boldsymbol{y}$ is often called the ***General Linear Hypothesis*** (GLH) form of the numerator SS. It is also called a ***Wald statistic.***

If $G'\boldsymbol{\beta}$ is estimable, and hence $\text{sp}(G) \subset \text{sp}(X')$, then $AG = AX'L = \mathbf{P}_XL$ for some L, and hence $\text{sp}(AG) \subset \text{sp}(X)$ and F_{AG} follows an F distribution. If $G'\boldsymbol{\beta}$ is not estimable, then it can happen that A could be chosen such that $XA' = \mathbf{P}_X$ but $\text{sp}(AG)$ contains members not in $\text{sp}(X)$, in which case F_{AG} would not follow an F distribution because the numerator and denominator SSs would not be independent. See Exercise 21 on p. 101.

To forestall that possibility, A such that $XA' = \mathbf{P}_X$ can be replaced with $\mathbf{P}_XA$, so that $\text{sp}(AG) \subset \text{sp}(X)$. Further, by Proposition 10.2, SS_{AG} is then the unique best SS that tests exclusively $G'A'X\boldsymbol{\beta} = \mathbf{0}$.

Under the GM model, if $\text{sp}(AG) \subset \text{sp}(X)$ and $\hat{\sigma}^2 = MSE$, then F_{AG} follows an F distribution. Its ncp is $\boldsymbol{\beta}'X'\mathbf{P}_{AG}X\boldsymbol{\beta}/\sigma^2$, which is 0 iff $G'A'X\boldsymbol{\beta} = \mathbf{0}$: F_{AG} tests exclusively $(X'AG)'\boldsymbol{\beta}$, while its intended target is $G'\boldsymbol{\beta}$. Proposition 10.5 establishes relations between F_H, F_{AG}, and their ncps.

Proposition 10.5. *Let G be a $(k+1) \times g$ matrix and X an $n \times (k+1)$ matrix. Let H be a matrix such that $\text{sp}(H) \subset \text{sp}(X)$ and $\text{sp}(X'H) = \text{sp}(X') \cap \text{sp}(G)$. Let A be a matrix such that $XA' = \mathbf{P}_X$.*

1. $\text{sp}(H) \subset \text{sp}(AG)$ *and* $\text{sp}(X'H) \subset \text{sp}(X'AG)$.

2. $\text{sp}(X'AG) = \text{sp}(G)$ *iff* $\text{sp}(G) \subset \text{sp}(X')$.

For any G, estimable or not, $\text{sp}(H) \subset \text{sp}(AG)$. That implies that $\mathbf{P}_{AG} - \mathbf{P}_H$ is a symmetric, idempotent matrix, and hence $\boldsymbol{y}'\mathbf{P}_{AG}\boldsymbol{y} \geqslant \boldsymbol{y}'\mathbf{P}_H\boldsymbol{y}$ for all $\boldsymbol{y} \in \Re^n$. Furthermore, $\text{sp}(X'H) \subset \text{sp}(X'AG)$, and so $H'X\boldsymbol{\beta} \neq \mathbf{0}$ causes the ncp of F_{AG} to be positive. If $G'\boldsymbol{\beta}$ is estimable, then $\text{sp}(G) = \text{sp}(X'H)$, $\text{sp}(AG) = \text{sp}(AX'H) = \text{sp}(\mathbf{P}_X H) = \text{sp}(H)$, and hence $\mathbf{P}_{AG} = \mathbf{P}_H$. If $G'\boldsymbol{\beta}$ is not estimable, in some cases (seemingly unusual) there exist matrices A such that $XA' = \mathbf{P}_X$ and $\text{sp}(X'AG) = \text{sp}(X'H)$. In those special cases F_{AG} tests the estimable part of $G'\boldsymbol{\beta}$. Then, by Proposition 10.1, either it is identical to SS_H or SS_H is better. Otherwise, it also tests other functions $\boldsymbol{g}'\boldsymbol{\beta}$ with $\boldsymbol{g}$ in $\text{sp}(X')$ but not in $\text{sp}(G)$. It is possible, then, that $H'X\boldsymbol{\beta} = 0$ but also $\delta^2_{AG} > 0$: then the magnitude of $F_{AG}(\boldsymbol{y})$ cannot be taken as evidence whether H_0 is false.

There is no impediment to constructing $F_{AG}(\boldsymbol{y})$ for any G, without regard to its estimability. However, while the function that it tests, $G'A'X\boldsymbol{\beta}$, is estimable, it is the same as $G'\boldsymbol{\beta}$ if $G'\boldsymbol{\beta}$ is estimable but not necessarily otherwise.

10.6 Rationales Leading to Tests

As noted in the Introduction to this chapter, there is not a neat, logical argument that leads to the conventional test statistic as the unique best one in terms of power. However, several rationales seem reasonable and appealing. One is the "extra" SSE due to imposing the conditions $G'\boldsymbol{\beta} = \mathbf{0}$ on the model, which leads to the RMFM SS described in Section 10.4. If any part of $G'\boldsymbol{\beta}$ is estimable, imposing the conditions always increases SSE; but if the increase is small, then apparently the restricted model fits about as well as the full model. It remains only to calibrate the increase in terms of a probability statement, and that follows neatly from the distribution of (10.1).

Other rationales are described next. From the previous section, H is a matrix such that $\mathbf{P}_H = \mathbf{P}_X - \mathbf{P}_{XN}$, and N is a matrix such that $\text{sp}(G)^{\perp} \subset \text{sp}(N) \subset [\text{sp}(X') \cap \text{sp}(G)]^{\perp}$.

10.6.1 Squared Distance from $\hat{\boldsymbol{y}}$ to $\mathcal{M}_0$

It is important to keep in mind that the only information in $\boldsymbol{y}$ about $\boldsymbol{\beta}$ comes indirectly through the mean vector. The development in this subsection is to demonstrate that full models and restricted models for the mean vector can be expressed directly in terms of linear subspaces, without expressing models in terms of $X\boldsymbol{\beta}$ and $\boldsymbol{\beta}$. Thus X and G are arbitrary, as long as they produce the same linear subspaces sp(X) and sp(G). This is sometimes called a *coordinate-free* formulation of linear models and linear hypotheses. While it is elegant, formulations of models and hypotheses are usually easier and more intuitively appealing in terms of X, $\boldsymbol{\beta}$, and G in practical settings. Here, it serves to emphasize that the parameterization $X\boldsymbol{\beta}$ is considerably arbitrary, and that any inference about $\boldsymbol{\beta}$ (which also is somewhat arbitrary) must come through the mean vector itself.

A concomitant objective here is to see that the RMFM SS can be viewed as the squared distance from $\hat{\boldsymbol{y}}$ in the linear subspace that is the full model to its nearest neighbor in the restricted model. Let the full model for the mean vector be $\mathcal{M}$, which is a linear subspace of $\Re^n$. Let $\mathcal{M}_0$, the restricted model, be a linear subspace of $\mathcal{M}$. Given a vector $\boldsymbol{m}$ in $\mathcal{M}$, the vector $\boldsymbol{m}_0$ in $\mathcal{M}_0$ that minimizes the squared Euclidean distance from $\boldsymbol{m}$ to $\boldsymbol{m}_0$, $(\boldsymbol{m}-\boldsymbol{m}_0)'(\boldsymbol{m}-\boldsymbol{m}_0)$, is $\hat{\boldsymbol{m}} = \mathbf{P}_{\mathcal{M}_0}\boldsymbol{m}$, and

$$(\boldsymbol{m} - \hat{\boldsymbol{m}})'(\boldsymbol{m} - \hat{\boldsymbol{m}}) = \boldsymbol{m}'(\mathbf{P}_{\mathcal{M}} - \mathbf{P}_{\mathcal{M}_0})\boldsymbol{m}.$$

Given that $\boldsymbol{m} \in \mathcal{M}$, this sum of squares is 0 iff $\boldsymbol{m} \in \mathcal{M}_0$. It is a natural gauge of how far $\boldsymbol{m}$ is from being in $\mathcal{M}_0$.

Given the response vector $\boldsymbol{y}$, the nearest member of $\mathcal{M}$ to $\boldsymbol{y}$ is $\hat{\boldsymbol{y}} = \mathbf{P}_{\mathcal{M}}\boldsymbol{y}$, and the squared distance from it to its nearest neighbor in $\mathcal{M}_0$ is

$$(\hat{\boldsymbol{y}} - \mathbf{P}_{\mathcal{M}_0}\hat{\boldsymbol{y}})'(\hat{\boldsymbol{y}} - \mathbf{P}_{\mathcal{M}_0}\hat{\boldsymbol{y}}) = \hat{\boldsymbol{y}}'(\mathbf{P}_{\mathcal{M}} - \mathbf{P}_{\mathcal{M}_0})\hat{\boldsymbol{y}} = \boldsymbol{y}'(\mathbf{P}_{\mathcal{M}} - \mathbf{P}_{\mathcal{M}_0})\boldsymbol{y}. \tag{10.8}$$

This is the RMFM SS for testing $\text{H}_0 : \boldsymbol{\mu} \in \mathcal{M}_0$, which is equivalent to $\text{H}_0 : H'\boldsymbol{\mu} = \mathbf{0}$, where $\mathbf{P}_H = \mathbf{P}_{\mathcal{M}} - \mathbf{P}_{\mathcal{M}_0}$. Its ncp is proportional to $\boldsymbol{\mu}'\mathbf{P}_H\boldsymbol{\mu}$, and so, for $\boldsymbol{\mu} \in \mathcal{M}$, it tests exclusively that $H'\boldsymbol{\mu} = \mathbf{0}$.

The squared distance from $\boldsymbol{y}$ itself (instead of $\hat{\boldsymbol{y}}$) to $\mathbf{P}_{\mathcal{M}_0}\boldsymbol{y}$ can be partitioned orthogonally into the sum of the denominator and numerator SSs of (10.1) as

$$\boldsymbol{y}'(\text{I} - \mathbf{P}_{\mathcal{M}_0})\boldsymbol{y} = \boldsymbol{y}'(\text{I} - \mathbf{P}_{\mathcal{M}})\boldsymbol{y} + \boldsymbol{y}'(\mathbf{P}_{\mathcal{M}} - \mathbf{P}_{\mathcal{M}_0})\boldsymbol{y}. \tag{10.9}$$

The two SSs in (10.9) can be viewed as squared distances from $\boldsymbol{y}$ to $\hat{\boldsymbol{y}}$ in the full model and from $\hat{\boldsymbol{y}}$ to $\mathbf{P}_{\mathcal{M}_0}\boldsymbol{y}$ in the restricted model. The test statistic is proportional to the ratio of these two.

Now suppose X, G, and N are given matrices such that sp(N) = sp$(G)^\perp$, $\mathcal{M}$ = sp(X), and $\mathcal{M}_0$ = sp(XN). (Note that sp(N) = $[\text{sp}(G) \cap \text{sp}(X')]^\perp$ would work as well.) Then for any matrices with these properties, the RMFM SS for testing $\text{H}_0 : G'\boldsymbol{\beta} = \mathbf{0}$ in the model $X\boldsymbol{\beta}$ is exactly (10.8).

10.6.2 The Likelihood-Ratio Test Statistic

As another rationale leading to the same test statistic, the likelihood-ratio test statistic is defined as the ratio of the suprema of the likelihood functions restricted by H_0 and unrestricted, respectively. It is an appealing statistic because it has good, known asymptotic properties.

Assuming that the response $\boldsymbol{Y}$ follows a multivariate normal distribution, the likelihood function at the realized response $\boldsymbol{y}$ is

$$L(\boldsymbol{\mu}, \sigma^2; \boldsymbol{y}) = (2\pi)^{-n/2}(\sigma^2)^{-n/2}\exp\{-(\boldsymbol{y}-\boldsymbol{\mu})'(\boldsymbol{y}-\boldsymbol{\mu})/(2\sigma^2)\}. \qquad (10.10)$$

We'll sketch the derivation of the supremum of L first for an arbitrary model for $\boldsymbol{\mu}$, an affine set, say $\boldsymbol{\alpha}_0 + \mathcal{S}$, where $\mathcal{S}$ is a linear subspace of $\Re^n$ and $\boldsymbol{\alpha}_0$ is a given n-vector.

Examining the log of the likelihood function as a function of σ^2 for fixed $\boldsymbol{\mu}$, it may be seen that this function is increasing for $\sigma^2 < (\boldsymbol{y}-\boldsymbol{\mu})'(\boldsymbol{y}-\boldsymbol{\mu})/n$ and decreasing for $\sigma^2 > (\boldsymbol{y}-\boldsymbol{\mu})'(\boldsymbol{y}-\boldsymbol{\mu})/n$, so, for any given $\boldsymbol{\mu}$, it is maximized at $\tilde{\sigma}^2(\boldsymbol{\mu}) = (\boldsymbol{y}-\boldsymbol{\mu})'(\boldsymbol{y}-\boldsymbol{\mu})/n$. Then the ***profile*** likelihood is

$$\log L(\boldsymbol{\mu}, \tilde{\sigma}^2(\boldsymbol{\mu}); \boldsymbol{y}) = -\frac{n}{2}\log(2\pi) - \frac{n}{2}\log[\tilde{\sigma}^2(\boldsymbol{\mu})] - \frac{n}{2};$$

and it is maximized with respect to $\boldsymbol{\mu}$ where $\boldsymbol{\mu}$ is such that $(\boldsymbol{y}-\boldsymbol{\mu})'(\boldsymbol{y}-\boldsymbol{\mu})$ is minimized. This occurs at

$$\hat{\boldsymbol{\mu}} = \boldsymbol{\alpha}_0 + \mathbf{P}_{\mathcal{S}}(\boldsymbol{y} - \boldsymbol{\alpha}_0).$$

Then

$$\tilde{\sigma}^2 = \tilde{\sigma}^2(\hat{\boldsymbol{\mu}}) \;=\; (\boldsymbol{y}-\boldsymbol{\alpha}_0)'(\mathrm{I} - \mathbf{P}_{\mathcal{S}})(\boldsymbol{y}-\boldsymbol{\alpha}_0)/n.$$

Now consider the full model $\mathcal{M} = \mathrm{sp}(X)$ and the restricted model $\mathcal{M}_0 = \mathrm{sp}(X) \cap \mathrm{sp}(H)^{\perp}$. As shown above, this is the same as $\{X\boldsymbol{\beta} : \boldsymbol{\beta} \in \Re^{k+1} \text{ and } G'\boldsymbol{\beta} = \mathbf{0}\}$. In the full model, the quadratic form in $\tilde{\sigma}^2$ is $Q = SSE$. In the restricted model, it is

$$Q_0 = \boldsymbol{y}'(\mathrm{I} - \mathbf{P}_{\mathcal{M}_0})\boldsymbol{y}.$$

To test the hypothesis $\mathrm{H}_0 : \boldsymbol{\mu} \in \mathcal{M}_0$, the likelihood ratio statistic is

$$LR(\boldsymbol{y}) = \left(\frac{Q}{Q_0}\right)^{n/2}.$$

From (10.9),

$$Q_0 \;=\; Q + \boldsymbol{y}'\mathbf{P}_H\boldsymbol{y}.$$

Then it can be seen that the likelihood-ratio statistic is a monotone decreasing function of $F_H(\boldsymbol{y})$, the RMFM test statistic.

10.6.3 $T^2_{\max}$, Scheffé's Test

Let A be a matrix such that $XA' = \mathbf{P}_X$. Given $\boldsymbol{y}$, $\hat{\boldsymbol{\beta}} = A'\boldsymbol{y}$ is a least-squares solution. Suppose our target of inference is $\mathrm{H}_0 : G'\boldsymbol{\beta} = \mathbf{0}$, where G is a given $(k+1) \times r$ matrix. Of course, $G'\boldsymbol{\beta} = \mathbf{0}$ iff $\boldsymbol{g}'\boldsymbol{\beta} = 0$ for every $\boldsymbol{g} \in \mathrm{sp}(G)$. For any such $\boldsymbol{g}$ in $\mathrm{sp}(G)$, to test that $\boldsymbol{g}'\boldsymbol{\beta} = 0$ the conventional Student's t statistic is

$$T_{\boldsymbol{g}} = \frac{\boldsymbol{g}'\hat{\boldsymbol{\beta}}}{\sqrt{\hat{\mathrm{V}}\mathrm{ar}(\boldsymbol{g}'\hat{\boldsymbol{\beta}})}}. \tag{10.11}$$

A simple rationale for testing all of $\mathrm{H}_0 : G'\boldsymbol{\beta} = \mathbf{0}$ is to ask what $\boldsymbol{g}$ in $\mathrm{sp}(G)$ would maximize $T^2_{\boldsymbol{g}}$ and then to use that maximum value as a test statistic.

For $\boldsymbol{g} \in \mathrm{sp}(G)$,

$$\begin{aligned}
T^2_{\boldsymbol{g}} &= \frac{(\boldsymbol{g}'\hat{\boldsymbol{\beta}})^2}{\hat{\mathrm{V}}\mathrm{ar}(\boldsymbol{g}'\hat{\boldsymbol{\beta}})} \\
&= \frac{(\boldsymbol{g}'A'\boldsymbol{y})^2}{\boldsymbol{g}'A'A\boldsymbol{g}\hat{\sigma}^2} \\
&= \frac{(\boldsymbol{g}'A'\mathbf{P}_{AG}\boldsymbol{y})^2}{\boldsymbol{g}'A'A\boldsymbol{g}\hat{\sigma}^2} \\
&\leqslant \frac{\boldsymbol{y}'\mathbf{P}_{AG}\boldsymbol{y}}{\hat{\sigma}^2}, \text{ by the Cauchy-Schwarz inequality,} \\
&= \nu_{AG}F_{AG}(\boldsymbol{y}),
\end{aligned} \tag{10.12}$$

with equality iff $A\boldsymbol{g}$ is proportional to $\mathbf{P}_{AG}\boldsymbol{y}$. $T^2_{\boldsymbol{g}}$ follows an F distribution with ncp proportional to $(\boldsymbol{g}'A'X\boldsymbol{\beta})^2$, and so it tests exclusively that $\boldsymbol{g}'A'X\boldsymbol{\beta} = 0$. Given $\boldsymbol{y}$, its maximum possible value over all vectors $\boldsymbol{g}$ in $\mathrm{sp}(G)$ is $\nu_{AG}F_{AG}(\boldsymbol{y})$. So this rationale to construct a statistic to test $G'\boldsymbol{\beta}$ leads to the GLH form $F_{AG}(\boldsymbol{y})$ aimed at testing $G'\boldsymbol{\beta}$.

If the target were in fact $G'A'X\boldsymbol{\beta}$, Scheffé (1959, pp. 70-72) showed the inequality (10.12) and concluded (p. 71), "From this relationship of the S-method [based on $T^2_{\boldsymbol{g}}$] to the F-test springs perhaps its chief usefulness: Whenever a hypothesis [$\mathrm{H}_0 : G'A'X\boldsymbol{\beta} = \mathbf{0}$] is rejected by the F-test we can investigate the different estimable functions [$\boldsymbol{g}'A'X\boldsymbol{\beta}$ with $\boldsymbol{g} \in \mathrm{sp}(G)$] ... to find out which ones are responsible for rejecting [H_0]."

10.7 Summary

Of these rationales that lead to test statistics, the squared distance from $\hat{\boldsymbol{y}}$ to the restricted model and the likelihood-ratio test both lead to the RMFM SS and $F_H(\boldsymbol{y})$, they test exclusively the estimable part of $G'\boldsymbol{\beta}$, and they do so without any effort needed to verify or identify estimable functions. The others, equivalent to the GLH form $F_{AG}(\boldsymbol{y})$, can be used for any G, and they test the estimable part of $G'\boldsymbol{\beta}$, but, unless all of $G'\boldsymbol{\beta}$ is estimable, they may also simultaneously test other unintended estimable functions, and those generally depend on the choice of A such that $XA' = \mathbf{P}_X$.

Whether F_H or F_{AG} is used, it makes sense to identify the estimable part of $G'\boldsymbol{\beta}$. If the RMFM SS is used, and not all of $G'\boldsymbol{\beta}$ is estimable, then the hypothesis that is tested exclusively is $\mathrm{H}_{0*} : H'X\boldsymbol{\beta} = \mathbf{0}$, which is different from $\mathrm{H}_0 : G'\boldsymbol{\beta} = \mathbf{0}$. Recognizing this difference can be important for interpretation of the results. On the other hand, the GLH form tests the estimable part of $G'\boldsymbol{\beta}$ plus some, and it would be sensible to identify the estimable part and provide results for it. That requires a little extra work, and when it is done, the result is the same as choosing the RMFM SS to begin with.

10.8 Illustration

The purpose of this section is to illustrate the elements and constructions covered in this chapter in a simple setting. Verify all assertions and computations.

Let

$$X = \begin{pmatrix} 1 & 1 & 0 & 0 \\ 1 & 0 & 1 & 0 \\ 1 & 0 & 0 & 1 \end{pmatrix},$$

$$G = \begin{pmatrix} 0 & 0 \\ 1 & 0 \\ 1 & 1 \\ 1 & 0 \end{pmatrix},$$

and $\boldsymbol{c}_0 = (100, 0)'$. Note that $\mathrm{sp}(X) = \Re^3$, and so $\mathbf{P}_X = \mathrm{I}_3$.

Denote the coefficient vector by $\boldsymbol{\beta} = (\beta_0, \beta_1, \beta_2, \beta_3)'$. With the full model $\mathcal{S} = \mathrm{sp}(X)$ and the hypothesis $\mathrm{H}_0 : G'\boldsymbol{\beta} = \boldsymbol{c}_0$, the restricted model is

$$\mathcal{M}_0 = \{X\boldsymbol{\beta} : \boldsymbol{\beta} \in \Re^4, \beta_1 + \beta_2 + \beta_3 = 100, \beta_2 = 0\}.$$

Denote the columns of G by $\boldsymbol{g}_1$ and $\boldsymbol{g}_2$. Verify that neither $\boldsymbol{g}_1$ nor $\boldsymbol{g}_2$ is in $\mathrm{sp}(X')$, but $\boldsymbol{g}_* = \boldsymbol{g}_1 - 3\boldsymbol{g}_2 = (0, 1, -2, 1)' = X'\boldsymbol{h}$, with $\boldsymbol{h} = (1, -2, 1)'$, is in $\mathrm{sp}(X')$. Thus $G'\boldsymbol{\beta}$ is not estimable, and it can be shown that the estimable part of $\mathrm{sp}(G)$ is $\mathrm{sp}(G) \cap \mathrm{sp}(X') = \mathrm{sp}(\boldsymbol{g}_*)$.

Suppose that the mean vector $\boldsymbol{\mu}$ in this model is known to be

$$\boldsymbol{\mu}_0 = \begin{pmatrix} 110 \\ 75 \\ 140 \end{pmatrix}.$$

Let $\boldsymbol{\beta}_0 = (75, 35, 0, 65)'$, and verify that $G'\boldsymbol{\beta}_0 = \boldsymbol{c}_0$ and $\boldsymbol{\mu}_0 = X\boldsymbol{\beta}_0$. From the results established in Equation (7.2), p. 52,

$$\{G'\boldsymbol{\beta} : X\boldsymbol{\beta} = \boldsymbol{\mu}_0\} \quad = \quad \{\boldsymbol{c}_0\} + \{G'\boldsymbol{d} : \boldsymbol{d} \in \mathrm{sp}[(\mathrm{I} - \mathbf{P}_{X'})G]\}.$$

For any non-zero $\boldsymbol{d}$ in $\text{sp}[(\text{I}-\mathbf{P}_{X'})G]$, $X(\boldsymbol{\beta}+\boldsymbol{d}) = X\boldsymbol{\beta}$ and $G'(\boldsymbol{\beta}+\boldsymbol{d}) \neq G'\boldsymbol{\beta}$ because $G'\boldsymbol{d} \neq \mathbf{0}$. Thus if there exists a non-zero $\boldsymbol{d}$ in $\text{sp}[(\text{I}-\mathbf{P}_{X'})G]$, we can identify another $\boldsymbol{\beta}$ such that $X\boldsymbol{\beta} = \boldsymbol{\mu}_0$ but $G'\boldsymbol{\beta} \neq G'\boldsymbol{\beta}_0$. Given $\boldsymbol{\mu}_0$, it is impossible to determine whether $G'\boldsymbol{\beta} = \boldsymbol{c}_0$ from the relation $X\boldsymbol{\beta} = \boldsymbol{\mu}_0$.

Verify that

$$\mathbf{P}_{X'} = \begin{pmatrix} \frac{3}{4} & \frac{1}{4}\mathbf{1}_3' \\ \frac{1}{4}\mathbf{1}_3 & \text{I}_3 - \frac{1}{4}\mathbf{1}_3\mathbf{1}_3' \end{pmatrix}$$

and that $\text{sp}[(\text{I}-\mathbf{P}_{X'})G] = \text{sp}(\boldsymbol{d})$, with $\boldsymbol{d} = (-1,1,1,1,)'$. Let $\boldsymbol{\beta}_1 = \boldsymbol{\beta}_0 + 10\boldsymbol{d} = (65,45,10,75)'$, and verify that $X\boldsymbol{\beta}_1 = \boldsymbol{\mu}_0$. But $G'\boldsymbol{\beta}_1 = (130,10)' \neq \boldsymbol{c}_0$. This ilustrates, in this example, that if $G'\boldsymbol{\beta}$ is not estimable, then the mean vector does not uniquely determine $G'\boldsymbol{\beta}$.

Now let's look at the RMFM SS for $\text{H}_0 : G'\boldsymbol{\beta} = \boldsymbol{c}_0$. The full model is $\text{sp}(X) = \Re^3$, and $\mathbf{P}_X = \text{I}_3$. For the restricted model, find that N such that $\text{sp}(N) = \text{sp}(G)^\perp$ is

$$N = \begin{pmatrix} 1 & 0 \\ 0 & 1 \\ 0 & 0 \\ 0 & -1 \end{pmatrix}, \text{ and so}$$

$$XN = \begin{pmatrix} 1 & 1 \\ 1 & 0 \\ 1 & -1 \end{pmatrix},$$

$$\mathbf{P}_{XN} = \frac{1}{6}\begin{pmatrix} 5 & 2 & -1 \\ 2 & 2 & 2 \\ -1 & 2 & 5 \end{pmatrix},$$

and thus

$$\mathbf{P}_X - \mathbf{P}_{XN} = \frac{1}{6}\begin{pmatrix} 1 & -2 & 1 \\ -2 & 4 & -2 \\ 1 & -2 & 1 \end{pmatrix}.$$

Finally, verify that $\mathbf{P}_X - \mathbf{P}_{XN} = \mathbf{P}_{\boldsymbol{h}} = \frac{1}{6}\boldsymbol{h}\boldsymbol{h}'$, where $\boldsymbol{h} = (1,-2,1)'$, as given above, and $X'\boldsymbol{h} = \boldsymbol{g}_*$.

The RMFM SS is $(\boldsymbol{y} - X\boldsymbol{\beta}_0)'\mathbf{P}_{\boldsymbol{h}}(\boldsymbol{y} - X\boldsymbol{\beta}_0)$; its ncp is

$$\delta^2_{\boldsymbol{h}}(X\boldsymbol{\beta}/\sigma) = [X(\boldsymbol{\beta} - \boldsymbol{\beta}_0)]'\mathbf{P}_{\boldsymbol{h}}[X(\boldsymbol{\beta} - \boldsymbol{\beta}_0)]/\sigma^2,$$

and it is 0 iff

$$\begin{aligned} \boldsymbol{h}'X(\boldsymbol{\beta} - \boldsymbol{\beta}_0) &= 0 \\ &= \boldsymbol{g}_*'\boldsymbol{\beta} - \boldsymbol{g}_*'\boldsymbol{\beta}_0 \\ &= \beta_1 - 2\beta_2 + \beta_3 - 100. \end{aligned}$$

It tests exclusively that $\beta_1 - 2\beta_2 + \beta_3 = 100$.

In summary, if we set out for inference about the proposition $\text{H}_0 : G'\boldsymbol{\beta} = G'\boldsymbol{\beta}_0$, the RMFM SS implicitly addresses instead $\text{H}_{0*} : \boldsymbol{g}_*'\boldsymbol{\beta} = \boldsymbol{g}_*'\boldsymbol{\beta}_0$. The intended null set $\{\boldsymbol{\beta} \in \Re^{k+1} : G'\boldsymbol{\beta} = \boldsymbol{c}_0\} = \{\boldsymbol{\beta}_0\} + \text{sp}(G)^\perp$ becomes instead the less restricted set $\{\boldsymbol{\beta}_0\} + \text{sp}(\boldsymbol{g}_*)^\perp$.

We turn now to the GLH form of the test statistic for inference about $G'\boldsymbol{\beta}$. Let A be a matrix such that $XA' = \mathbf{P}_X$.

Consider first the homogeneous version of the GLH SS that is aimed at $\text{H}_0 : G'\boldsymbol{\beta} = \mathbf{0}$, that is, with $\check{\boldsymbol{\beta}} = A'\boldsymbol{y}$,

$$(G'\check{\boldsymbol{\beta}})'[\hat{\text{V}}\text{ar}(G'\check{\boldsymbol{\beta}})]^{-}(G'\check{\boldsymbol{\beta}})\hat{\sigma}^2 \;=\; \boldsymbol{y}'\mathbf{P}_{AG}\boldsymbol{y}.$$

It tests exclusively that $\mathbf{P}_{AG}X\boldsymbol{\beta} = \mathbf{0}$ or, equivalently, that $G'A'X\boldsymbol{\beta} = \mathbf{0}$.

It was noted in Section 10.5 that this SS tests the estimable part of $G'\boldsymbol{\beta}$, which is $\boldsymbol{g}_*'\boldsymbol{\beta}$ with $\boldsymbol{g}_* = (0, 1, -2, 1)'$. We shall see here that it tests also another estimable function $\boldsymbol{g}_2'\boldsymbol{\beta}$, but not one that is contained in $\text{sp}(G)$, and that this other estimable function depends on the choice of solution A' to $XA' = \mathbf{P}_X$.

Shown next are three matrices A' such that (verify) $XA' = \mathbf{P}_X$. The first is $A_0' = TQ'\mathbf{P}_X$, from GS on X; the second is $A_1' = A_0' + (\text{I} - \mathbf{P}_{X'})Z$ with $Z = 10 \cdot \mathbf{1}_4\mathbf{1}_3'$; and the third is $A_2' = X^+\mathbf{P}_X = (X'X)^+X'\mathbf{P}_X = (X'X)^+X'$, where M^+ denotes the Moore-Penrose pseudoinverse of the matrix M.

$$A_0' \;=\; \begin{pmatrix} 0 & 0 & 1 \\ 1 & 0 & -1 \\ 0 & 1 & -1 \\ 0 & 0 & 0 \end{pmatrix},$$

$$A_1' \;=\; \begin{pmatrix} -5 & -5 & -4 \\ 6 & 5 & 4 \\ 5 & 6 & 4 \\ 5 & 5 & 5 \end{pmatrix},$$

$$A_2' \;=\; \frac{1}{4}\begin{pmatrix} 1 & 1 & 1 \\ 3 & -1 & -1 \\ -1 & 3 & -1 \\ -1 & -1 & 3 \end{pmatrix}.$$

From these,

$$A_0G \;=\; \begin{pmatrix} 1 & 0 \\ 1 & 1 \\ -2 & -1 \end{pmatrix}, \; X'A_0G = \begin{pmatrix} 0 & 0 \\ 1 & 0 \\ 1 & 1 \\ -2 & -1 \end{pmatrix} = G_0,$$

$$A_1G \;=\; \begin{pmatrix} 16 & 5 \\ 16 & 6 \\ 13 & 4 \end{pmatrix}, \; X'A_1G = \begin{pmatrix} 45 & 15 \\ 16 & 5 \\ 16 & 6 \\ 13 & 4 \end{pmatrix} = G_1,$$

$$A_2G \;=\; \frac{1}{4}\begin{pmatrix} 1 & -1 \\ 1 & 3 \\ 1 & -1 \end{pmatrix}, \; X'A_2G = \frac{1}{4}\begin{pmatrix} 3 & 1 \\ 1 & -1 \\ 1 & 3 \\ 1 & -1 \end{pmatrix} = G_2,$$

and

$$\mathbf{P}_{A_0G} = \frac{1}{3}\begin{pmatrix} 2 & 1 & 1 \\ 1 & 2 & 1 \\ 1 & 1 & 2 \end{pmatrix},$$

$$\mathbf{P}_{A_1G} = \frac{1}{453}\begin{pmatrix} 257 & 14 & 224 \\ 14 & 452 & -16 \\ 224 & -16 & 197 \end{pmatrix}, \text{ and}$$

$$\mathbf{P}_{A_2G} = \frac{1}{2}\begin{pmatrix} 1 & 0 & 1 \\ 0 & 2 & 0 \\ 1 & 0 & 1 \end{pmatrix}.$$

The numerator SS based on A_j is $SS_j = (\boldsymbol{y} - X\boldsymbol{b}_0)'\mathbf{P}_{A_jG}(\boldsymbol{y} - X\boldsymbol{b}_0)$. That the $\mathbf{P}_{A_jG}$s are different shows that, for the same $\boldsymbol{y}$, the numerator SSs are different.

It is clear that $\boldsymbol{g}_* = G_0\binom{1}{-3} = G_1\binom{1}{-3} = G_2\binom{1}{-3}$, so $\boldsymbol{g}_*$ is contained in $\mathrm{sp}(G_j)$, $j = 0, 1, 2$. In each case, there is another estimable function in addition to $\boldsymbol{g}_*$ in $\mathrm{sp}(G_j)$. To find it, one way is GS on $(\boldsymbol{g}_*, G_j)$, so that the first column of Q is proportional to $\boldsymbol{g}_*$ and the second, $\boldsymbol{q}_2$, is orthogonal to it. The resulting vectors are proportional to these:

$$\boldsymbol{g}_{20} = \begin{pmatrix} 0 \\ 1 \\ 0 \\ -1 \end{pmatrix}, \quad \boldsymbol{g}_{21} = \begin{pmatrix} 30 \\ 11 \\ 10 \\ 9 \end{pmatrix}, \quad \boldsymbol{g}_{22} = \begin{pmatrix} 3 \\ 1 \\ 1 \\ 1 \end{pmatrix}.$$

Let $G_{*j} = (\boldsymbol{g}_*, \boldsymbol{g}_{2j})$, so that $\mathrm{sp}(G_{*j}) = \mathrm{sp}(G_j)$, $j = 0, 1, 2$.

What is tested instead of $G'\boldsymbol{\beta}$ depends on which A_j is used; instead of $G'\boldsymbol{\beta}$, $G'_{*j}\boldsymbol{\beta}$ is tested exclusively, where $G_{*j} = (\boldsymbol{g}_*, \boldsymbol{g}_{2j})$. Each $G'_{*j}\boldsymbol{\beta}$ is estimable, and the corresponding ncp is 0 iff $G'_{*j}\boldsymbol{\beta} = \mathbf{0}$. As the three linear subspaces $\mathrm{sp}(G_j) = \mathrm{sp}(G_{*j})$ are different, the null sets $\{\boldsymbol{\beta} : G'_{*j}\boldsymbol{\beta} = \mathbf{0}\}$ are distinct. Thus what the GLH SS tests depends on the choice of A such that $XA' = \mathbf{P}_X$, and in any case it does not test exclusively $\mathrm{H}_0 : G'\boldsymbol{\beta} = \mathbf{0}$.

10.9 Exercises

1. Let $\boldsymbol{\alpha}_0$ be a vector in $\Re^n$ and let $\mathcal{S}$ be a linear subspace in $\Re^n$. Let $\mathcal{A} = \boldsymbol{\alpha}_0 + \mathcal{S}$. Show that:

   (a) If λ is a scalar and $\boldsymbol{z}_1$ and $\boldsymbol{z}_2$ are vectors in $\mathcal{A}$, then $(1 - \lambda)\boldsymbol{z}_1 + \lambda\boldsymbol{z}_2$ is in $\mathcal{A}$. That is, $\mathcal{A}$ is closed under affine combinations.

   (b) $\mathcal{A} = \mathcal{S}$ iff $\boldsymbol{\alpha}_0 \in \mathcal{S}$.

   (c) For any vector $\boldsymbol{z} \in \mathcal{A}$, $\mathcal{A} = \{\boldsymbol{z}\} + \mathcal{S}$.

2. Let $\boldsymbol{\mu}_0$ and $\boldsymbol{\eta}_0$ be vectors, and let $\mathcal{S}$ and $\mathcal{S}_0$ be linear subspaces, all in $\Re^n$. Let $\mathcal{M} = \{\boldsymbol{\mu}_0\} + \mathcal{S}$ and $\mathcal{M}_0 = \{\boldsymbol{\eta}_0\} + \mathcal{S}_0$.

   (a) $\mathcal{M}_0 \subset \mathcal{M}$ iff $\boldsymbol{\mu}_0 - \boldsymbol{\eta}_0 \in \mathcal{S}$ and $\mathcal{S}_0 \subset \mathcal{S}$.

   (b) If $\mathcal{M}_0 \subset \mathcal{M}$, then $\mathcal{M} = \{\boldsymbol{\eta}_0\} + \mathcal{S}$.

3. Let $\mathcal{S}$ be a linear subspace of $\Re^n$. Let H be a matrix such that $\text{sp}(H) \subset \mathcal{S}$, let $\boldsymbol{\eta}_0$ be a particular vector in $\mathcal{S}$, and let

$$\mathcal{M}_0 = \{\boldsymbol{\mu} \in \mathcal{S} : H'(\boldsymbol{\mu} - \boldsymbol{\eta}_0) = \mathbf{0}\}.$$

   (a) $\mathcal{M}_0 = \{\boldsymbol{\eta}_0\} + \mathcal{S}_0$ where $\mathcal{S}_0 = \mathcal{S} \cap \text{sp}(H)^\perp$.

   (b) $\mathcal{S} = \mathcal{S}_0 + \mathcal{S}_0^\perp \cap \mathcal{S}$ and $\mathcal{S} = \text{sp}(H) + \text{sp}(H)^\perp \cap \mathcal{S}$.

   (c) If $\mathcal{S}_1$ is a linear subspace of $\mathcal{S} \cap \mathcal{S}_0^\perp$, and if $\mathcal{S} = \mathcal{S}_0 + \mathcal{S}_1$, then $\mathcal{S}_1 = \mathcal{S} \cap \mathcal{S}_0^\perp$.

   (d) $\text{sp}(H) = \mathcal{S} \cap \mathcal{S}_0^\perp$.

   (e) $\mathcal{S} \cap \mathcal{S}_0^\perp$ is the orthogonal complement of $\mathcal{S}_0$ in $\mathcal{S}$. That is, it is the set of all vectors in $\mathcal{S}$ that are orthogonal to all vectors in $\mathcal{S}_0$.

   (f) $\mathbf{P}_H = \mathbf{P}_{\mathcal{S}} - \mathbf{P}_{\mathcal{S}_0}$.

   (g) If $\boldsymbol{\mu} \in \mathcal{S}$, then $\boldsymbol{\mu} \in \mathcal{M}_0$ iff $\mathbf{P}_H(\boldsymbol{\mu} - \boldsymbol{\eta}_0) = \mathbf{0}$. And $\mathbf{P}_H(\boldsymbol{\mu} - \boldsymbol{\eta}_0) = \mathbf{0}$ iff $(\boldsymbol{\mu} - \boldsymbol{\eta}_0)'\mathbf{P}_H(\boldsymbol{\mu} - \boldsymbol{\eta}_0) = 0$.

   (h) $\mathbf{P}_{\mathcal{S}_0^\perp} = \mathbf{P}_{\mathcal{S}^\perp} + \mathbf{P}_H$.

4. Let $\mathcal{M}_0 = \{X\boldsymbol{b} : \boldsymbol{b} \in \Re^{k+1}$ and $G'\boldsymbol{b} = \mathbf{0}\}$. Show that

$$\{\boldsymbol{b} : X\boldsymbol{b} \in \mathcal{M}_0\} = [\text{sp}(X') \cap \text{sp}(G)]^\perp.$$

   This means that $\mathcal{M}_0$ contains mean vectors $\boldsymbol{m}$ such that $\boldsymbol{m} = X\boldsymbol{b}$ with $G'\boldsymbol{b} \neq \mathbf{0}$ unless $\text{sp}(G) \subset \text{sp}(X')$. Whether $G'\boldsymbol{\beta} = \mathbf{0}$ cannot be determined from $X\boldsymbol{\beta}$ unless $G'\boldsymbol{\beta}$ is estimable.

5. Let X and G be matrices, X with $k+1$ columns and G with $k+1$ rows; and let $\boldsymbol{c}_0$ be a vector in $\text{sp}(G')$. Show that there exists a vector $\boldsymbol{\eta}_0$ and a linear subspace $\mathcal{S}_0$ such that

$$\{X\boldsymbol{\beta} : \boldsymbol{\beta} \in \Re^p \text{ and } G'\boldsymbol{\beta} = \boldsymbol{c}_0\} = \{\boldsymbol{\eta}_0\} + \mathcal{S}_0.$$

6. Assume that $\boldsymbol{Y} \sim \mathbf{N}(X\boldsymbol{\beta}, \sigma^2\mathrm{I})$. Consider the F-statistic for testing $\mathrm{H}_0 : G'\boldsymbol{\beta} = \boldsymbol{c}_0$. Assume that $\text{sp}(G) \subset \text{sp}(X')$, and let H be a matrix with $\text{sp}(H) \subset \text{sp}(X)$ such that $X'H = G$. Let $\boldsymbol{\beta}_0$ be a $(k+1)$-vector such that $G'\boldsymbol{\beta}_0 = \boldsymbol{c}_0$, and let $\boldsymbol{\eta}_0 = X\boldsymbol{\beta}_0$. Prove the following assertions.

   (a) $(\boldsymbol{Y} - \boldsymbol{\eta}_0)'\mathbf{P}_H(\boldsymbol{Y} - \boldsymbol{\eta}_0)/\sigma^2 \sim \chi^2_{\nu_1}(\delta^2_H)$, with $\nu_1 = \text{tr}(\mathbf{P}_H)$ and δ^2_H as given in Equation (10.2).

(b) $(\boldsymbol{Y}-\boldsymbol{\eta}_0)'(\mathrm{I}-\mathbf{P}_X)(\boldsymbol{Y}-\boldsymbol{\eta}_0)/\sigma^2 \sim \chi^2_{\nu_2}(0)$, with $\nu_2 = \mathrm{tr}(\mathrm{I}-\mathbf{P}_X)$.

(c) These two quadratic forms are independent.

(d) $\delta^2_H(X\boldsymbol{\beta}-\boldsymbol{\eta}_0) = 0$ iff $G'\boldsymbol{\beta} = \boldsymbol{c}_0$.

7. Given the $n \times (k+1)$ matrix X, assume that the n-variate random variable $\boldsymbol{Y}$ has mean vector $\mathrm{E}(\boldsymbol{Y}) = \boldsymbol{\mu} = X\boldsymbol{\beta}$, $\boldsymbol{\beta} \in \Re^{k+1}$, and variance-covariance matrix $\mathrm{Var}(\boldsymbol{Y}) = \sigma^2\mathrm{I}$, $\sigma^2 > 0$. Let $\mathcal{S} = \mathrm{sp}(X)$.

Let G be a matrix such that $\mathrm{sp}(G) \subset \mathrm{sp}(X')$. Let $\boldsymbol{c}_0$ be a vector in $\mathrm{sp}(G')$, and let $\boldsymbol{b}_0$ be a vector such that $G'\boldsymbol{b}_0 = \boldsymbol{c}_0$. Let $\boldsymbol{\eta}_0 = X\boldsymbol{b}_0$.

Let $\hat{\boldsymbol{\beta}} = \hat{\boldsymbol{\beta}}(\boldsymbol{y})$ be a function of $\boldsymbol{y}$ such that for any $\boldsymbol{y} \in \Re^n$, $X\hat{\boldsymbol{\beta}}(\boldsymbol{y}) = \mathbf{P}_X\boldsymbol{y} = \hat{\boldsymbol{y}}$. That is, $\hat{\boldsymbol{\beta}}(\boldsymbol{y})$ is a least-squares solution.

Show that:

(a) For any matrix M such that $X'M = G$,

$$\{X\boldsymbol{\beta} \in \mathrm{sp}(X) : G'\boldsymbol{\beta} = \boldsymbol{c}_0\} = \{\boldsymbol{\eta}_0\} + \mathcal{S}_0,$$

where $\mathcal{S}_0 = \{\boldsymbol{\mu} \in \mathrm{sp}(X) : M'\boldsymbol{\mu} = \mathbf{0}\} = \mathrm{sp}(X) \cap \mathrm{sp}(M)^{\perp}$.

(b) If M_1 and M_2 are matrices such that $X'M_1 = X'M_2$, then $\mathrm{sp}(X) \cap \mathrm{sp}(M_1)^{\perp} = \mathrm{sp}(X) \cap \mathrm{sp}(M_2)^{\perp}$.

(c) There exists a matrix H with $\mathrm{sp}(H) \subset \mathrm{sp}(X)$ such that $X'H = G$.

(d) For any $\boldsymbol{y} \in \Re^n$, $G'\hat{\boldsymbol{\beta}}(\boldsymbol{y}) = H'\hat{\boldsymbol{y}} = H'\boldsymbol{y}$.

(e) For any $\boldsymbol{\beta} \in \Re^{k+1}$, if $\mathrm{E}(\boldsymbol{Y}) = X\boldsymbol{\beta}$, then $\mathrm{E}[G'\hat{\boldsymbol{\beta}}(\boldsymbol{Y})] = G'\boldsymbol{\beta}$.

(f) $\mathrm{Var}[G'\hat{\boldsymbol{\beta}}(\boldsymbol{Y})] = \sigma^2 H'H$.

(g) $(G'\hat{\boldsymbol{\beta}} - \boldsymbol{c}_0)'[\mathrm{Var}(G'\hat{\boldsymbol{\beta}})]^-(G'\hat{\boldsymbol{\beta}} - \boldsymbol{c}_0)$

$$\begin{aligned} &= (G'\hat{\boldsymbol{\beta}} - \boldsymbol{c}_0)'(H'H)^-(G'\hat{\boldsymbol{\beta}} - \boldsymbol{c}_0)/\sigma^2 \\ &= \delta^2_H[(\boldsymbol{y}-\boldsymbol{\eta}_0)/\sigma]. \end{aligned}$$

(h) $\boldsymbol{\beta}$ satisfies $\mathbf{P}_H(X\boldsymbol{\beta}-\boldsymbol{\eta}_0) = \mathbf{0}$ iff $G'\boldsymbol{\beta} = \boldsymbol{c}_0$.

(i) With δ^2_H defined by (10.2),

$$\delta^2_H(X\hat{\boldsymbol{\beta}}-\boldsymbol{\eta}_0) = (\boldsymbol{y}-\boldsymbol{\eta}_0)'\mathbf{P}_H(\boldsymbol{y}-\boldsymbol{\eta}_0).$$

8. **Inference on a single estimable function $\boldsymbol{g}'\boldsymbol{\beta}$.** Let $\boldsymbol{g}$ be a non-zero $(k+1)$-vector in $\mathrm{sp}(X')$, and let $\boldsymbol{h}$ be an n-vector such that $X'\boldsymbol{h} = \boldsymbol{g}$; this means that $\boldsymbol{g}'\boldsymbol{\beta}$ is an estimable function of $\boldsymbol{\beta}$. Let c_0 be a given, arbitrary scalar. Let $\hat{\boldsymbol{\beta}}$ be a $(k+1)$-vector, a function of $\boldsymbol{y}$ such that for any $\boldsymbol{y}$, $\mathbf{P}_X\boldsymbol{y} = X\hat{\boldsymbol{\beta}}$: $\hat{\boldsymbol{\beta}}(\boldsymbol{y})$ is a least-squares solution. Let $\hat{\boldsymbol{y}} = \mathbf{P}_X\boldsymbol{y}$. Assume that $\boldsymbol{Y} \sim \mathbf{N}_n(X\boldsymbol{\beta}, \sigma^2\mathrm{I})$, $(\boldsymbol{\beta}, \sigma^2) \in \Re^{k+1} \times \Re_+$.

(a) *The least-squares estimate of an estimable function is unique.* Show: If $\boldsymbol{b}_1$ and $\boldsymbol{b}_2$ are $(k+1)$-vectors such that $\hat{\boldsymbol{y}} = X\boldsymbol{b}_1 = X\boldsymbol{b}_2$, then $\boldsymbol{g}'\boldsymbol{b}_1 = \boldsymbol{g}'\boldsymbol{b}_2$.

(b) *The least-squares estimate of an estimable function is linear in* $\boldsymbol{y}$. Show: $\boldsymbol{g}'\hat{\boldsymbol{\beta}} = \boldsymbol{h}'\hat{\boldsymbol{y}}$. Further, if $\boldsymbol{h}$ is chosen as the unique vector in $\mathrm{sp}(X)$ such that $X'\boldsymbol{h} = \boldsymbol{g}$, then $\boldsymbol{g}'\hat{\boldsymbol{\beta}} = \boldsymbol{h}'\boldsymbol{y}$.

(c) Show: $\mathrm{Var}(\boldsymbol{g}'\hat{\boldsymbol{\beta}}) = \boldsymbol{h}'\mathbf{P}_X\boldsymbol{h}\sigma^2$. If $\boldsymbol{h} \in \mathrm{sp}(X)$, then $\mathrm{Var}(\boldsymbol{g}'\hat{\boldsymbol{\beta}}) = \boldsymbol{h}'\boldsymbol{h}\sigma^2$.

(d) *LS estimators of estimable functions are BLUE.* Show: If $\boldsymbol{\ell}$ is an n-vector of constants such that $\mathrm{E}(\boldsymbol{\ell}'\boldsymbol{Y}) = \boldsymbol{g}'\boldsymbol{\beta}$ for all $\boldsymbol{\beta} \in \Re^{(k+1)}$ and all $\sigma^2 > 0$, then

$$\mathrm{Var}(\boldsymbol{\ell}'\boldsymbol{Y}) \geqslant \mathrm{Var}(\boldsymbol{g}'\hat{\boldsymbol{\beta}}).$$

(e) Show that

$$T = \frac{\boldsymbol{g}'\hat{\boldsymbol{\beta}} - c_0}{\sqrt{\hat{\sigma}^2\boldsymbol{h}'\mathbf{P}_X\boldsymbol{h}}}$$

follows a Student's t distribution. Identify its degrees of freedom and non-centrality parameter. Here, $\hat{\sigma}^2 = MSE = \boldsymbol{y}'(\mathrm{I}-\mathbf{P}_X)\boldsymbol{y}/\mathrm{tr}(\mathrm{I}-\mathbf{P}_X)$.

(f) Show that (10.1) becomes in this case

$$F = T^2.$$

(g) Show that the non-centrality parameter for this F-statistic increases with $|\boldsymbol{g}'\boldsymbol{\beta} - c_0|$, that is, for both $\boldsymbol{g}'\boldsymbol{\beta} < c_0$ and $> c_0$.

(h) Using (10.5), describe a $100(1-\alpha)\%$ confidence interval on $\boldsymbol{g}'\boldsymbol{\beta}$.

9. Let X be a matrix with n rows, and let H be a matrix such that $\mathrm{sp}(H) \subset \mathrm{sp}(X)$. Let $\boldsymbol{\eta}_0$ be a vector in $\mathrm{sp}(X)$. Think of X, H, and $\boldsymbol{\eta}_0$ as given, fixed throughout this exercise.

   Prove the following propositions.

   (a) In order that P be an orthogonal projection matrix such that

   $$\{\boldsymbol{\mu} \in \Re^n : P(\boldsymbol{\mu} - \boldsymbol{\eta}_0) = \mathbf{0}\} = \{\boldsymbol{\mu} \in \mathrm{sp}(X) : H'(\boldsymbol{\mu} - \boldsymbol{\eta}_0) = \mathbf{0}\},$$

   it is necessary and sufficient that $P = \mathbf{P}_H + \mathbf{P}_{X^\perp}$, where $\mathbf{P}_{X^\perp} = \mathrm{I} - \mathbf{P}_X$ is the orthogonal projection matrix onto $\mathrm{sp}(X)^\perp$.

   (b) In order that P be an orthogonal projection matrix such that

   $$\begin{aligned}\{\boldsymbol{\mu} \in \mathrm{sp}(X) &: P(\boldsymbol{\mu} - \boldsymbol{\eta}_0) = \mathbf{0}\}\\ &= \{\boldsymbol{\mu} \in \mathrm{sp}(X) : H'(\boldsymbol{\mu} - \boldsymbol{\eta}_0) = \mathbf{0}\}\end{aligned}$$

   it is necessary and sufficient that $P = \mathbf{P}_H + \mathbf{P}_{\mathcal{T}}$, where $\mathcal{T}$ is a linear subspace of $\mathrm{sp}(X)^\perp$.

   (c) Let $G = X'H$. With $P = \mathbf{P}_H + \mathbf{P}_{\mathcal{T}}$ as described in (9b), show that $G'\boldsymbol{\beta} = H'\boldsymbol{\eta}_0$ iff $P(X\boldsymbol{\beta} - \boldsymbol{\eta}_0) = \mathbf{0}$. That is, show that

   $$\{\boldsymbol{\beta} \in \Re^q : G'\boldsymbol{\beta} = H'\boldsymbol{\eta}_0\} = \{\boldsymbol{\beta} \in \Re^q : P(X\boldsymbol{\beta} - \boldsymbol{\eta}_0) = \mathbf{0}\}.$$

(d) With $P = \mathbf{P}_{\mathcal{H}} + \mathbf{P}_{\mathcal{T}}$ as described in (9b), suppose $\boldsymbol{Y} \sim \mathbf{N}(X, \sigma^2 \mathrm{I})$.

i. Show that $(\boldsymbol{Y} - \boldsymbol{\eta}_0)' P (\boldsymbol{Y} - \boldsymbol{\eta}_0)/\sigma^2$ follows a chi-squared distribution, and specify its degrees of freedom and non-centrality parameter.

ii. Does

$$F_* = \frac{(\boldsymbol{Y} - \boldsymbol{\eta}_0)' P (\boldsymbol{Y} - \boldsymbol{\eta}_0)/\mathrm{tr}(P)}{\boldsymbol{Y}' \mathbf{P}_{X^\perp} \boldsymbol{Y}/\mathrm{tr}(\mathbf{P}_{X^\perp})}$$

follow an F distribution? Justify your answer.

iii. Find a statistic F_a that has an F distribution and is a one-to-one function of F_*. What are its degrees of freedom and non-centrality parameter?

10. Let $L(\boldsymbol{\mu}, \sigma^2; \boldsymbol{y})$ be the likelihood function (10.10), where the domain of $\boldsymbol{\mu}$ is $\{\boldsymbol{\alpha}_0\} + \mathcal{A}$, where $\boldsymbol{\alpha}_0$ is a given vector and $\mathcal{A}$ is a linear subspace, both in $\Re^n$.

(a) Show that, given $\boldsymbol{\mu}$, L is maximized with respect to σ^2 at $\tilde{\sigma}^2(\boldsymbol{\mu}) = (\boldsymbol{y} - \boldsymbol{\mu})'(\boldsymbol{y} - \boldsymbol{\mu})/n$.

(b) Show that $\log L(\boldsymbol{\mu}, \tilde{\sigma}^2(\boldsymbol{\mu}); \boldsymbol{y})$ is maximized with respect to $\boldsymbol{\mu}$ at

$$\hat{\boldsymbol{\mu}} = \{\boldsymbol{\alpha}_0\} + \mathbf{P}_{\mathcal{A}}(\boldsymbol{y} - \boldsymbol{\alpha}_0).$$

(c) Show that

$$\tilde{\sigma}^2(\hat{\boldsymbol{\mu}}) = (\boldsymbol{y} - \boldsymbol{\alpha}_0)' \mathbf{P}_{\mathcal{A}^\perp} (\boldsymbol{y} - \boldsymbol{\alpha}_0)/n.$$

11. Let V be a given $n \times n$ symmetric, positive-definite matrix. Suppose $\boldsymbol{Y} \sim \mathbf{N}(X\boldsymbol{\beta}, \sigma^2 V)$, where σ^2 is an unknown positive constant and $\boldsymbol{\beta}$ is an unknown $(k+1)$-vector. Find the MLEs of $X\boldsymbol{\beta}$ and σ^2.

12. Given matrices X and G, show how to get H such that $\mathrm{sp}(H) \subset \mathrm{sp}(X)$ and $\mathrm{sp}(X'H) = \mathrm{sp}(G) \cap \mathrm{sp}(X')$.

13. Given matrices X and G, let H be a matrix such that $\mathrm{sp}(H) \subset \mathrm{sp}(X)$ and $\mathrm{sp}(X'H) = \mathrm{sp}(G) \cap \mathrm{sp}(X')$. Let A be a matrix such that $XA' = \mathbf{P}_X$. Prove that $\mathrm{sp}(H) \subset \mathrm{sp}(AG)$. This implies that $\mathbf{P}_{AG} - \mathbf{P}_H$ is idempotent, and hence that $\delta_H^2(X\boldsymbol{\beta}) \leqslant \delta_{AG}^2(X\boldsymbol{\beta})$ for all $\boldsymbol{\beta} \in \Re^{k+1}$. Prove also that if $\mathrm{sp}(G) \subset \mathrm{sp}(X')$, then $\mathrm{sp}(H) = \mathrm{sp}(AG)$.

14. Let A be a matrix such that $XA' = \mathbf{P}_X$. Prove that $X'AG = G$ iff $\mathrm{sp}(G) \subset \mathrm{sp}(X')$. This establishes that $G'A'\boldsymbol{y}$ is an unbiased estimator of $G'\boldsymbol{\beta}$ iff $G'\boldsymbol{\beta}$ is estimable.

15. Given matrices X and G, let N be a matrix such that $\mathrm{sp}(N) = \mathrm{sp}(G)^\perp$.

(a) Prove: If H is a matrix with $\mathrm{sp}(H) \subset \mathrm{sp}(X)$ and $\mathrm{sp}(X'H) = \mathrm{sp}(G) \cap \mathrm{sp}(X')$, then $\mathbf{P}_H = \mathbf{P}_X - \mathbf{P}_{XN}$ (that is, $\mathrm{sp}(H) = \mathrm{sp}(X) \cap \mathrm{sp}(XN)^\perp$).

(b) Prove: If H is a matrix such that $\mathbf{P}_H = \mathbf{P}_X - \mathbf{P}_{XN}$, then $\mathrm{sp}(X'H) = \mathrm{sp}(G) \cap \mathrm{sp}(X')$.

(c) Prove: If H is a matrix such that $\mathrm{sp}(H) \subset \mathrm{sp}(X)$ and $\mathrm{sp}(X'H) = \mathrm{sp}(G) \cap \mathrm{sp}(X')$, then $\{X\boldsymbol{b} : \boldsymbol{b} \in \Re^{k+1} \text{ and } G'\boldsymbol{b} = \mathbf{0}\} = \{X\boldsymbol{b} : \boldsymbol{b} \in \Re^{k+1} \text{ and } H'X\boldsymbol{b} = \mathbf{0}\}$.

16. Let Q be a symmetric idempotent matrix. In the model $X\boldsymbol{\beta}$, the numerator SS $\boldsymbol{y}'Q\boldsymbol{y}$ tests exclusively that $QX\boldsymbol{\beta} = \mathbf{0}$. Let $H = \mathbf{P}_X Q$. Then $SS_H = \boldsymbol{y}'\mathbf{P}_H\boldsymbol{y}$ also tests exclusively that $H'X\boldsymbol{\beta} = QX\boldsymbol{\beta} = \mathbf{0}$. In fact, because $\mathrm{sp}(H) \subset \mathrm{sp}(X)$, SS_H is the RMFM SS for testing $QX\boldsymbol{\beta} = \mathbf{0}$.

    Prove: $X'(\mathbf{P}_H - Q)X$ is nnd, and therefore $\delta^2_H(X\boldsymbol{\beta}) \geqslant \delta^2_Q(X\boldsymbol{\beta})$ for all $X\boldsymbol{\beta} \in \mathrm{sp}(X)$.

17. In the model $X\boldsymbol{\beta}$, suppose that $\mathrm{sp}(H) \subset \mathrm{sp}(X)$ and that both SS_H and SS_L test exclusively $G'\boldsymbol{\beta}$. Prove that if $\mathrm{sp}(L) \subset \mathrm{sp}(X)$, then $\mathrm{sp}(L) = \mathrm{sp}(H)$ and hence $\mathbf{P}_L = \mathbf{P}_H$ and $SS_L \equiv SS_H$.

18. Let M_1 and M_2 be matrices, both with r rows, such that $\mathrm{sp}(M_1) \subset \mathrm{sp}(M_2)$. Let $\mathbb{K}$ be an $n \times r$ matrix. Let $X_1 = \mathbb{K}\mathbf{P}_{M_1}$ and $X_2 = \mathbb{K}\mathbf{P}_{M_2}$. Let Q be an $n \times n$ symmetric idempotent matrix.

    In the model $X_2\boldsymbol{\eta} = \mathbb{K}\mathbf{P}_{M_2}\boldsymbol{\eta}$, the numerator SS $SS_Q = \boldsymbol{y}'Q\boldsymbol{y}$ tests $(\mathbf{P}_{M_2}\mathbb{K}'Q)'\boldsymbol{\eta} = \mathbf{0}$. Let $G_2 = \mathbf{P}_{M_2}\mathbb{K}'Q$. Then SS_Q tests $G_2'\mathbf{P}_{M_2}\boldsymbol{\eta} = \mathbf{0}$ in model 2.

    Prove: In model 1, with $X_1 = \mathbb{K}\mathbf{P}_{M_1}\boldsymbol{\eta}$, SS_Q tests $G_2'\mathbf{P}_{M_1}\boldsymbol{\eta} = \mathbf{0}$.

19. This is one formulation of the ***one-way Analysis of Variance*** (ANOVA) model.

    Suppose independent random samples are taken from a populations, which correspond to levels $1, \ldots, a$ of a *treatment factor* named "A." A real-valued response Y is to be observed from each sampled unit. Denote the response of the j-th unit from the i-th population by Y_{ij}, $i = 1, \ldots, a$, $j = 1, \ldots, n_i$. All sample sizes n_i are positive. Let $n = \sum_i n_i$. Let $\boldsymbol{Y}_i = (Y_{i1}, \ldots, Y_{in_i})'$ and $\boldsymbol{Y} = (\boldsymbol{Y}_1', \ldots, \boldsymbol{Y}_a')'$. Denote the realized value of $\boldsymbol{Y}$ by $\boldsymbol{y} = (y_{ij})$, indexed and partitioned in the same way as $\boldsymbol{Y}$.

    Suppose that, in the respective populations, the population mean of Y is η_i and the population variance is σ^2. Then $\mathrm{E}(\boldsymbol{Y}_i) = \eta_i \mathbf{1}_{n_i}$ and $\mathrm{Var}(\boldsymbol{Y}_i) = \sigma^2 \mathrm{I}_{n_i}$.

    A common representation of the ***cell means*** is $\eta_i = \beta_0 + \beta_i$, $i = 1, \ldots, a$. This follows from the conceptualization that all subjects come initially from a population in which the mean response is β_0. Then the i, j-th subject receives level i of factor A, which has the effect of adding β_i to the subject's initial untreated response.

    Let $\boldsymbol{\eta} = (\eta_i) = (\eta_1, \ldots, \eta_a)' \in \Re^a$. Then, in terms of $\boldsymbol{\beta} = (\beta_0, \beta_1, \ldots, \beta_a)'$, $\boldsymbol{\eta} = M\boldsymbol{\beta}$, with $M = (\mathbf{1}_a, \mathrm{I}_a)$. Partition $\boldsymbol{\beta}$ as $(\beta_0, \boldsymbol{\beta}_1')'$.

Let $\mathbb{K} = \text{Diag}(\mathbf{1}_{n_1}, \ldots, \mathbf{1}_{n_a})$. See Table 14.1, p. 138, for an example. $\mathbb{K}\boldsymbol{\eta}$ repeats each η_i in n_i rows.

Let $\boldsymbol{\mu} = \text{E}(\boldsymbol{Y})$ denote the mean vector of $\boldsymbol{Y}$. Then $\boldsymbol{\mu} = \mathbb{K}\boldsymbol{\eta} = \mathbb{K}M\boldsymbol{\beta}$. This is a multiple regression model $X\boldsymbol{\beta}$ with $X = \mathbb{K}M$.

Let $\boldsymbol{e}_a$ denote the a-th column of I_a. Define the matrix C_1 to be the first $a-1$ columns of $\text{I}_a - \boldsymbol{e}_a\mathbf{1}'_a$. Let $C = \binom{\mathbf{0}'}{C_1}$.

Denote cell sample means by $\bar{y}_{i\cdot} = (1/n_i)\sum_{j=1}^{n_i} y_{ij}$, and denote the overall sample mean by $\bar{y}_{\cdot\cdot} = (1/n)\sum_{i=1}^{a}\sum_{j=1}^{n_i} y_{ij}$.

In addition to proving the assertions that follow, work through them numerically in terms of an unbalanced model with $a = 3$, $n_1 = 2$, $n_2 = 1$, $n_3 = 4$ and $\boldsymbol{y} = (50, 22, 111, 65, 53, 54, 101)'$.

(a) Show that $\text{sp}(M) = \Re^a$ and hence that $\text{sp}(X) = \text{sp}(\mathbb{K})$, and that therefore $\mathbf{P}_X = \mathbf{P}_{\mathbb{K}}$.

(b) Show that $\mathbb{K}'\mathbb{K} = N = \text{diag}(n_i)$, so that $(\mathbb{K}'\mathbb{K})^{-1} = N^{-1} = \text{diag}(1/n_i)$, and hence that $\mathbf{P}_{\mathbb{K}} = \mathbb{K}N^{-1}\mathbb{K}'$.

(c) Let $\bar{\boldsymbol{y}} = (\bar{y}_{1\cdot}, \ldots, \bar{y}_{a\cdot})'$. Show that $\bar{\boldsymbol{y}} = N^{-1}\mathbb{K}'\boldsymbol{y}$ and $\mathbf{P}_X\boldsymbol{y} = \mathbb{K}\bar{\boldsymbol{y}}$.

(d) Show that $SSE = \boldsymbol{y}'(\text{I} - \mathbf{P}_X)\boldsymbol{y} = \sum_{i=1}^{a}\sum_{j=1}^{n_i}(y_{ij} - \bar{y}_{i\cdot})^2$. What are its df and non-centrality parameter?

(e) Find a matrix A such that $XA' = \mathbf{P}_X$ and hence that $\hat{\boldsymbol{\beta}}(\boldsymbol{y}) = A'\boldsymbol{y}$ is a linear least-squares solution for $\boldsymbol{\beta}$.

(f) Let $\boldsymbol{g} = (g_0, \boldsymbol{g}_1')' \in \Re^{a+1}$. Show that $\boldsymbol{g}'\boldsymbol{\beta}$ is estimable in the model $X\boldsymbol{\beta}$ iff $g_0 = \mathbf{1}'_a\boldsymbol{g}_1$. Is β_0 estimable? Is $\boldsymbol{\beta}_1$ estimable?

(g) Show that $C'\boldsymbol{\beta} = \mathbf{0}$ iff $\beta_1 = \cdots = \beta_a$.

(h) Show that $C'\boldsymbol{\beta}$ is estimable in the model $X\boldsymbol{\beta}$.

(i) Show that, if A_1 and A_2 are matrices such that $XA_1' = XA_2' = \mathbf{P}_X$, then $A_1C = A_2C$. With $A = (\mathbf{0}, \mathbb{K}N^{-1})$, then, $AC = \mathbb{K}N^{-1}C_1$.

(j) The GLH numerator SS for testing $\text{H}_0 : C'\boldsymbol{\beta} = \mathbf{0}$ is $\boldsymbol{y}'\mathbf{P}_{AC}\boldsymbol{y}$, where $A'\boldsymbol{y}$ is a linear least-squares solution. Show that $\text{sp}(AC) = \text{sp}(\mathbb{K}) \cap \text{sp}(\mathbb{K}\mathbf{1}_a)^{\perp}$ and therefore $\mathbf{P}_{AC} = \mathbf{P}_{\mathbb{K}} - \mathbf{P}_{\mathbf{1}_n}$.

(k) Show that

$$\boldsymbol{y}'\mathbf{P}_{AC}\boldsymbol{y} = \sum_{i=1}^{a}\sum_{j=1}^{n_i}(\bar{y}_{i\cdot} - \bar{y}_{\cdot\cdot})^2 = \sum_{i=1}^{a} n_i(\bar{y}_{i\cdot} - \bar{y}_{\cdot\cdot})^2.$$

What are its df and ncp?

(l) Show that the restricted model $\{X\boldsymbol{\beta} : \boldsymbol{\beta} \in \Re^{a+1} \text{ and } C'\boldsymbol{\beta} = \mathbf{0}\}$ is $\text{sp}(\mathbf{1}_n)$.

(m) Show that the RMFM SS for testing $\text{H}_0 : C'\boldsymbol{\beta} = \mathbf{0}$ is $\boldsymbol{y}'\mathbf{P}_{AC}\boldsymbol{y}$.

(n) Let $M_C = (\mathbf{1}_a, C_1)$. Show that $\text{sp}(M_C) = \Re^a$ and hence that $\text{sp}(X) = \text{sp}(\mathbb{K}M_C)$. Show that $\text{sp}(C_1) = \text{sp}(S_a) = \text{sp}(\mathbf{1}_a)^\perp$. ($S_a$ is defined on p. 25.)

(o) Show that the RMFM SS for testing H_0 is the same as the extra SSE due to deleting the columns $\mathbb{K}C_1$ from $\mathbb{K}M_C = \mathbb{K}(\mathbf{1}_a, C_1)$.

(p) Show that if $a = 2$, then

$$\boldsymbol{y}'\mathbf{P}_{AC}\boldsymbol{y} = \frac{(\bar{y}_{1\cdot} - \bar{y}_{2\cdot})^2}{\frac{1}{n_1} + \frac{1}{n_2}}.$$

The next two sets of exercises are about properties of the GLH form of numerator SS and its independence from SSE.

20. Let X be a non-zero $n \times (k+1)$ matrix. Let $\boldsymbol{g}$ be a $(k+1)$-vector.

(a) There exists a matrix A such that $XA' = \mathbf{P}_X$.

(b) If $\boldsymbol{g} \in \text{sp}(X')$, then $A\boldsymbol{g} \in \text{sp}(X)$.

(c) There exists a matrix A_* such that $\text{sp}(A_*) \subset \text{sp}(X)$ and $XA_*' = \mathbf{P}_X$.

(d) If $\text{sp}(A) \subset \text{sp}(X)$, then $A\boldsymbol{g} \in \text{sp}(X)$.

21. Let $X = \begin{pmatrix} 1 & 0 & 0 \\ 0 & 1 & 1 \\ 0 & 0 & 0 \end{pmatrix}$, $\boldsymbol{g} = (g_1, g_2, g_3)' \in \Re^3$, and $\boldsymbol{y} = (y_1, y_2, y_3)' \in \Re^3$.

(a) $\boldsymbol{g}'\boldsymbol{\beta}$ is estimable in the model $X\boldsymbol{\beta}$ iff $g_2 = g_3$.

(b) $\mathbf{P}_X = \begin{pmatrix} \mathrm{I}_2 & \mathbf{0} \\ \mathbf{0}' & 0 \end{pmatrix}$ and $\mathbf{P}_X\boldsymbol{y} = \begin{pmatrix} y_1 \\ y_2 \\ 0 \end{pmatrix}$.

(c) Let $c(\boldsymbol{y})$ be an arbitrary real-valued function of $\boldsymbol{y}$. The function $\hat{\boldsymbol{b}}(\boldsymbol{y})$ is a least-squares solution (that is, $X\hat{\boldsymbol{b}}(\boldsymbol{y}) = \mathbf{P}_X\boldsymbol{y}$ for all $\boldsymbol{y}$) iff

$$\hat{\boldsymbol{b}}(\boldsymbol{y}) = \begin{pmatrix} y_1 \\ y_2 - c(\boldsymbol{y}) \\ c(\boldsymbol{y}) \end{pmatrix} = \mathbf{P}_X\boldsymbol{y} + c(\boldsymbol{y}) \begin{pmatrix} 0 \\ -1 \\ 1 \end{pmatrix}$$

for all $\boldsymbol{y} \in \Re^3$.

(d) Let $\boldsymbol{g}$ be a given 3-vector. There exists a single vector $\boldsymbol{\ell}$ such that, for every least-squares solution $\hat{\boldsymbol{b}}(\boldsymbol{y})$, $\boldsymbol{g}'\hat{\boldsymbol{b}}(\boldsymbol{y}) = \boldsymbol{\ell}'\boldsymbol{y}$ for all $\boldsymbol{y} \in \Re^3$ iff $\boldsymbol{g} \in \text{sp}(X')$.

(e) $XA' = \mathbf{P}_X$ iff

$$A = \begin{pmatrix} 1 & z_1 & -z_1 \\ 0 & z_2 & 1 - z_2 \\ 0 & z_3 & -z_3 \end{pmatrix} \text{ for some } \boldsymbol{z} = \begin{pmatrix} z_1 \\ z_2 \\ z_3 \end{pmatrix} \in \Re^3.$$

Then $\text{sp}(A) \notin \text{sp}(X)$ iff $z_3 \neq 0$.

(f) For any such A,

$$A'\boldsymbol{y} = \begin{pmatrix} y_1 \\ \boldsymbol{z}'\boldsymbol{y} \\ y_2 - \boldsymbol{z}'\boldsymbol{y} \end{pmatrix}.$$

(g) For any such A,

$$A_* = \mathbf{P}_X A = \begin{pmatrix} 1 & z_1 & -z_1 \\ 0 & z_2 & 1 - z_2 \\ 0 & 0 & 0 \end{pmatrix}.$$

(h) For any such A,

$$A\boldsymbol{g} = \begin{pmatrix} g_1 + z_1(g_2 - g_3) \\ g_3 + z_2(g_2 - g_3) \\ z_3(g_2 - g_3) \end{pmatrix}.$$

If $\boldsymbol{g}'\boldsymbol{\beta}$ is estimable, then $g_2 = g_3$ and $A\boldsymbol{g} \in \text{sp}(X)$. If $\boldsymbol{g}'\boldsymbol{\beta}$ is not estimable (hence $g_2 \neq g_3$) and$z_3 \neq 0$, then $A\boldsymbol{g} \notin \text{sp}(X)$. In that case $\boldsymbol{g}'\check{\boldsymbol{b}}(\boldsymbol{y})$ is not independent of $SSE = \boldsymbol{y}'(\mathrm{I} - \mathbf{P}_X)\boldsymbol{y}$.

(i) For any such A, $\check{\boldsymbol{b}}(\boldsymbol{y}) = A'\boldsymbol{y}$ is a linear least-squares solution. That is, for each $\boldsymbol{y} \in \Re^3$, $X\check{\boldsymbol{b}}(\boldsymbol{y}) = \mathbf{P}_X\boldsymbol{y}$.

(j) Consider solutions to the normal equation:

$$\begin{aligned} X'X\boldsymbol{b} &= \begin{pmatrix} 1 & 0 & 0 \\ 0 & 1 & 1 \\ 0 & 1 & 1 \end{pmatrix}\boldsymbol{b} = \begin{pmatrix} b_1 \\ b_2 + b_3 \\ b_2 + b_3 \end{pmatrix} \\ &= X'\boldsymbol{y} = \begin{pmatrix} y_1 \\ y_2 \\ y_2 \end{pmatrix}. \end{aligned}$$

Show that every such solution takes the same form as least-squares solutions, that is,

$$\check{\boldsymbol{b}}(\boldsymbol{y}) = \begin{pmatrix} y_1 \\ y_2 \\ 0 \end{pmatrix} + c(\boldsymbol{y}) \begin{pmatrix} 0 \\ -1 \\ 1 \end{pmatrix}.$$

11

Restricted Linear Models

Formulations of linear models for the mean vector that incorporate restrictions on its parameters are centrally important in the theory and practice of statistical methods. Their essential role in inference is evident from Chapter 10. Further coverage in this chapter includes the equivalence of the algebraic formulation and the traditional derivation through stationary points of a Lagrangian function. Relations to full-rank reparameterizations are derived. It is shown that restricted least squares and ordinary least squares are essentially equivalent, and that imposing restrictions is equivalent to deleting a subset of predictor variables from the full model. Penalized least squares provides a means to impose conditions a little at a time. A method is derived to provide full-model and restricted-model results together in a single model.

11.1 Introduction

The models for the mean vector of the response vector $\boldsymbol{y}$ considered here are all affine sets in $\Re^n$, so they can be represented as $\boldsymbol{m}_0 + \mathcal{S}$, where $\boldsymbol{m}_0$ is an n-vector and $\mathcal{S}$ is a linear subspace of $\Re^n$, both known. The simplest and most commonly used form is the multiple regression model $\{X\boldsymbol{\beta} : \boldsymbol{\beta} \in \Re^{k+1}\} = \mathrm{sp}(X)$. Other forms result when linear restrictions are placed on $\boldsymbol{\beta}$, as in $\{X\boldsymbol{\beta} : \boldsymbol{\beta} \in \Re^{k+1} \text{ and } R'\boldsymbol{\beta} = \boldsymbol{r}_0\}$. It was shown in Chapter 8 that all such models can be expressed equivalently as multiple regression models.

The assemblage of model, methods, and computations entailed in fitting the multiple regression model $X\boldsymbol{\beta}$ to the realized response $\boldsymbol{y}$ is often called ordinary least squares (OLS). When applied to models with linear restrictions on $\boldsymbol{\beta}$, it is sometimes called ***restricted least squares*** (RLS). One objective of this chapter is to demonstrate the fundamental equivalence of these two, that is, that RLS is OLS.

Restricted models are central to the construction of statistics to test propositions like $\mathrm{H}_0 : G'\boldsymbol{\beta} = \boldsymbol{r}_0$, as detailed in Chapter 10. In addition, in some settings, they reflect known structural relations among the regression coefficients.

Several accounts of linear models describe methods so that least-squares solutions are unique when columns of X are linearly dependent. One is to replace X by X_{f} such that $\mathrm{sp}(X) = \mathrm{sp}(X_{\mathrm{f}})$ and the columns of X_{f} are linearly independent, and hence there is only one least-squares solution $\hat{\boldsymbol{b}}_{\mathrm{f}}$ such that $X_{\mathrm{f}}\hat{\boldsymbol{b}}_{\mathrm{f}} = \mathbf{P}_X\boldsymbol{y}$. For example, X_{f} can be comprised of a linearly independent subset of the columns of X that span the same subspace. See Scheffé (1959, p. 17) and Seber and Lee (2003, p. 376), for example.

Another work-around is to impose linear restrictions that leave the model unchanged but serve to identify exactly one least-squares solution. In this way, full-rank formulations can be used on the transformed model. These methods are described below as ***full-rank reparameterizations***. In models for analysis of variance effects, this device is so customary that the conditions often are described as the "usual" conditions. See Section 16.7.

To simplify the presentation here, only homogeneous restrictions, $R'\boldsymbol{b} = \mathbf{0}$, will be considered. Non-zero right-hand sides and offsets, so that the model is represented as $\{\boldsymbol{m}_0 + X\boldsymbol{b} : \boldsymbol{b} \in \Re^{k+1} \text{ and } R'\boldsymbol{b} = \boldsymbol{r}_0\}$, require only translating $\boldsymbol{y}$ to $\boldsymbol{z} = \boldsymbol{y} - \boldsymbol{m}_0 - X\boldsymbol{b}_0$, where $\boldsymbol{b}_0$ is a solution to $R'\boldsymbol{b} = \boldsymbol{r}_0$, reducing the model for the mean of $\boldsymbol{z}$ to $\{X\boldsymbol{b} : \boldsymbol{b} \in \Re^{k+1} \text{ and } R'\boldsymbol{b} = \mathbf{0}\}$.

11.2 RLS

The model considered here for the mean of the response vector $\boldsymbol{y}$ is $\mathcal{M} = \{X\boldsymbol{b} : \boldsymbol{b} \in \Re^{k+1} \text{ and } R'\boldsymbol{b} = \mathbf{0}\}$, where X $(n \times (k+1))$ and R $((k+1) \times r)$ are given. The results needed here were established in Chapter 7, mainly in Exercises 24 and 27.

Let N be a matrix such that $\mathrm{sp}(N) = \mathrm{sp}(R)^{\perp}$. Then $\mathcal{M}$ can also be expressed as $\mathrm{sp}(XN)$, a multiple regression model. The least-squares fit of $XN\boldsymbol{c}$ to $\boldsymbol{y}$ is $\hat{\boldsymbol{y}} = XN\hat{\boldsymbol{c}} = \mathbf{P}_{XN}\boldsymbol{y}$. You might say that $\hat{\boldsymbol{b}} = N\hat{\boldsymbol{c}}$ is an RLS solution under the restriction $R'\boldsymbol{b} = \mathbf{0}$. At the same time, $\hat{\boldsymbol{c}}$ is an OLS solution in the model $\mathrm{sp}(XN)$. The extra steps required by the restriction are computing N and XN.

In many textbooks on linear models, least-squares fitting under linear restrictions is framed in terms of finding stationary points of a Lagrangian function

$$q(\boldsymbol{b}, \boldsymbol{\lambda}) = (\boldsymbol{y} - X\boldsymbol{b})'(\boldsymbol{y} - X\boldsymbol{b}) + 2\boldsymbol{\lambda}'R'\boldsymbol{b}.$$

See C. R. Rao (1973, p. 232), for example. Equating all the partial derivatives of q with respect to $\boldsymbol{b}$ and $\boldsymbol{\lambda}$ to 0 results in the equations

$$\begin{aligned} X'X\boldsymbol{b} + R\boldsymbol{\lambda} &= X'\boldsymbol{y}, \\ R'\boldsymbol{b} &= \mathbf{0}. \end{aligned} \tag{11.1}$$

Proposition 11.1 establishes that solutions to these equations are the same as $\hat{\boldsymbol{b}} = N\hat{\boldsymbol{c}}$ and $X\hat{\boldsymbol{b}} = \mathbf{P}_{XN}\boldsymbol{y}$, as described in the previous paragraph.

Proposition 11.1. *Vectors* $\boldsymbol{b}$ *and* $\boldsymbol{\lambda}$ *satisfy (11.1) iff* $R'\boldsymbol{b} = \mathbf{0}$, $X\boldsymbol{b} = \mathbf{P}_{XN}\boldsymbol{y}$, *and* $X'(\boldsymbol{y} - X\boldsymbol{b}) = R\boldsymbol{\lambda}$.

A quick search online shows that extant accounts of RLS develop explicit expressions for $\hat{\boldsymbol{b}}$ by algebraically manipulating (11.1) under the assumption that columns of both X and R are linearly independent (some assume instead that the columns of $\binom{X}{R'}$ are linearly independent). There is no need for such conditions. Computations can be made explicit in several ways. For example, find N as Q_2 from GS on (R, I). Then find $\hat{\boldsymbol{c}}$ as $TQ'\boldsymbol{y}$, where T and Q come from GS on XN, and thence $\hat{\boldsymbol{b}} = N\hat{\boldsymbol{c}}$. Residual SS is $SSE = \boldsymbol{y}'(\mathrm{I} - \mathbf{P}_{XN})\boldsymbol{y}$, with $\mathbf{P}_{XN} = QQ'$ from the same GS on XN. It has $n - \nu_{XN} = \mathrm{tr}(\mathrm{I}_n - \mathbf{P}_{XN})$ degrees of freedom, where ν_{XN} is the number of columns of Q. Standard errors of the RLS coefficient estimators $\hat{\boldsymbol{b}}$ can be computed from MSE and the diagonal entries of $(1/\sigma^2)\mathrm{Var}(\hat{\boldsymbol{b}}) = NTT'N'$.

Note that the restriction $R'\boldsymbol{b} = \mathbf{0}$ does not decrease SSE, comparing the models $\mathrm{sp}(X)$ and $\mathrm{sp}(XN)$:

$$SSE_{XN} - SSE_X = \boldsymbol{y}'(\mathbf{P}_X - \mathbf{P}_{XN})\boldsymbol{y},$$

which is $\geqslant 0$ because $\mathrm{sp}(XN) \subset \mathrm{sp}(X)$. Similarly, the restriction "shrinks" the estimated mean, because $||\mathbf{P}_{XN}\boldsymbol{y}||^2 = \boldsymbol{y}'\mathbf{P}_{XN}\boldsymbol{y} \leqslant ||\mathbf{P}_X\boldsymbol{y}||^2$. Further, the restriction reduces variances of least-squares estimators of linear functions of the mean vector, because

$$\mathrm{Var}(\boldsymbol{\ell}'\mathbf{P}_X\boldsymbol{y}) = \sigma^2\boldsymbol{\ell}'\mathbf{P}_X\boldsymbol{\ell} \geqslant \sigma^2\boldsymbol{\ell}'\mathbf{P}_{XN}\boldsymbol{\ell} = \mathrm{Var}(\boldsymbol{\ell}'\mathbf{P}_{XN}\boldsymbol{y}).$$

If the mean vector is $X\boldsymbol{\beta} \in \mathrm{sp}(X)$, then $\mathrm{E}(\mathbf{P}_X\boldsymbol{y}) = X\boldsymbol{\beta}$, and $\mathrm{E}(\mathbf{P}_{XN}\boldsymbol{y}) = \mathbf{P}_{XN}X\boldsymbol{\beta} = X\boldsymbol{\beta} - (\mathbf{P}_X - \mathbf{P}_{XN})X\boldsymbol{\beta}$. The bias of $\mathbf{P}_{XN}\boldsymbol{y}$ in estimating $X\boldsymbol{\beta}$ is $(\mathbf{P}_X - \mathbf{P}_{XN})X\boldsymbol{\beta} = \mathbf{P}_H X\boldsymbol{\beta}$, which is identically $\mathbf{0}$ iff the estimable part of $R'\boldsymbol{\beta}$ is $\mathbf{0}$, that is, $H'X\boldsymbol{\beta} = \mathbf{0}$.

This theme, that imposing restrictions improves inference if in fact the restriction is true, is recurrent in statistics. The problem is, of course, that whether the restriction is true is never known with certainty (otherwise, it would be incorporated directly into the model), and so it is not possible to know whether imposing it actually does any good.

An alternative to imposing restrictions is to penalize the least-squares criterion for departures from $R'\boldsymbol{b} = \mathbf{0}$. This is described in Section 11.5.

11.3 Full-Rank Reparameterization

Let M be a matrix with linearly independent columns such that $\mathrm{sp}(M) = \mathrm{sp}(X')$. Then the columns of XM are linearly independent, and $\mathrm{sp}(XM) = \mathrm{sp}(X)$. It can be shown that, for the equivalent models $X\boldsymbol{\beta} = XM\boldsymbol{\gamma}$, $\boldsymbol{\gamma}$ is a linear function of $X\boldsymbol{\beta}$ and so it is estimable in $X\boldsymbol{\beta}$. In addition, every estimable linear function of $\boldsymbol{\beta}$ is a linear function of $\boldsymbol{\gamma}$. Inference about an estimable function $G'\boldsymbol{\beta}$ translates to inference about $G'M\boldsymbol{\gamma}$ in the XM model.

Another approach starts by identifying an exhaustive set of non-estimable functions. Let R be a matrix such that $\text{sp}(X') \cap \text{sp}(R) = \{\mathbf{0}\}$ and $\text{sp}(X') + \text{sp}(R) = \Re^{k+1}$. Then the condition $R'\boldsymbol{b} = \mathbf{0}$ does not reduce the model, that is, $\{X\boldsymbol{b} : \boldsymbol{b} \in \Re^{k+1} \text{ and } R'\boldsymbol{b} = \mathbf{0}\} = \text{sp}(X)$. Further, the columns of $\binom{X}{R'}$ are linearly independent, and therefore for each $\boldsymbol{m} \in \text{sp}(X)$, there exists exactly one vector $\boldsymbol{b}_m$ such that $X\boldsymbol{b}_m = \boldsymbol{m}$ and $R'\boldsymbol{b}_m = \mathbf{0}$. Under these non-estimable conditions, the least-squares estimate of $\boldsymbol{\beta}$ is unique. It is established in the exercises that these computations can be had by regressing $\binom{\boldsymbol{y}}{\mathbf{0}}$ on $\binom{X}{R'}$, so that

$$\mathbf{P}_{\binom{X}{R'}}\begin{pmatrix}\boldsymbol{y}\\ \mathbf{0}\end{pmatrix} = \begin{pmatrix}X\\ R'\end{pmatrix}(X'X + RR')^{-1}X'\boldsymbol{y} = \begin{pmatrix}\hat{\boldsymbol{y}}\\ \mathbf{0}\end{pmatrix}, \tag{11.2}$$

resulting in $\check{\boldsymbol{b}} = (X'X + RR')^{-1}X'\boldsymbol{y}$ such that $X\check{\boldsymbol{b}} = \mathbf{P}_X\boldsymbol{y}$ and $R'\check{\boldsymbol{b}} = \mathbf{0}$.

This construction, imposing the non-estimable conditions $R'\boldsymbol{b} = \mathbf{0}$, with $\text{sp}(X') + \text{sp}(R) = \Re^{k+1}$, is equivalent to a full-rank reparameterization, as follows. Let M be a matrix whose columns are linearly independent and $\text{sp}(M) = \text{sp}(R)^{\perp}$. Then the columns of XM are linearly independent and $\text{sp}(XM) = \text{sp}(X)$. Regressing $\boldsymbol{y}$ on XM, let $\check{\boldsymbol{c}} = [(XM)'(XM)]^{-1}(XM)'\boldsymbol{y}$. Then $(XM)\check{\boldsymbol{c}} = \mathbf{P}_{XM}\boldsymbol{y} = \mathbf{P}_X\boldsymbol{y}$. With $\check{\boldsymbol{b}} = M\check{\boldsymbol{c}}$, then, $X\check{\boldsymbol{b}} = \mathbf{P}_X\boldsymbol{y}$ and $R'\check{\boldsymbol{b}} = \mathbf{0}$.

It is not clear that converting the model to full column rank gains any simplicity or efficiency of formulation or in the computations required for estimation and statistical inference as described in the previous chapters.

11.4 RLS as OLS

Regressing $\boldsymbol{y}$ on XN produces $\hat{\boldsymbol{c}}$ such that $XN\hat{\boldsymbol{c}} = \mathbf{P}_{XN}\boldsymbol{y}$, and from it the least-squares solution under the conditions $R'\boldsymbol{b} = \mathbf{0}$ appears as $\hat{\boldsymbol{b}} = N\hat{\boldsymbol{c}}$. Next we show an alternative formulation that produces $\hat{\boldsymbol{b}}$ directly. It is based on the relationship that $\text{sp}(X|H) = \text{sp}(XN)$, and it illustrates a device that has been described in multiple settings for multiple purposes. D. M. Allen (1974) calls it ***data augmentation***. It is to augment the model matrix and the vector of observed responses with additional rows. In some applications, the additional rows serve to add penalties to residual SS. Here they serve to impose the conditions $R'\boldsymbol{b} = \mathbf{0}$ on the coefficient vector.

Let H be a matrix such that $\mathbf{P}_H = \mathbf{P}_X - \mathbf{P}_{XN}$, and let $X|H = (\mathrm{I} - \mathbf{P}_H)X$. We have seen that then $\text{sp}(X|H) = \text{sp}(XN)$ and $\text{sp}(X'H) = \text{sp}(X') \cap \text{sp}(R)$.

It can be shown that $\text{sp}(R) \cap \text{sp}[(X|H)'] = \{\mathbf{0}\}$, which implies that none of $R'\boldsymbol{b}$ is estimable in the model $\text{sp}(X|H)$, and hence imposing $R'\boldsymbol{b} = \mathbf{0}$ does not reduce the model. In addition, by Exercise 24, p. 52, this implies that

$$\mathbf{P}_{\binom{X|H}{R'}} = \begin{pmatrix} \mathbf{P}_{X|H} & 0 \\ 0 & \mathbf{P}_{R'} \end{pmatrix}.$$

It follows that, for any $(n+r)$-vector $\binom{\boldsymbol{y}}{\mathbf{0}}$, there exists a vector $\hat{\boldsymbol{b}}$ such that

$$\begin{aligned} \mathbf{P}_{\binom{X|H}{R'}} \begin{pmatrix} \boldsymbol{y} \\ \mathbf{0} \end{pmatrix} &= \begin{pmatrix} \mathbf{P}_{X|H}\boldsymbol{y} \\ \mathbf{P}_{R'}\mathbf{0} \end{pmatrix} \\ &= \begin{pmatrix} X|H \\ R' \end{pmatrix} \hat{\boldsymbol{b}} \\ &= \begin{pmatrix} (X|H)\hat{\boldsymbol{b}} \\ \mathbf{0} \end{pmatrix} \end{aligned}$$

with $R'\hat{\boldsymbol{b}} = \mathbf{0}$. Further, since

$$\mathrm{sp}(R) \supset \mathrm{sp}(X'H),$$

$R'\hat{\boldsymbol{b}} = \mathbf{0} \Longrightarrow H'X\hat{\boldsymbol{b}} = \mathbf{0} \iff \mathbf{P}_H X\hat{\boldsymbol{b}} = \mathbf{0}$, and therefore $(X|H)\hat{\boldsymbol{b}} = X\hat{\boldsymbol{b}}$. Thus the OLS regression of $\binom{\boldsymbol{y}}{\mathbf{0}}$ on $\binom{X|H}{R'}$ yields $X\hat{\boldsymbol{b}}$ in the restricted model $\mathcal{M}$ as

$$X\hat{\boldsymbol{b}} = \mathbf{P}_{X|H}\boldsymbol{y}.$$

Note further that, because $\mathrm{sp}(X|H) = \mathrm{sp}(XN)$, where $\mathrm{sp}(N) = \mathrm{sp}(R)^{\perp}$, regressing $\boldsymbol{y}$ on XN (also OLS) gives $\mathbf{P}_{XN}\boldsymbol{y} = XN\hat{\boldsymbol{c}} = X\hat{\boldsymbol{b}}$, with $R'\hat{\boldsymbol{b}} = R'N\hat{\boldsymbol{c}} = \mathbf{0}$.

Customary output from multiple regression programs includes coefficient estimates and their standard errors, the ANOVA table showing SSE and its degrees of freedom, and predicted values and their standard errors. Applied to $\binom{\boldsymbol{y}}{\mathbf{0}}$ regressed on $\binom{X|H}{R'}$, the coefficient estimates are $\hat{\boldsymbol{b}}$ satisfying $R'\hat{\boldsymbol{b}} = \mathbf{0}$, and SSE is correct for the restricted model $\{X\boldsymbol{b} : \boldsymbol{b} \in \Re^{k+1} \text{ and } R'\boldsymbol{b} = \mathbf{0}\}$.

A generic OLS program assumes that the variance-covariance matrix of the input response vector is proportional to an identity matrix, and the standard errors of estimates that it provides are based on this assumption. When $\binom{\boldsymbol{y}}{\mathbf{0}}$ is regressed on $\binom{X|H}{R'}$, though, the variance-covariance matrix should be $\mathrm{Var}\binom{\boldsymbol{Y}}{\mathbf{0}} = \sigma^2 \begin{pmatrix} \mathrm{I}_n & 0 \\ 0 & 0_{r\times r} \end{pmatrix}$ but it is assumed by the program to be $\sigma^2\mathrm{I}_{n+r}$ instead.

Consequences of this mis-assumption can be seen in terms of using GS for the computations. GS on $\binom{X|H}{R'}$ yields $Q_1 = \binom{Q_{11}}{Q_{21}}$ and T such that $\mathrm{sp}(Q_1) = \mathrm{sp}\binom{X|H}{R'}$, $Q_1 = \binom{(X|H)T}{R'T}$, and $Q_1'Q_1 = Q_{11}'Q_{11} + Q_{21}'Q_{21} = \mathrm{I}$. A least-squares solution is

$$\hat{\boldsymbol{b}} = TQ_1' \begin{pmatrix} \boldsymbol{y} \\ \mathbf{0} \end{pmatrix} = TQ_{11}'\boldsymbol{y}. \tag{11.3}$$

Its variance-covariance matrix is $\sigma^2 TQ'_{11}Q_{11}T'$, but the OLS program considers it to be

$$\text{Var}(\hat{\boldsymbol{b}}) = \sigma^2 T(Q'_{11}Q_{11} + Q'_{21}Q_{21})T' = \sigma^2 TT'. \tag{11.4}$$

Further, as shown above,

$$\mathbf{P}_{\binom{X|H}{R'}} = \begin{pmatrix} \mathbf{P}_{X|H} & 0 \\ 0 & \mathbf{P}_{R'} \end{pmatrix}, \tag{11.5}$$

so the program returns

$$SSE = \begin{pmatrix} \boldsymbol{y} \\ \mathbf{0} \end{pmatrix}' \left(\mathrm{I}_{n+r} - \mathbf{P}_{\binom{X|H}{R'}}\right) \begin{pmatrix} \boldsymbol{y} \\ \mathbf{0} \end{pmatrix} = \boldsymbol{y}'(\mathrm{I}_n - \mathbf{P}_{X|H})\boldsymbol{y}, \tag{11.6}$$

which is correct, but it returns its degrees of freedom as $df_E = \text{tr}(\mathrm{I}_n - \mathbf{P}_{X|H}) + \text{tr}(\mathrm{I}_r - \mathbf{P}_{R'})$, which is correctly equal to $\text{tr}(\mathrm{I}_n - \mathbf{P}_{X|H})$ iff $\text{tr}(\mathbf{P}_{R'}) = r$, that is, iff the r columns of R are linearly independent. While that can always be engineered, in many cases replacing R by a full-rank substitute requires extra steps.

The upshot is that regressing $\binom{\boldsymbol{y}}{\mathbf{0}}$ on $\binom{X|H}{R'}$ produces a correct least-squares solution $\hat{\boldsymbol{b}}$ such that $R'\hat{\boldsymbol{b}} = \mathbf{0}$, and it produces the correct SSE. But the standard errors and test statistics it produces are not correct. All results would be correct if the fact that $\text{Var}\binom{\boldsymbol{Y}}{\mathbf{0}} = \sigma^2 \begin{pmatrix} \mathrm{I}_n & 0 \\ 0 & 0_{r\times r} \end{pmatrix}$ instead of $\sigma^2 \begin{pmatrix} \mathrm{I}_n & 0 \\ 0 & \mathrm{I}_r \end{pmatrix}$ were built into the program. This could be done, for instance, by providing a designation of rows in the response as artificial with variances all 0.

This kind of construction can be extended further to see the restricted estimates along with the estimates for the unrestricted model $\text{sp}(X) = \text{sp}(X|H) + \text{sp}(H)$ by regressing $\binom{\boldsymbol{y}}{\mathbf{0}}$ on $\begin{pmatrix} X|H & H \\ R' & 0 \end{pmatrix}$. Then the RMFM SS for testing $\mathrm{H}_0 : R'\boldsymbol{\beta} = \mathbf{0}$ is the additional SSE due to omitting the columns $\binom{H}{\mathbf{0}}$. Because $\binom{X|H}{R'}$ and $\binom{H}{0}$ are orthogonal, this RMFM SS is $\binom{\boldsymbol{y}}{\mathbf{0}}'\mathbf{P}_{\binom{H}{0}}\binom{\boldsymbol{y}}{\mathbf{0}} = \boldsymbol{y}'\mathbf{P}_H\boldsymbol{y}$.

It is possible to perform all these computations using only an OLS program, in several steps on several different response vectors and model matrices, but just to make the point that it can be done.

11.5 Penalized Least Squares

An alternative to imposing conditions outright is to add a penalty to the least-squares criterion that increases it for departures from $R'\boldsymbol{b} = \mathbf{0}$. This can be done by regressing $\binom{\boldsymbol{y}}{\mathbf{0}}$ on $\binom{X}{\sqrt{k}R'}$, so that the LS criterion becomes $(\boldsymbol{y} - X\boldsymbol{b})'(\boldsymbol{y} - X\boldsymbol{b}) + k\boldsymbol{b}'RR'\boldsymbol{b}$. At $k = 0$ this is OLS. As k increases, it would seem that the penalty would dominate and force the optimizing solution toward $R'\boldsymbol{b} = \mathbf{0}$. However, as we have seen, if $\mathrm{sp}(R) \cap \mathrm{sp}(X') = \{\mathbf{0}\}$, then there is a solution to $R'\boldsymbol{b} = \mathbf{0}$ among LS solutions such that $X\boldsymbol{b} = \mathbf{P}_X\boldsymbol{y}$, and so the penalty has no additional effect. On the other hand, if some of $R'\boldsymbol{b}$ is estimable, then the penalty does affect minimizing solutions.

Regressing $\binom{\boldsymbol{y}}{\mathbf{0}}$ on $\binom{X}{\sqrt{k}R'}$ results in

$$\tilde{\boldsymbol{y}}_k = H_k\boldsymbol{y} = X\hat{\boldsymbol{b}}_k \tag{11.7}$$

where $H_k = X(X'X + kRR')^{-}X'$ and $\hat{\boldsymbol{b}}_k = (X'X + kRR')^{-}X'\boldsymbol{y}$. Under the conditions that $\mathrm{E}(\boldsymbol{Y}) = X\boldsymbol{\beta}$ and $\mathrm{Var}(\boldsymbol{Y}) = \sigma^2\mathrm{I}_n$, $\mathrm{E}(\tilde{\boldsymbol{Y}}_k) = H_kX\boldsymbol{\beta}$ and $\mathrm{Var}(\tilde{\boldsymbol{Y}}_k) = \sigma^2H_k^2$.

The details of what happens to H_k as k increases are established in LaMotte et al. (2020): both H_k and H_k^2 decrease monotonically (in the Löwner sense) to $\mathbf{P}_{XN}$, and $\nu_k = \mathrm{tr}(H_k)$ decreases monotonically from $\mathrm{tr}(\mathbf{P}_X)$ to $\mathrm{tr}(\mathbf{P}_{XN})$. Given $\boldsymbol{y}$, $\tilde{\boldsymbol{y}}_k$ then shrinks monotonically (in its squared norm) to the RLS estimate $\mathbf{P}_{XN}\boldsymbol{y}$ for the restricted model $\{X\boldsymbol{b} : \boldsymbol{b} \in \Re^{k+1} \text{ and } R'\boldsymbol{b} = \mathbf{0}\} = \mathrm{sp}(XN)$. Its variance-covariance matrix shrinks monotonically from $\sigma^2\mathbf{P}_X$ to $\sigma^2\mathbf{P}_{XN}$. Finally, it can be shown by the results just cited that $R'\hat{\boldsymbol{b}}_k$ shrinks monotonically to $\mathbf{0}$.

Figure 2.1, p. 14, depicts this setting. Some sources call ν_k the ***effective degrees of freedom*** of the model, in analogy to $\mathrm{tr}(\mathbf{P}_X)$. For $k \geqslant 0$, both H_k and ν_k are differentiable, decreasing functions of k. The effect of the magnitude of k on the resulting fit depends on scaling of X and R, among other things. Some sources suggest that useful values of k to examine are those such that ν_k goes by integer values from $\mathrm{tr}(\mathbf{P}_X)$ to $\mathrm{tr}(\mathbf{P}_{XN})$. Because ν_k is monotone and differentiable in k, the non-linear equation $\nu_k = j$ can be solved readily for k by a Newton-Raphson procedure.

Penalized least-squares has many and varied applications. Allen (1974) showed that ridge regression could be viewed as penalized least squares, with $R = \mathrm{I}_{k+1}$. It can be used to view effects of imposing constraints "a little at a time" on goodness-of-fit measures, such as residual SS corresponding to ν_k. It can be used to see incremental effects of smoothing of semiparametric fits, like spline functions: there, R represents, say, second differences, so that increasing k smooths the model toward a straight-line fit. By translating $\boldsymbol{y}$ to $\boldsymbol{y} - X\boldsymbol{b}_0$, where $R'\boldsymbol{b}_0 = \boldsymbol{r}_0$, the effect of increasing k is to shrink $\tilde{\boldsymbol{y}}_k$ to $\mathbf{P}_{XN}\boldsymbol{y} + X\boldsymbol{b}_0$ and $R'\hat{\boldsymbol{b}}_k$ to $\boldsymbol{r}_0$.

If $\binom{\boldsymbol{y}}{\mathbf{0}}$ is fed into an OLS program, the program assumes that its variance-covariance matrix is $\sigma^2\mathrm{I}_{n+r}$ when in fact it should be $\sigma^2\begin{pmatrix} \mathrm{I}_n & 0 \\ 0 & 0_{r\times r} \end{pmatrix}$. Suppose for a moment that its variance-covariance matrix is instead $\sigma^2\begin{pmatrix} \mathrm{I}_n & 0 \\ 0 & \frac{1}{k}\mathrm{I}_r \end{pmatrix}$. Transform from $\binom{\boldsymbol{y}}{\mathbf{0}}$ to

$$\begin{pmatrix} \mathrm{I}_n & 0 \\ 0 & \frac{1}{k}\mathrm{I}_r \end{pmatrix}^{-1/2} \begin{pmatrix} \boldsymbol{y} \\ \mathbf{0} \end{pmatrix} = \begin{pmatrix} \boldsymbol{y} \\ \mathbf{0} \end{pmatrix}, \tag{11.8}$$

so that the model matrix for the transformed random variable is

$$\begin{pmatrix} \mathrm{I}_n & 0 \\ 0 & \frac{1}{k}\mathrm{I}_r \end{pmatrix}^{-1/2} \begin{pmatrix} X \\ R' \end{pmatrix} = \begin{pmatrix} X \\ \sqrt{k}R' \end{pmatrix}. \tag{11.9}$$

The transformed response then has variance-covariance matrix $\sigma^2\mathrm{I}_{n+r}$, and so it follows a GM model. This leads to the same estimates as shown above. In this heuristic, k can be regarded as indexing the certainty of the assertion that $R'\boldsymbol{\beta} = \mathbf{0}$, it being small for k close to 0 (variance σ^2/k large), and ranging to certain as k increases without bound (variance small). These extremes correspond to $X\hat{\boldsymbol{b}}_0 = \mathbf{P}_X\boldsymbol{y}$ and $X\hat{\boldsymbol{b}}_\infty = \mathbf{P}_{XN}\boldsymbol{y}$.

11.6 Example

The purpose of the numerical example shown here is to illustrate the relations and constructions discussed in the previous sections. The data set is artificial, generated as responses y under combinations of three levels each of factors A and B along with a covariate x. The model comprises an intercept, six dummy variables indicating levels of A and B, respectively, and a term for the covariate. The data set is shown in Table 11.1.

Columns (1)–x comprise the matrix X. Denote their coefficients in the model $X\boldsymbol{\beta}$ for the mean of the response vector $\boldsymbol{y}$ by

$$\boldsymbol{\beta} = (\mu_0, \alpha_1, \alpha_2, \alpha_3, \beta_1, \beta_2, \beta_3, \delta)'.$$

Define the matrix R for the restrictions $R'\boldsymbol{\beta} = \mathbf{0}$ by

$$R = \begin{pmatrix} 0 & 0 & 0 & 0 & 0 \\ 0 & 0 & 0 & 0 & -1 \\ 1 & 0 & 0 & 0 & 0 \\ 0 & 0 & 0 & 0 & 1 \\ 0 & 1 & 0 & 1 & 0 \\ 0 & 1 & 1 & -2 & 0 \\ 0 & 1 & 0 & 1 & 0 \\ 0 & 0 & 0 & 0 & 0 \end{pmatrix}. \tag{11.10}$$

The restrictions are: $\alpha_2 = 0$, $\beta_1 + \beta_2 + \beta_3 = 0$, $\beta_2 = 0$, $\beta_1 - 2\beta_2 + \beta_3 = 0$, and $-\alpha_1 + \alpha_3 = 0$. The estimable part of the conjunction of these $r = 5$ restrictions comprises the two restrictions $\alpha_1 - \alpha_3 = 0$ and $\beta_1 - 2\beta_2 + \beta_3 = 0$. Note that R has column rank 4, as the fourth column is a linear combination of the second and third.

TABLE 11.1: Data for the example. The response y is simulated at combinations of levels of factors A and B, each at three levels, and values of a covariate x.

(1)	A1	A2	A3	B1	B2	B3	x	y
1	1	0	0	1	0	0	28	50.0
1	1	0	0	0	1	0	21	22.2
1	1	0	0	0	1	0	77	111.7
1	1	0	0	0	0	1	54	65.3
1	0	1	0	1	0	0	34	59.2
1	0	1	0	1	0	0	26	60.2
1	0	1	0	1	0	0	59	101.3
1	0	1	0	0	1	0	12	51.0
1	0	1	0	0	0	1	69	109.5
1	0	1	0	0	0	1	42	70.4
1	0	0	1	1	0	0	62	64.8
1	0	0	1	0	1	0	86	94.8
1	0	0	1	0	1	0	53	60.3
1	0	0	1	0	0	1	46	63.6
1	0	0	1	0	0	1	93	108.2
1	0	0	1	0	0	1	31	37.0

Let N be a matrix such that $\text{sp}(N) = \text{sp}(R)^{\perp}$:

$$N = \begin{pmatrix} 1 & 0 & 0 & 0 \\ 0 & 1 & 0 & 0 \\ 0 & 0 & 0 & 0 \\ 0 & 1 & 0 & 0 \\ 0 & 0 & 1 & 0 \\ 0 & 0 & 0 & 0 \\ 0 & 0 & -1 & 0 \\ 0 & 0 & 0 & 1 \end{pmatrix}. \tag{11.11}$$

A matrix H such that $\mathbf{P}_H = \mathbf{P}_X - \mathbf{P}_{XN}$ can be found as Q_2 from $\text{GS}(XN, X) \to (Q_1, Q_2)$. Recall that $X|H = (\text{I} - \mathbf{P}_H)X$, and $\text{sp}(X|H)$ is the same as the restricted model, $\{X\boldsymbol{\beta} : \boldsymbol{\beta} \in \Re^8\}$. Define $X_+ = \binom{X|H}{R'}$ and $\boldsymbol{y}_+ = \binom{\boldsymbol{y}}{\mathbf{0}_5}$.

Table 11.2 shows results of multiple regression procedures for six models. The first and last models are for the regression of y on (1), A1, ..., B3, x without any restrictions on the coefficients. The other four models impose the restrictions $R'\boldsymbol{\beta} = \mathbf{0}$ shown in (11.10). So the first and sixth models are equivalent, and the second through fifth models are equivalent. The second model was fit using SAS proc reg with the restrictions imposed. Its results are correct. As noted above, the restricted model is equivalent to $\text{sp}(X|H)$: regressing y directly on $X|H$ gives the results shown in the fifth model. The models $\{X\boldsymbol{b} : \boldsymbol{b} \in \Re^8$ and $R'\boldsymbol{b} = \mathbf{0}\}$ and $\text{sp}(X|H)$ are the same, and so SSE

TABLE 11.2: Summary of results of regressions y on several models, data of Table 11.1.

vbls.	No restr.		SAS proc reg w/ restr.		XN	
x_j	Est.	Std. Err.	Est.	Std. Err.	Est.	Std. Err.
(1)	-1.28043	8.31183	30.01191	5.92119	30.01191	5.92119
A1	10.46125	6.15692	-23.85282	5.14904	-23.85282	5.14904
A2	29.25342	5.86513	0	0	0	0
A3	0	0	-23.85282	5.14904	-23.85282	5.14904
B1	-0.26526	5.73288.	0.44968	2.96100	0.44968	2.96100
B2	0.86560	5.61360	0	0	0	0
B3	0	0	-0.44968	2.96100	-0.44968	2.96100
x	1.17228	0.10867	1.12016	0.10895	1.12016	0.10895
SSE/df_E	808.32534	10	1054.67184	12	1054.67184	12

vbls.	X_+		$X\|H$		$(X\|H, H//R', 0)$	
x_j	Est.	Std. Err.	Est.	Std. Err.	Est.	Std. Err.
(1)	30.01191	11.18907	153.50592	8.71486	30.01191	10.64887
A1	-23.85282	11.31402	-226.87260	21.94673	-23.85282	10.76780
A2	0	9.00714	-90.82205	11.08169	0	8.57229
A3	-23.85282	11.24559	0	0	-23.85282	10.70267
B1	0.44968	4.49709	25.16559	7.04406	0.44968	4.27997
B2	0	3.84066	0	0	0	3.65524
B3	-0.44968	4.31410	0	0	-0.44968	4.10582
x	1.12016	0.11362	0	0	1.12016	0.10813
H_1	0	0	0	0	15.59541	8.57229
H_2	0	0	0	0	-1.76911	8.57229
SSE/df_E	1054.67184	13	1054.67184	12	808.32534	11

and df_E are the same for the second and fifth models. The difference is that the coefficient estimates in the fifth model are not constrained by $R'\boldsymbol{b} = \mathbf{0}$, which are non-estimable in the model $X|H$, while they are so constrained in the second. For the discussion here, the second model is the gold standard for the restricted model.

The same results as the second model are obtained in the third model. In it, y is regressed on XN. However, the coefficient estimates are $\hat{\boldsymbol{c}}$ for the model $XN\boldsymbol{c}$, and so additional computations must be done to get $\hat{\boldsymbol{b}} = N\hat{\boldsymbol{c}}$ and its standard errors.

For the fourth model, $\boldsymbol{y}_+ = \binom{\boldsymbol{y}}{\mathbf{0}_5}$ is regressed on $X_+ = \binom{X|H}{R'}$. This gives the correct coefficient estimates and SSE for the restricted model, the same as the second model. But the standard errors of the coefficient estimates are wrong, as noted in (11.4), and the degrees of freedom for SSE are wrong because the column dimension of R is 5 while its rank is 4.

In the sixth model, $\binom{\boldsymbol{y}}{\mathbf{0}}$ is regressed on $\begin{pmatrix} X|H & H \\ R' & 0 \end{pmatrix}$. The model is equivalent to the full model, so its SSE is the same as in the first model. Its estimates of the coefficients of the columns of X are the restricted-model estimates. As noted in (11.4), standard errors and df_E could be fixed easily so that correct restricted-model results could be shown within the full model.

11.7 Exercises

Let X be a given $n \times (k+1)$ matrix, and let R be a given $(k+1) \times r$ matrix. Let H be a matrix such that $\text{sp}(H) \subset \text{sp}(X)$ and $\text{sp}(X'H) = \text{sp}(X') \cap \text{sp}(R)$. Define $X|H = (\text{I} - \mathbf{P}_H)X$. Let N be a matrix such that $\text{sp}(N) = \text{sp}(R)^{\perp}$. Most of the results in these exercises follow from results established in Exercise 27, p. 53.

1. (a) Show that

$$\text{sp}[(X|H)'] \cap \text{sp}(X'H) = \{\mathbf{0}\}.$$

   (b) Show that

$$\text{sp}[(X|H)'] \cap \text{sp}(R) = \{\mathbf{0}\}.$$

   (c) Let $\boldsymbol{z} \in \Re^n$. Show that

$$\mathbf{P}_{\binom{X|H}{R'}} \begin{pmatrix} \boldsymbol{z} \\ \mathbf{0} \end{pmatrix} = \begin{pmatrix} \mathbf{P}_{X|H}\boldsymbol{z} \\ \mathbf{0} \end{pmatrix} = \begin{pmatrix} (X|H)\boldsymbol{b} \\ R'\boldsymbol{b} \end{pmatrix}.$$

   This specifies a vector $\boldsymbol{b}$ such that $\mathbf{P}_{X|H}\boldsymbol{z} = (X|H)\boldsymbol{b}$ with $R'\boldsymbol{b} = \mathbf{0}$. In addition, $R'\boldsymbol{b} = \mathbf{0} \Longrightarrow \mathbf{P}_H X\boldsymbol{b} = \mathbf{0}$ and hence $\mathbf{P}_{X|H}\boldsymbol{z} = X\boldsymbol{b}$ with $R'\boldsymbol{b} = \mathbf{0}$.

(d) Show that $\text{sp}(XN) = \{X\boldsymbol{b} : \boldsymbol{b} \in \Re^{k+1} \text{ and } R'\boldsymbol{b} = \mathbf{0}\}$.

(e) Show that $\text{sp}(X|H) = \text{sp}(XN)$.

(f) Show that $\text{sp}(X|H) + \text{sp}(H) = \text{sp}(X)$ and $\text{sp}(X|H) \cap \text{sp}(H) = \mathbf{0}$.

2. Let

$$\mathcal{M} = \{\boldsymbol{\mu}_0 + X\boldsymbol{\beta} : \boldsymbol{\beta} \in \Re^{k+1} \text{ and } R'\boldsymbol{\beta} = \boldsymbol{r}_0\}.$$

(a) Show that, for any given vector $\boldsymbol{\eta}_0$ in $\mathcal{M}$,

$$\mathcal{M} = \{\boldsymbol{\eta}_0 + XN\boldsymbol{\gamma} : \boldsymbol{\gamma} \in \Re^q\} = \{\boldsymbol{\eta}_0\} + \text{sp}(XN).$$

(b) Let $\boldsymbol{g}$ be a q-vector. The definition $\star$ of estimability says that $\boldsymbol{g}'\boldsymbol{\beta}$ is estimable in this model iff there exists an n-vector $\boldsymbol{\ell}$ and a constant c such that

$$\boldsymbol{g}'\boldsymbol{\beta} = \boldsymbol{\ell}'\boldsymbol{\mu} + c$$

for all vectors $\boldsymbol{\beta}$ such that $\boldsymbol{\mu} = \boldsymbol{\mu}_0 + X\boldsymbol{\beta}$ is in $\mathcal{M}$.
Show that $\boldsymbol{g}'\boldsymbol{\beta}$ is estimable in $\mathcal{M}$ iff $\boldsymbol{g} \in \text{sp}(X') + \text{sp}(R) = \text{sp}\binom{X}{R'}'$.

(c) With the model expressed as $\mathcal{M} = \{\boldsymbol{\eta}_0 + XN\boldsymbol{\gamma} : \boldsymbol{\gamma} \in \Re^q\}$, $(N'\boldsymbol{g})'\boldsymbol{\gamma}$ is estimable iff $N'\boldsymbol{g} \in \text{sp}(N'X')$. Show that this is true iff $\boldsymbol{g} \in \text{sp}(X') + \text{sp}(R)$.

(d) From these two parts, it is evident that $\boldsymbol{g}'\boldsymbol{\beta}$ is estimable in $\mathcal{M}$ iff $\boldsymbol{g} \in \text{sp}\binom{X}{R'}'$.

(e) Show that $\text{sp}\binom{X}{R'}' = \text{sp}\binom{X|H}{R'}'$.
[Hint: $X'\boldsymbol{\ell} + R\boldsymbol{z} = X'(\text{I} - \mathbf{P}_H)\boldsymbol{\ell} + X'\mathbf{P}_H\boldsymbol{\ell} + R\boldsymbol{z}$. Show that $X'\mathbf{P}_H\boldsymbol{\ell} \in \text{sp}(R)$.]

3. One use of linear constraints on $\boldsymbol{\beta}$ is to make linear functions $\boldsymbol{p}'\boldsymbol{\beta}$ ***identifiable*** from the mean vector $X\boldsymbol{\beta}$. The notion of identifiability is that, if $\boldsymbol{\beta}_1$ and $\boldsymbol{\beta}_2$ yield the same mean vector $X\boldsymbol{\beta}_1 = X\boldsymbol{\beta}_2$, then the value of $\boldsymbol{p}'\boldsymbol{\beta}_1$ should be the same as the value of $\boldsymbol{p}'\boldsymbol{\beta}_2$. That is, the mean vector determines the value of $\boldsymbol{p}'\boldsymbol{\beta}$. A little more formally, a function $\boldsymbol{p}'\boldsymbol{\beta}$ is identifiable iff $X\boldsymbol{\beta}_1 = X\boldsymbol{\beta}_2 \Longrightarrow \boldsymbol{p}'\boldsymbol{\beta}_1 = \boldsymbol{p}'\boldsymbol{\beta}_2$ (equivalently, $\boldsymbol{p}'\boldsymbol{\beta}_1 \neq \boldsymbol{p}'\boldsymbol{\beta}_2 \Longrightarrow X\boldsymbol{\beta}_1 \neq X\boldsymbol{\beta}_2$).

In the model $\mathcal{M} = \{X\boldsymbol{\beta} : \boldsymbol{\beta} \in \Re^{k+1}\}$, show that a vector $\boldsymbol{p}$ is such that $\boldsymbol{p}'\boldsymbol{\beta}$ is identifiable iff $\boldsymbol{p} \in \text{sp}(X')$. That is, the properties, estimability and identifiability, are equivalent.

4. How do restrictions on the model affect the set of estimable functions? Consider $\mathcal{M} = \{X\boldsymbol{\beta} : \boldsymbol{\beta} \in \Re^{k+1}\}$ and $\mathcal{M}_0 = \{X\boldsymbol{\beta} : \boldsymbol{\beta} \in \Re^{k+1} \text{ and } R'\boldsymbol{\beta} = R'\boldsymbol{b}_0\}$, where X, R, and $\boldsymbol{b}_0$ are given. The set of vectors $\boldsymbol{p}$ such that $\boldsymbol{p}'\boldsymbol{\beta}$ is estimable in $\mathcal{M}$ is $\text{sp}(X')$. Identify the set of estimable functions in $\mathcal{M}_0$.

5. In the model specified by $X\boldsymbol{\beta}$ with

$$X = \begin{pmatrix} 1 & 1 & 0 & 0 \\ 1 & 0 & 1 & 0 \\ 1 & 0 & 0 & 1 \end{pmatrix},$$

and $\boldsymbol{\beta} = (\beta_0, \beta_1, \beta_2, \beta_3)'$, $\beta_1 + \beta_2 + \beta_3$ is not estimable, nor is β_1, but $\beta_1 - \beta_3$ is estimable. Show that in the model $\{X\boldsymbol{\beta} : \boldsymbol{\beta} \in \Re^4 \text{ and } \beta_1 + \beta_2 + \beta_3 = 0\}$, every linear function of $\boldsymbol{\beta}$ is estimable.

6. Suppose that $\text{sp}(X') \neq \Re^{(k+1)}$.

   (a) Show that the dimension of $\text{sp}(X)$, and hence the column rank of X, is less than $k + 1$.

   (b) With R such that $\text{sp}(X') + \text{sp}(R) = \Re^{k+1}$, show that the columns of $\binom{X}{R'}$ are linearly independent.

   (c) Prove: If $\text{sp}(X') + \text{sp}(R) = \Re^{k+1}$, and if N is a matrix with linearly independent columns such that $\text{sp}(N) = \text{sp}(R)^{\perp}$, then the columns of XN are linearly independent.

   (d) Describe how to compute R such that $\text{sp}(X') \cap \text{sp}(R) = \{\mathbf{0}\}$ and $\text{sp}(X') + \text{sp}(R) = \Re^{k+1}$.

   (e) Describe how to compute N with linearly independent columns such that $\text{sp}(N) = \text{sp}(R)^{\perp}$.

12

Special Hypotheses: Outliers, Prediction, Inverse Prediction, Collapsibility

This chapter develops linear hypotheses and their sums of squares that are particular to testing responses as outliers, constructing prediction intervals on future responses, and inferring what conditions, in terms of settings of predictor variables, might have led to a given observed value of the response. These all can be considered within the general framework of collapsibility.

12.1 Introduction

Given a linear model for the mean vector, we have developed tools with which we can address any set of questions about $\boldsymbol{\mu}$ that can be formulated as a linear hypothesis in the form $\text{H}_0 : L_0'\boldsymbol{\mu} = \boldsymbol{c}_0$. In models built around $X\boldsymbol{\beta}$, with or without linear restrictions on $\boldsymbol{\beta}$, for a hypothesis $\text{H}_0 : G'\boldsymbol{\beta} = \boldsymbol{c}_0$, we can identify the subset of functions $H'X\boldsymbol{\beta} = M'G'\boldsymbol{\beta}$ of $\boldsymbol{\beta}$ that could possibly cause violations of H_0, and we can construct the conventional F statistic for testing the estimable part of H_0, namely $\text{H}_{0*} : H'X\boldsymbol{\beta} = M'\boldsymbol{c}_0$.

Given a linear hypothesis, we have the tools to handle it. Often that is the easy part. Formulating the linear hypothesis itself may require some care. In a regression model $\boldsymbol{\mu} = \beta_0\mathbf{1} + X_1\boldsymbol{\beta}_1 + X_2\boldsymbol{\beta}_2$, questions about the factor effects in $X_1\boldsymbol{\beta}_1$ would seem naturally to be formulated in terms of linear functions of $\boldsymbol{\beta}_1$. The question, whether there are no such effects, is simply $\text{H}_0 : \boldsymbol{\beta}_1 = \mathbf{0}$. It can be that simple. On the other hand, it can happen that $\boldsymbol{\beta}_1$ is not estimable, or that effects in $X_1\boldsymbol{\beta}_1$ cannot be separated from effects in $X_2\boldsymbol{\beta}_2$. In such cases, more work is required to identify the questions of interest that can be addressed in terms of the model.

Formulations of linear hypotheses for addressing ***outliers***, ***prediction***, and ***inverse prediction*** share a common approach, as presented in the next section. They can be viewed in the broader context of ***collapsibility***. The notion of collapsibility is simple enough. As described by Clogg et al. (1992), in the model $\boldsymbol{\mu} = X_1\boldsymbol{\beta}_1 + X_2\boldsymbol{\beta}_2$, $\boldsymbol{\beta}_1$ is collapsible with respect to $\boldsymbol{\beta}_2$ if it is the same in the full model $\mathrm{sp}(X_1, X_2)$ as it is in the submodel $\mathrm{sp}(X_1)$.

12.2 Outliers

An outlier is a value of the response that is so far off the pattern of the rest of the data that it seems farfetched to believe that it came from the same regime. However, keep in mind that extreme values can occur without violating the model. Discarding or disregarding an observation simply because it's extreme may lead to understating variability and hence to spurious inferences. Our objective here is to describe inferential tools that can be used to assess the extent to which an observation is an outlier. What to do with an apparent outlier is a delicate subject that requires a deeper discussion than we shall undertake here. In summary, though, a good rule is, don't discard or disregard an observation unless you can substantiate that something occurred – a lab accident, equipment failure, for example – that caused this observation to be different.

Whether an observed value is an outlier is a question about y, a realized value of a random variable, not directly a question about the parameters $\boldsymbol{\beta}$. Methods of statistical inference address questions about parameters, not observed values. In order to address whether an observed value is an outlier, then, we must formulate a corresponding question in terms of parameters.

LaMotte and Volaufova (1999) describe a rationale for examining a single observation in the data set. For convenience, suppose it's the last. Then the observed response vector and the X matrix are

$$\boldsymbol{y} = \begin{pmatrix} \boldsymbol{y}_{(-n)} \\ y_n \end{pmatrix} \text{ and } X = \begin{pmatrix} X_{(-n)} \\ \boldsymbol{x}'_n \end{pmatrix},$$

where the subscript $(-n)$ indicates that the n-th row has been omitted. The original model, one that doesn't permit the possibility that the last observation came from a different regime, is

$$\mathrm{E}(\boldsymbol{Y}) = X\boldsymbol{\beta} = \begin{pmatrix} X_{(-n)}\boldsymbol{\beta} \\ \boldsymbol{x}'_n\boldsymbol{\beta} \end{pmatrix}.$$

To include the possibility that the model for the n-th observation is different, let μ_* denote the population mean of the population from which the n-th subject was drawn; μ_* can be any real number, not tied to $\boldsymbol{x}_n$ or $\boldsymbol{\beta}$. Then the model becomes

$$\mathrm{E}(\boldsymbol{Y}) = \begin{pmatrix} X_{(-n)}\boldsymbol{\beta} \\ \mu_* \end{pmatrix} = \begin{pmatrix} X_{(-n)} & \mathbf{0}_{n-1} \\ \mathbf{0}' & 1 \end{pmatrix} \begin{pmatrix} \boldsymbol{\beta} \\ \mu_* \end{pmatrix}.$$

Verify that

$$\mathrm{sp}(X) \subset \mathrm{sp}\begin{pmatrix} X_{(-n)} & \mathbf{0} \\ \mathbf{0}' & 1 \end{pmatrix}.$$

The question, whether y_n comes from the same model as the rest of the observations, can be formulated as the hypothesis $\mathrm{H}_0 : \mu_* = \boldsymbol{x}_n'\boldsymbol{\beta}$. It can be shown (in the exercises) that sp(X) is a proper subset of $\mathrm{sp}\begin{pmatrix} X_{(-n)} & \mathbf{0} \\ \mathbf{0}' & 1 \end{pmatrix}$ iff $\mu_* - \boldsymbol{x}_n'\boldsymbol{\beta}$ is an estimable function in this augmented model. Then, if $\mu_* - \boldsymbol{x}_n'\boldsymbol{\beta}$ is estimable, the difference in SSE for the two models gives the numerator sum of squares for the F-statistic to test H_0. It is in the sense of testing whether $\mu_* - \boldsymbol{x}_n'\boldsymbol{\beta} = 0$ that we can test whether y_n is an outlier with respect to the model $\mathrm{E}(\boldsymbol{Y}) = X\boldsymbol{\beta}$.

This approach can be simplified a bit. The full model shown above, $\mathrm{sp}\begin{pmatrix} X_{(-n)} & \mathbf{0} \\ \mathbf{0}' & 1 \end{pmatrix}$, is the same as $\mathrm{sp}(X, \boldsymbol{e}_n)$, where $\boldsymbol{e}_n$ is the n-th column of I_n. It can be seen that the coefficient δ of $\boldsymbol{e}_n$ in this model is the same as $\mu_* - \boldsymbol{x}_n'\boldsymbol{\beta}$ in the model above, and so the test statistic can be found directly in output for the regression of $\boldsymbol{y}$ on $(X, \boldsymbol{e}_n)$. Apparently, we can accomplish the same for any observation, say the i-th, with the model $(X, \boldsymbol{e}_i)$. In this context, the vectors $\boldsymbol{e}_i$ are sometimes called ***unique event dummies***.

12.3 Prediction

The approach we have taken regarding outliers leads nicely to questions of prediction. Given a data set $\boldsymbol{y}, X$, suppose we would like to predict the value y_* of the response that would be observed if an additional, independent observation were taken at the specific conditions $\boldsymbol{x}_*$. That the observed value will be a particular, specified value has probability zero, so it makes more sense to seek a range of values for y_* that are plausible. One way to accomplish that is to find the range of values of y_* that, if observed, would not be outliers. This can be formalized as a ***prediction interval*** on y_* at $\boldsymbol{x}_*$. Now $X_{(-n)}$ becomes X, $\boldsymbol{y}_{(-n)}$ becomes $\boldsymbol{y}$, y_n becomes y_*, and $\boldsymbol{x}_n$ becomes $\boldsymbol{x}_*$. The setup can be formulated as

$$\mathrm{E}\begin{pmatrix} \boldsymbol{Y} \\ Y_* \end{pmatrix} = \begin{pmatrix} X & 0 \\ \boldsymbol{x}_*' & 1 \end{pmatrix}\begin{pmatrix} \boldsymbol{\beta} \\ \delta \end{pmatrix},$$

with the response vector $\begin{pmatrix} \boldsymbol{y} \\ y_* \end{pmatrix}$. Conceptually, the prediction interval is the range of values of y_* for which the test statistic for $\mathrm{H}_0 : \delta = 0$ is less than the α-level critical value. This can be worked out to be the range between

$$\boldsymbol{x}_*'\hat{\boldsymbol{\beta}} \pm t_{1-\alpha/2,\nu_E}\sqrt{\hat{\sigma}^2 + \hat{\mathrm{V}}\mathrm{ar}(\boldsymbol{x}_*'\hat{\boldsymbol{\beta}})}.$$

Note that $\hat{\delta} = y_* - \boldsymbol{x}_*'\hat{\boldsymbol{\beta}}$. [Some details are required here. In particular, in order that δ be estimable, it is necessary that $\boldsymbol{x}_* \in \text{sp}(X')$.]

12.4 Inverse Prediction

Another set of questions that can be formulated in this context is called *inverse prediction* or ***calibration***. To keep notation simple, consider a setting with only one predictor variable, so that $X = (\mathbf{1}_n, \boldsymbol{x})$, where $\boldsymbol{x}$ is an n-vector and, to avoid pathologies, not proportional to $\mathbf{1}_n$. Let $\boldsymbol{y}$ denote the vector of observed values of the response variable. Now we are given a specific value y_* of the response, not corresponding to any observation in the data set, and we would like to know a plausible range of values of the predictor variable that might have led to y_*.

One setting where this question arises is in estimating the post-mortem interval from the size of insect larvae found on the body. Suppose we have ***training data*** from controlled experiments in which sizes y are measured on larvae of known age x, providing the data $\boldsymbol{y}$ and $\boldsymbol{x}$. Suppose further that the relation between size (dry weight, for example) and age is linear, so that the expected value of size at age x is $\text{E}(Y) = \beta_0 + x\beta_1$. The mystery specimen collected at the crime scene has size y_*, and we would like to estimate somehow what its age is. We can re-frame the question as, "If the age of this larva were x_*, would y_* be an outlier?" We have addressed this question by testing $\text{H}_0 : \mu_* - (\beta_0 + x_*\beta_1) = 0$. We can test this for any x_*, so we can (not going into the computational details here) find the set of all values x_* for which H_0 would not be rejected against a two-sided alternative hypothesis at the α level of significance. LaMotte and Wells (1995) show an example with this approach. The result is a $100(1 - \alpha)\%$ confidence set on x_*. See Lehmann (1959, p. 79) for a proof of this assertion. Depending on the model and the data, this set may or may not be an interval of values.

The settings in which we have developed these three topics – testing an observed response as an outlier, prediction intervals, and inverse prediction – are similar, but they differ in one important respect. In testing the observed value y_i as an outlier, the rest of the data set, $\boldsymbol{y}_{(i)}$ and $X_{(i)}$, becomes the training data. In prediction intervals and inverse prediction, we imagine taking an additional observation, and so the data set we have in hand is the training data.

12.5 Collapsibility

The notion of *collapsibility* in multiple regression models was examined by Clogg et al. (1992) and extended by LaMotte (1999). Within a model of the form $\mathcal{M} = \{X\boldsymbol{\beta} + Z\boldsymbol{\gamma} : \boldsymbol{\beta} \in \Re^q,\ \boldsymbol{\gamma} \in \Re^r\}$, Clogg et al. (1992) addressed the question, whether $\boldsymbol{\beta}_1$ and $\boldsymbol{\beta}_2$ are the same in the models "$X\boldsymbol{\beta}_1$ (reduced model)" and "$X\boldsymbol{\beta}_2 + Z\boldsymbol{\gamma}$ (full model)." They say that "Z is collapsible with respect to the $\boldsymbol{Y}$–X relationship whenever $\boldsymbol{\delta}[= \boldsymbol{\beta}_1 - \boldsymbol{\beta}_2] = \mathbf{0}$."

This objective is not altogether clear. Often the mean vector does not uniquely specify the coefficient vector, and in such cases it is not clear what $\boldsymbol{\beta}_1$ and $\boldsymbol{\beta}_2$ are. Furthermore, it can happen that $\mathrm{sp}(X) \cap \mathrm{sp}(Z)$ contains more than $\mathbf{0}$, so that the meanings of the coefficient vectors are even more ambiguous. To make any sense at all, X and Z must be such that, for any $\boldsymbol{z} \in \mathrm{sp}(X) + \mathrm{sp}(Z)$, there is exactly one vector $\boldsymbol{z}_1$ in $\mathrm{sp}(X)$ and exactly one vector $\boldsymbol{z}_2$ in $\mathrm{sp}(Z)$ such that $\boldsymbol{z} = \boldsymbol{z}_1 + \boldsymbol{z}_2$. In this case, it can be shown that $\mathrm{sp}(X) + \mathrm{sp}(Z)$ is the direct sum of $\mathrm{sp}(X)$ and $\mathrm{sp}(Z)$. Then $\boldsymbol{z}_1$ is said to be the projection of $\boldsymbol{z}$ onto $\mathrm{sp}(X)$ *along* $\mathrm{sp}(Z)$. This is denoted $\mathbf{P}_{X||Z}\boldsymbol{z}$. Similarly, $\boldsymbol{z}_2$ is the projection of $\boldsymbol{z}$ onto $\mathrm{sp}(Z)$ along $\mathrm{sp}(X)$.

In these terms, the collapsibility question may be presented as asking whether $\boldsymbol{\mu}$ is in the set of vectors in $\mathcal{M}$ such that the function $\mathbf{P}_{X||Z}\boldsymbol{\mu}$ is the same as $\mathbf{P}_X\boldsymbol{\mu}$. This can be formulated as the null hypothesis $\mathrm{H}_0 : (\mathbf{P}_{X||Z} - \mathbf{P}_X)\boldsymbol{\mu} = \mathbf{0}$. The restricted model is then $\mathcal{M}_0 = \{\boldsymbol{\mu} \in \mathcal{M} : (\mathbf{P}_{X||Z} - \mathbf{P}_X)\boldsymbol{\mu} = \mathbf{0}\}$.

This sort of question can be generalized as asking whether a given linear function $L'\boldsymbol{\mu}$ is the same in two different subsets (linear subspaces or affine sets) of $\mathcal{M}$. In the current setting, let $L = (\mathbf{P}_{X||Z})'$. The question is whether $L'\boldsymbol{\mu}$ is the same on $\mathrm{sp}(X, Z)$ and $\mathrm{sp}(X)$, that is, whether $\mathbf{P}_{X||Z}(\mathbf{P}_{(X,Z)} - \mathbf{P}_X)\boldsymbol{\mu} = \mathbf{0}$.

Although any linear hypothesis can be regarded as a collapsibility hypothesis, these sorts of questions are clearly relevant when assessing the influence of one or a few individual cases on a model. The two models in question are the model $\mathrm{sp}(X, \boldsymbol{e}_i)$, extended to accommodate the i-th observation as a special exemption, and $\mathrm{sp}(X)$. Assuming that $\boldsymbol{g}'\boldsymbol{\beta} = \boldsymbol{\ell}'X\boldsymbol{\beta}$ is estimable, and that $\boldsymbol{e}_i \notin \mathrm{sp}(X)$, whether $\boldsymbol{g}'\boldsymbol{\beta}$ is the same with and without the i-th observation becomes $\mathrm{H}_0 : \boldsymbol{\ell}'(\mathbf{P}_{(X,\boldsymbol{e}_i)} - \mathbf{P}_X)\boldsymbol{\mu} = 0$. LaMotte (1999) showed, and it is shown in exercises, that for any estimable function $\boldsymbol{g}'\boldsymbol{\beta}$, the test statistic is the same as the i-th externally-Studentized residual.

12.6 Exercises

1. Let A be an $m \times p$ matrix and let $\boldsymbol{a}$ be a p-vector.

   (a) Show: $\mathrm{sp}\begin{pmatrix} A & \mathbf{0}_m \\ \mathbf{0}_p' & 1 \end{pmatrix} = \mathrm{sp}\begin{pmatrix} A & \mathbf{0} \\ \boldsymbol{a}' & 1 \end{pmatrix}$. In particular, show that, for any p-vector $\boldsymbol{b}$ and scalar d,

$$\begin{pmatrix} A & \mathbf{0}_m \\ \mathbf{0}_p' & 1 \end{pmatrix}\begin{pmatrix} \boldsymbol{b} \\ d \end{pmatrix} = \begin{pmatrix} A & \mathbf{0} \\ \boldsymbol{a}' & 1 \end{pmatrix}\begin{pmatrix} \boldsymbol{b} \\ d - \boldsymbol{a}'\boldsymbol{b} \end{pmatrix}.$$

(b) Show: $\begin{pmatrix} \mathbf{0}_m \\ 1 \end{pmatrix} \notin \text{sp}\begin{pmatrix} A \\ \boldsymbol{a}' \end{pmatrix}$ iff $\boldsymbol{a} \in \text{sp}(A')$.

(c) Show that

$$\begin{pmatrix} -\boldsymbol{a} \\ 1 \end{pmatrix} \in \text{sp}\begin{pmatrix} A' & \mathbf{0}_p \\ \mathbf{0}'_m & 1 \end{pmatrix}$$

iff $\boldsymbol{a} \in \text{sp}(A')$.

(d) Show that

$$\begin{pmatrix} \mathbf{0}_p \\ 1 \end{pmatrix} \in \text{sp}\begin{pmatrix} A' & \boldsymbol{a} \\ \mathbf{0}'_m & 1 \end{pmatrix}$$

iff $\boldsymbol{a} \in \text{sp}(A')$.

(e) Show that the orthogonal projection of the $(m+1)$-vector $(\boldsymbol{y}', y_{m+1})'$ onto sp $\begin{pmatrix} A \\ \mathbf{0}' \end{pmatrix}$ is

$$\mathbf{P}_{\begin{pmatrix} A \\ \mathbf{0}' \end{pmatrix}} \begin{pmatrix} \boldsymbol{y} \\ y_{m+1} \end{pmatrix} = \begin{pmatrix} \mathbf{P}_A & \mathbf{0} \\ \mathbf{0}' & 0 \end{pmatrix} \begin{pmatrix} \boldsymbol{y} \\ y_{m+1} \end{pmatrix} = \begin{pmatrix} \mathbf{P}_A \boldsymbol{y} \\ 0 \end{pmatrix}.$$

2. In these exercises you will develop relations between least-squares statistics for the full data set and after omitting one observation. The full data set is $\boldsymbol{y}$, X. Assume that $\boldsymbol{e}_i \notin \text{sp}(X)$. Refer to $(X, \boldsymbol{e}_i)$ as the *augmented* model, and denote its coefficient vector by $(\boldsymbol{\beta}'_{(i)}, \delta_{(i)})'$. Let $\hat{\beta}$ denote a $(k+1)$-vector such that $\hat{\boldsymbol{y}} = \mathbf{P}_X \boldsymbol{y} = X\hat{\boldsymbol{\beta}}$, and let $\hat{\delta}_{(i)}$ and $\hat{\boldsymbol{\beta}}_{(i)}$ be a scalar and $(k+1)$-vector such that $\mathbf{P}_{(X,\boldsymbol{e}_i)}\boldsymbol{y} = X\hat{\boldsymbol{\beta}}_{(i)} + \boldsymbol{e}_i\hat{\delta}_{(i)}$. $\hat{\delta}_{(i)}$ will be unique because $\delta_{(i)}$ is estimable in the augmented model. Depending on X, there may be multiple vectors $\boldsymbol{b}$ such that $\mathbf{P}_{(X,\boldsymbol{e}_i)}\boldsymbol{y} = X\boldsymbol{b} + \boldsymbol{e}_i\hat{\delta}_{(i)}$; $\hat{\boldsymbol{\beta}}_{(i)}$ is a specific (but arbitrary) one, and it remains the same vector throughout this discussion.

(a) Show that, if $\boldsymbol{e}_i \notin \text{sp}(X)$, then $\text{sp}(X, \boldsymbol{e}_i) = \text{sp}(X) \oplus \text{sp}(\boldsymbol{e}_i)$. As a direct sum, this means that $X\boldsymbol{b}_1 + \boldsymbol{e}_i d_1 = X\boldsymbol{b}_2 + \boldsymbol{e}_i d_2$ implies that $X\boldsymbol{b}_1 = X\boldsymbol{b}_2$ and $d_1 = d_2$.

(b) Show: $\delta_{(i)} = (\mathbf{0}', 1)(\boldsymbol{\beta}'_{(i)}, \delta_{(i)})'$ is estimable in the model $X\boldsymbol{\beta}_{(i)} + \boldsymbol{e}_i\delta_{(i)}$ iff $\boldsymbol{e}_i \notin \text{sp}(X)$. Further, show that if $\delta_{(i)}$ is estimable in this model, then so is $\boldsymbol{x}'_i\boldsymbol{\beta}_{(i)}$.

(c) Show that $\mathbf{P}_{\boldsymbol{e}_i} = \boldsymbol{e}_i\boldsymbol{e}'_i$, and that hence $\boldsymbol{y}|\boldsymbol{e}_i = (\mathrm{I} - \mathbf{P}_{\boldsymbol{e}_i})\boldsymbol{y}$ is $\boldsymbol{y}$ with its i-th entry replaced by 0, and that $X|\boldsymbol{e}_i$ is X with its i-th row replaced by a row of zeros.

(d) For simplicity, consider here that $i = n$, that i refers to the last observation. Let $X_{(-n)}$ and $\boldsymbol{y}_{(-n)}$ denote the first $n-1$ rows of X and $\boldsymbol{y}$, respectively, that is, $\boldsymbol{y}_{(-n)}$ and $X_{(-n)}$ denote the data set after omitting the last observation. Show that

$$X|\boldsymbol{e}_n = (\mathrm{I} - \mathbf{P}_{\boldsymbol{e}_n})X = \begin{pmatrix} X_{(-n)} \\ \mathbf{0}' \end{pmatrix},$$

and hence (by 1e) that

$$\mathbf{P}_{(X|\boldsymbol{e}_n)}\boldsymbol{y} = \begin{pmatrix} \mathbf{P}_{X_{(-n)}}\boldsymbol{y}_{(-n)} \\ 0 \end{pmatrix}.$$

(e) Continuing with $i = n$, show that

$$\mathbf{P}_{X_{(-n)}}\boldsymbol{y}_{(-n)} = X_{(-n)}\hat{\boldsymbol{\beta}}_{(n)}$$

and $\hat{\delta}_{(n)} = y_n - \boldsymbol{x}_n'\hat{\boldsymbol{\beta}}_{(n)}$. Thus fitting the augmented model sp$(X, \boldsymbol{e}_n)$ to $\boldsymbol{y}$ gives the same vector of predicted values for the first n observations as fitting sp$(X_{(-n)})$ to $\boldsymbol{y}_{(-n)}$; and for the n-th observation, it gives y_n as the predicted value.
[Hint:

$$\begin{aligned} \mathbf{P}_{(X,\boldsymbol{e}_n)}\boldsymbol{y} &= X\hat{\boldsymbol{\beta}}_{(n)} + \boldsymbol{e}_n\hat{\delta}_{(n)} \\ &= \mathbf{P}_{\boldsymbol{e}_n}\boldsymbol{y} + \mathbf{P}_{(X|\boldsymbol{e}_n)}\boldsymbol{y} \\ &= y_n\boldsymbol{e}_n + (\mathrm{I} - \boldsymbol{e}_n\boldsymbol{e}_n')X\boldsymbol{b} \end{aligned}$$

for some $(k+1)$-vector $\boldsymbol{b}$. This implies that $X\hat{\boldsymbol{\beta}}_{(n)} = X\boldsymbol{b}$ and $\hat{\delta}_{(n)} = y_n - \boldsymbol{x}_n'\hat{\boldsymbol{\beta}}_{(n)}$: Why?]

It has become customary to denote $\boldsymbol{x}_n'\hat{\boldsymbol{\beta}}_{(n)}$ as $\hat{y}_{n(n)}$; it is the estimate of the mean of Y at $\boldsymbol{x}_n$ based only on the first $n-1$ observations. Analogous notation $\hat{y}_{i(i)} = \boldsymbol{x}_i'\hat{\boldsymbol{\beta}}_{(i)}$ is used in reference to deleting the i-th observation.

That $X_{(-n)}\hat{\boldsymbol{\beta}}_{(n)} = \mathbf{P}_{X_{(-n)}}\boldsymbol{y}_{(-n)}$ means that, for the first $n-1$ observations, the results of fitting the model $(X, \boldsymbol{e}_n)$ to $\boldsymbol{y}$ are the same as if we deleted the n-th observation and fit the remaining observations, that is, fit $X_{(-n)}$ to $\boldsymbol{y}_{(-n)}$.

(f) Expressing $\mathbf{P}_{(X,\boldsymbol{e}_i)}$ as $\mathbf{P}_X + \mathbf{P}_{(\boldsymbol{e}_i|X)}$, show that

$$\begin{aligned} \mathbf{P}_{(X,\boldsymbol{e}_i)}\boldsymbol{y} &= X\hat{\boldsymbol{\beta}}_{(i)} + \boldsymbol{e}_i\hat{\delta}_{(i)} \\ &= \mathbf{P}_X\boldsymbol{y} + \mathbf{P}_{(\boldsymbol{e}_i|X)}\boldsymbol{y} \\ &= X\hat{\boldsymbol{\beta}} - \frac{r_i}{1-h_{ii}}\mathbf{P}_X\boldsymbol{e}_i + \frac{r_i}{1-h_{ii}}\boldsymbol{e}_i, \end{aligned}$$

where $r_i = y_i - \hat{y}_i$ and $h_{ii} = \boldsymbol{e}_i'\mathbf{P}_X\boldsymbol{e}_i$ is the i-th diagonal entry of $\mathbf{P}_X$. Therefore,

$$X\hat{\boldsymbol{\beta}}_{(i)} = X\hat{\boldsymbol{\beta}} - \frac{r_i}{1-h_{ii}}\mathbf{P}_X\boldsymbol{e}_i \text{ and } \hat{\delta}_{(i)} = y_i - \hat{y}_{i(i)} = \frac{r_i}{1-h_{ii}}.$$

(g) Find an expression for $\text{Var}(\hat{\delta}_{(i)})$.

(h) Let SSE denote residual sum of squares for the original model $\text{sp}(X)$, and let $SSE_{(i)}$ denote residual sum of squares for the augmented model $\text{sp}(X, \boldsymbol{e}_i)$. Show that

$$SSE_{(i)} = \boldsymbol{y}'(\text{I} - \mathbf{P}_{(X,\boldsymbol{e}_i)})\boldsymbol{y} = SSE - \frac{r_i^2}{1 - h_{ii}}.$$

(i) Show that $SSE_{(i)}$ is also residual sum of squares for the regression of $\boldsymbol{y}_{(-i)}$ on $X_{(-i)}$.

13

On Methods of Model-Building

Throughout the preceding chapters, the model matrix X has been assumed to be fixed and known. That is rarely true in practice.

Take the Kentucky Utilities data set, for example, from which Table 2.1 was extracted. The company gathered data over the summer months in hopes of improving their ability to anticipate peak load on the system up to about a day in advance. In the full data set, date, day of week, time of day, seven weather-descriptive measurements, and instantaneous load on the system were recorded every 3 hours throughout June, July, and August. It was reasonable to think that load would be greater on warmer days, lower at night than during the workday, higher on weekdays than on weekends, that is, that all the factors they recorded might affect load. Other than that, there was no theoretical or physical model for load in terms of these factors. This is typical of many applications of multiple regression analysis.

There are many textbooks, other publications, and web sites, both in statistics and in other disciplines that use statistical methods, that devote great numbers of pages to methods and techniques of model-building. The purpose of this chapter is to describe some of the considerations that are addressed in such processes. Broadly, they entail formulating predictor variables from the basic condition variables and then identifying those that are important and winnowing out those that don't seem to help. Further diagnostics aim to identify observations that don't fit the pattern of the others in order to reduce their aberrant influence on the resulting model and inferences based on it. Some methods drop the restriction to unbiased linear estimators, raising questions of admissibility among that broader field.

At a very basic level, the modeler must formulate functional forms of the relations between the response and the predictor variables, jointly and separately. In the Kentucky Utilities example, nothing is known about the relation between load and temperature other than, *ceteris paribus*, load will generally be greater at greater ambient temperatures. One approach then would be to start with a first-order approximation, a linear term, and to do the same for the other quantitative weather variables. For a categorical factor, like day of week, different levels of load for each can be accommodated by indicator variables. Time-of-day is quantitative and cyclic, so its effects might be represented by cyclic terms, like sines and cosines; or effects of its eight different levels might be included simply with indicator variables. Additional indicator variables of time-of-day by day-of-week combinations might be included in the model to account for the possibility that time-of-day effects on load differ with day-of-week. Other functional forms for the effects of the weather variables might work better than linear terms. For example, the effect of temperature on load might be flatter for temperatures below a certain point and above another point and increasing more or less linearly in between. The relation between load and relative humidity, as another example, might be increasing but curvilinear, so that both a linear and a quadratic term in relative humidity might be appropriate.

These kinds of considerations arise in practically every application that entails positing a model. It would be easy, in this fairly simple setting, to list several dozen predictor variables that might credibly be included in the model. Clearly then, at least two things can happen: predictors are included that don't really have any effects on the response; and predictors are not included that do.

The most common measure of goodness-of-fit for multiple regression models is R^2, which is the ratio of Regression SOS to Total SOS. We feel good when R^2 is over 90% and bad when it's below 60%. The fact is, though, that including another predictor variable in the model cannot decrease R^2, and usually it increases it, at least a little bit. For this reason, focusing on R^2 to see how we did, it's always tempting to add more predictors if we have some lying around. In the other direction, removing a predictor from the model reduces R^2, if anything, so there's a disincentive to do that. In terms of R^2, the more predictors in the model, the better. The same tendency holds for SSE: additional predictors cannot increase SSE, and usually they reduce it.

The costs of including predictors that don't have effects is not so instantly apparent in terms of statistical properties. Some accounts mention Occam's razor at this point, or appeal to our puritanical appreciation of parsimony.

But how can it be so wrong when it increases R^2? The statistical cost is described in the two paragraphs after Proposition 10.2: inclusion of extra predictors in the model reduces the non-centrality parameter of the test statistic for effects that are estimable in both models. In addition, including them decreases error df, depressing the power of the test further. On the other hand, they might be necessary to ensure that the distribution of SSE is central. There are costs and benefits, and there is no way to tell with certainty which is greater. In practice, such decisions often are based on the respective p-values of the additional predictors.

In most model-building applications, one or a few aberrant observations can have greatly disproportionate effects on the fitted model. Tests of individual cases as outliers were described in Chapter 12. They can be formulated as testing the regression coefficients of dummy variables that indicate the suspect cases. In that sense, omitting an outlier is equivalent to including a predictor variable. Then identifying and excluding outliers from the model is equivalent to including or not including predictor variables. Examining predictor variables for inclusion and suspect cases for exclusion are one and the same. Peixoto and LaMotte (1989) describe and demonstrate this unified approach to simultaneous predictor variable selection and outlier detection in an interesting example.

Including indicator variables for potential outliers increases the number of predictor variables. A broad approach would be to include one dummy variable for each observation. Then the pool of potential predictors would include those along with all the rest, which would far exceed the number of observations. Even then, though, it is at least conceptually possible to examine the efficacy of every possible combination of predictors from this big pool in order to identify a few good ones.

The paper just cited tried this with 58 observations in the data set, 9 predictor variables, and 16 dummy variables for 16 observations nominated as potential outliers. The predictor pool then comprised 25 predictors. The number of possible subsets of these is $2^{25} = 33,554,432$. Evaluating each would entail fitting that many models.

Instead, the authors employed the variable-selection algorithm created by Furnival and Wilson (1974) as it is implemented in SAS. The same algorithm seems to have been implemented by most of the widely-available statistical packages. It identifies subset regressions that have least SSE among all subsets of the same size. Taking advantage of the fact that SSE is a monotone set function, the algorithm orders the subsets it considers in such a way that it can, by comparisons with results from those already examined, possibly delete many potential subsets from consideration, knowing that there is none among them that could possibly be better than the best found so far. Being able to reduce the field in that way is not necessarily always possible, and if not then the algorithm would end up examining all subsets. In practice, though, it seems to accomplish its goal while examining only a very small proportion of the 2^k possible subset regression models.

Other variable-selection algorithms have been around since the 1940s. They are step-wise, at each step choosing only one predictor at a time to be added to or deleted from the current model. As such, they cannot guarantee any optimal property of the subsets they identify. Their sub-optimality has been demonstrated often. See Peixoto and LaMotte (1989) for one example.

The reach and utility of the very efficient Furnival-Wilson variable-selection algorithm would be extended greatly if it were programmed to handle groups of predictors as it now handles predictors individually. Instead of finding best subsets of $x_1, \ldots, x_k$, it would find best subsets of groups $G_1, \ldots, G_g$, where each group could be a subset of one or more of the pool of predictor variables. A group might include, for example, the indicators of levels of a categorical factor, so that they could be included or excluded together. In concept, different groups could share predictors in common. For example, for day of week, one group could include indicators for each of the seven. Another might combine these into three categories, for weekdays, Saturdays, and Sundays. Even when subsets overlap like that, the basic operating principle of the selection algorithm, that SSE is a monotone set function, holds also with subsets of groups of predictors, and the algorithm would still work.

Often when such a selection procedure has been used and a final model settled on, inference is undertaken as if the final model was fixed from the beginning. As the objective of the procedure is to improve the model, this is clearly data snooping or p-value shopping. Effects that would not have been found significant in the original model are more likely to be found significant in the carefully-selected model. This subject of post-selection inference has generated considerable discussion. The paper by Kuchibhotla et al. (2022) gives an excellent overview, and its bibliography is extensive.

Corrective measures to mitigate effects of outliers in fitted multiple regression models have been described and analyzed extensively in the statistics literature. See Belsley et al. (1980), Cook and Weisberg (1982), and Chatterjee and Hadi (1986). Several diagnostic statistics for "nominating" observations as outliers are widely available in statistical computing packages. They include Cook's D (Cook, 1977); DFBETAS, DFFITS, and COVRATIO (Belsley et al., 1980); and the Studentized residual and the externally Studentized residual. Bounds on each of these have been proposed to trigger nomination: for example, $4/n$ for the square of DFBETAS (Belsley et al., 1980) and $F_{.5;p,n-p}$ for Cook's D (Cook, 1977). Both Chatterjee and Hadi (1986) and LaMotte (1999) showed that all these statistics are one-to-one with the square of the externally Studentized residual, so knowing one is as good as knowing all. However, probabilities of exceeding their suggested bounds are not consistent. In an example described by LaMotte (1999), median probabilities over 91 observations ranged from 0.000 to 0.081 over a dozen of these diagnostic statistics. In that same example, 26 of the 91 observations were nominated by at least one statistic, with the covariance ratio leading all with 15.

The paper by Stein (1956) established that in certain conditions the sample mean is not an admissible estimator of the population mean. That is, he demonstrated another estimator (not linear) that has mean squared error less than or equal to that of the sample mean, no matter what the values of the population parameters are. As the sample mean in that setting is a least-squares estimator, this brought all least-squares estimators into question, suggesting that, in terms of squared-error loss, there might exist other estimators that are better.

We know that, among unbiased linear estimators, least-squares estimators are best. Thus the only place to look for something better among linear estimators is among biased estimators. Dropping the restriction to unbiased linear estimators might reveal some that are biased but have lesser mean squared error.

The first outpouring of such attempts began with the papers by Hoerl and Kennard (1970a,b) describing ridge regression. See also Alldredge and Gilb (1976).

The abstract at the beginning of Hoerl and Kennard (1970a) asserted: "It is shown how to augment $X'X$ to obtain biased estimates with smaller mean square error." What is then proven, as Theorem 4.3, is that for each given value γ_* of $\boldsymbol{\beta}/\sigma$ there exists a value of the ridge parameter k such that the ridge estimator is better than the least-squares estimator at γ_*. That is, it is possible that lesser MSE might be attained locally at γ_* with the right k, which is about as reassuring as getting the right time from a clock that is known only to be exactly right twice a day. Identifying such a k depends on the unknown parameters, and so improvement cannot be guaranteed.

The allure of such a free lunch was strong. At the time of this writing, a search on JSTOR for "ridge regression" found 13,764 results. It became recognized that the attractive properties of ridge estimators were shared by other classes of biased linear estimators, including variable selection, restricted least squares, and fractional rank estimators (Marquardt, 1970).

More recently the LASSO procedure for simultaneously fitting a model and identifying subsets of predictors has received much attention in the literature. Some of its appeal is that it automatically identifies predictors to be deleted, without obvious user intervention. Since its beginnings with Tibshirani (1996), its mathematical and statistical properties have been explored extensively. See Freijeiro-González et al. (2022) for a thorough review.

Part III

ANOVA Models: Linear Models for Effects of Categorical Factors

14

ANOVA: Introduction and Summary

This part of the book covers Analysis of Variance (ANOVA). This chapter is an overview of the subject, including some of its historical landmarks. It defines notation for a general framework for modeling the mean vector when subjects are observed under combinations of levels of multiple factors, through models that include arbitrary lists of ANOVA factor effects.

14.1 Introduction to ANOVA and Its Development

Analysis of variance (ANOVA) models are multiple regression models formulated for effects of factors that are categorical. They fit into the same general form that we've been dealing with up to this point. There is no need, therefore, to do more than to show how the X matrix can be defined in terms of factor-level combinations and to define the special sets of linear functions of the means, like main effects and interaction effects, that are associated with ANOVA. Inference about any effect or set of effects can then be based on the RMFM SS, with the restricted model defined by the absence of the effects in question. By virtue of Proposition 10.3, the resulting F-statistic tests exclusively the estimable part of those effects.

However, ANOVA notions have been immensely influential. Their development and the questions and controversies they have raised have generated hundreds of pages in books and articles in the statistical literature. Their terminology permeates applied statistics, constituting much of the language we use to communicate and teach with. This is not the place for an extensive review of the history and importance of ANOVA. But I shall present what I consider to be principal developments, historical context where I think it is helpful, and some topics that remain controversial and, in my opinion, incompletely understood. In the process, I'll try to unify terminology and notation, and I'll provide some results and relations that seem to me to be new (although I've learned from experience that practically nothing really is).

The simplest ANOVA model is for effects of a single categorical factor. It is presented in Exercise 19, p. 99. That provides a partial, basic preview, but single-factor ANOVA lacks the features that complicate models for joint effects of multiple factors.

In their original form, ANOVA computations for balanced settings for multiple factors are simple, all based on sums of squared deviations from the mean, of the form $SOS_z = \sum_i (z_i - \bar{z})^2$. The relations that they test among the factor-level-combination population means are readily apparent. They are presented in most introductory and intermediate textbooks on statistical methods. But those same textbooks either do not mention unbalanced settings or they say that they are beyond the level that the book intends to address, hinting that ANOVA methods exist there but that they're complicated. More-mathematical textbooks on linear models cover balanced models and unbalanced models in separate sections. Most do not go much beyond two-factor models. They deal with empty cells either in painstaking detail or not at all.

What is it about unbalanced multi-factor models that makes them difficult? After all, almost since their inception, it has been known that they could be formulated as multiple regression models. In principle, then, once a set of effects is defined, the model can be fit with and without it, and a suitable test statistic developed for it based on the difference in SSE.

Several considerations hindered this conversion, as I see it:

1. There was no universal agreement on how factor effects should be defined in unbalanced models.

2. There was a preference for explicit, closed-form formulas for SSs, and preferably SSs like those in balanced models.

3. There was a preference for computations that could be accomplished conveniently with the computational devices that were readily available in the 1920s through 1940s.

4. For multiple regression models, there was a preference to get extra SSE by deleting subsets of predictor variables to form restricted models, instead of the general construction shown in Chapter 10.

5. There was the recognition that those extra SSEs didn't always test the same hypotheses in balanced and unbalanced models.

6. Models in terms of dummy variables have containment relations so that, for example, omitting terms for main effects in a model that includes interaction effects yields no change in SSE.

Fisher presented ANOVA in terms of "variances," which are sums of squared deviations from the average, SOSs, in balanced models. ANOVA methods soon became widely used in applications. Habits and customs formed around them. They were taught in statistical methods courses for non-statisticians, presented as recipes and calculating formulas. Of necessity to make their use broadly accessible, they were presented with only heuristic justification: these sums of squares tend to be greater the more and greater are the differences among the underlying population means. Very little in the way of a general theoretical framework was given, which might have indicated how snags that appeared in unbalanced models could be resolved.

Reflecting these considerations, F. Yates's 1934 paper on the Method of Weighted Squares of Means (MWSM) presented a formula for a SS to test factor A main effects in two-factor, all-cells-filled models that permit AB interaction effects. However, the formula was a little bit intimidating and opaque, and it was not clear how it might be extended to models with more factors or empty cells. Other simpler, but approximate, methods continued to be proposed, even up into recent years. But the MWSM came to be regarded as the gold standard for unbalanced models. SAS, in Goodnight (1976), validated its Type III SSs by saying that "[w]hen no missing cells exist ... Type III SS will coincide with Yate's [*sic*] weighted squares of means technique."

As early as the 1930s, methods were devised to perform ANOVA in unbalanced models as if they were balanced, and such efforts continued to appear up into the 1950s. Snedecor and Cox (1935) proposed that the cell sample sizes n_{ij} be replaced by what they would be if they were "proportional subclass numbers" in computing ANOVA sums of squares. Modifications like this were advocated in Anderson and Bancroft (1952) and Snedecor and Cochran (1980).

Yates's MWSM and the simplifications of proportional subclass numbers helped keep alive the hope that multi-factor ANOVA models in unbalanced settings might be analyzed with tractable adaptations of balanced-model methods. Then they could be covered in general-audience methods courses without imperiling enrollments by requiring a good working familiarity with multiple regression analysis as a prerequisite. Chapters 19 and 20, on Yates's MWSM and proportional subclass numbers, are included here for completeness and because these two topics are historically significant.

A paper appeared in 1973 (Francis, 1973) that reflected the understanding of ANOVA at that time, after almost a half-century of development and widespread use. It presents and compares SSs produced by four "canned program[s] to perform an analysis of variance on a two-way factorial design with unequal number of observations in the cells." The four programs produced three different values for each main effect sum of squares. The author noted these differences and explained them, but did not suggest which one was "correct" in any sense. Perhaps it is surprising, but that remains about the current state of understanding today. The example shown in Tables 2.6–2.8 give an expanded illustration of the current state.

Hocking (2013) gives the most complete account of these questions in terms of cell-means models. He stops short, though, of saying that any one type of SS is the best one to use. On the same question, Smith and Cribbie (2014) "recommend use of the Type II sums of squares for analysis of main effects because when no interaction is present it tests meaningful hypotheses and is the most statistically powerful alternative."

An internet search on "unbalanced anova" turns up many accounts from diverse points of view. The main impression one gets from these is that it is very busy and very complicated. One, for example, claims to demonstrate that results can depend on the types of "contrasts" used to code factor levels in Type III analysis. Several suggest schemes to remedy unbalancedness. Most build discussion around examples using particular statistical computing packages, including Minitab, SAS, Stata, and R. Few give good, satisfying accounts of the underlying mathematics or statistical theory.

The discussion and controversy over ANOVA effects and how to test them continues to this day. Here, to address these questions, we will define factor effects and show how to construct models that include any sets of them. Then numerator SSs for any effects in the model can be had as RMFM SSs, which automatically test exclusively their estimable parts.

Most of the pieces of this general resolution of questions raised by ANOVA models have been noted before. Searle (1966) noted that imposing a single non-estimable condition did not affect SSE, a hint that imposing any condition increases SSE only due to its estimable part. General formulations of factor effects in terms of orthogonal projection matrices have appeared off and on for several decades, as has the binary scheme for listing them (as (1), A, B, AB, C, AC, and so on). Hocking (2013) used these notions extensively.

For example, it is widely known that formulating factor effects in terms of contrasts avoids the containment relations inherent in the dummy-variable formulation. The relation of such formulations to imposing "the usual non-estimable conditions" on dummy-variable models has been noted in particular settings, like two-factor models. Orthogonal projection matrices have been used to define factor effects. Something like what is called here the "ANOVA Identity" has been noted in several forms. Hocking (2013, p. 416), for example, shows it explicitly as the "sums of squares identity."

An objective of this part of the book is to provide a general formulation of models that include effects of categorical factors in terms of the classical main effects and interaction effects, for any number of factors and any list of effects. That provides the mathematical framework for establishing general properties.

Chapter 15 defines ANOVA effects. Chapter 16 develops a general formulation of models that include ANOVA effects, and it extends it to include potential effects of covariates and covariate-by-factor interaction effects. Chapter 17 gives a formulation of Type III SSs and proves their basic properties.

Several fundamental questions about the theory and practice of analysis of variance took a long time to be resolved, and some are not resolved yet. I have the impression that practitioners, textbook writers, and instructors think of these issues, if at all, like gravity: we know we don't fully understand why it works, but we've adapted to it, and we know how to live with it. For many,

"Use Type III" is good enough for government work, everyone I know agrees, no need to agonize. Some things are widely believed and assumed to have been established – by someone, somewhere. As we shall see, some of those have not in fact been established, and some are not true.

For the most part, when ANOVA is covered in introductory methods textbooks and courses, it is introduced in the same basic form that Fisher first described a century ago, with emphasis on the sums of squares (Fisher's "variances"). I think it is reasonable to question whether that kind of introduction is useful to anyone anymore. The balanced-model computations of sums of squares might be useful if you're stranded on a desert isle equipped with only a stick to write in the sand. Otherwise, you'll be using a general-purpose linear models procedure in a statistical computing package that produces test statistics for the effects that you specify in the model. For that, what's important is understanding that those effects provide a systematic hierarchical structure, from simpler to more complex, for categorizing and assessing differences among multiple population means. The definitions of the effects and their properties as a group are important, not so much the sums of squares. But, as it has transpired, it was the definitions of the balanced-model sums of squares that came to define the effects. As for how numerator SSs are computed to test these effects in general settings, balanced or unbalanced, all cells filled or not, it is enough to know that they are computed as numerator SSs are described generally in Chapter 10.

14.2 Notation for the Multi-Factor Setting

Each of n experimental subjects is observed under a combination of levels of f ***factors***. The factors are indexed by $k = 1, \ldots, f$, and factor k has a_k distinct ***levels***, or ***classes***, indexed $1, \ldots, a_k$. A ***factor-level combination*** (FLC) comprises one level from each factor. Denote it by $\boldsymbol{\ell} = (\ell_1, \ldots, \ell_f)$, where $\ell_k \in \{1, \ldots, a_k\}$ denotes the level of factor k in this FLC. There are $a_1 \times \cdots \times a_f = a_*$ distinct FLCs possible. Denote the set of all FLCs by $\mathcal{L}$. FLCs are also called ***cells*** or ***subclasses***.

In particular settings and examples, factors may be named A, B, and so on, their numbers of levels by a, b, and so on, and FLCs by $i, j, \ldots$. Denote the number of subjects observed under FLC $\boldsymbol{\ell}$ by $n_{\boldsymbol{\ell}}$. Then $n = \sum_{\boldsymbol{\ell} \in \mathcal{L}} n_{\boldsymbol{\ell}}$. The $n_{\boldsymbol{\ell}}$s are called ***cell frequencies*** or ***subclass numbers***. The $\boldsymbol{\ell}$-th cell is ***empty*** iff $n_{\boldsymbol{\ell}} = 0$.

A univariate response y is observed from each subject. The n realized responses are the entries in the n-vector $\boldsymbol{y}$. Entries in $\boldsymbol{y}$ may be indexed by a single index, like y_s, $s = 1, \ldots, n$; or they may be indexed within each FLC, as $y_{\boldsymbol{\ell},s}$, $s = 1, \ldots, n_{\boldsymbol{\ell}}$. Let $\boldsymbol{\ell}_s$ denote the FLC under which the s-th response was observed, with $s = 1, \ldots, n$.

TABLE 14.1: $\mathbb{K}$ for the unbalanced setting from Table 2.6 with the $(3, 3)$ cell empty.

			Columns of $\mathbb{K}$								
			1	1	1	2	2	2	3	3	3
s	ℓ_1	ℓ_2	1	2	3	1	2	3	1	2	3
1	1	1	1	0	0	0	0	0	0	0	0
2	1	2	0	1	0	0	0	0	0	0	0
3	1	2	0	1	0	0	0	0	0	0	0
4	1	3	0	0	1	0	0	0	0	0	0
5	1	3	0	0	1	0	0	0	0	0	0
6	1	3	0	0	1	0	0	0	0	0	0
7	2	1	0	0	0	1	0	0	0	0	0
8	2	1	0	0	0	1	0	0	0	0	0
9	2	1	0	0	0	1	0	0	0	0	0
10	2	2	0	0	0	0	1	0	0	0	0
11	2	3	0	0	0	0	0	1	0	0	0
12	2	3	0	0	0	0	0	1	0	0	0
13	3	1	0	0	0	0	0	0	1	0	0
14	3	1	0	0	0	0	0	0	1	0	0
15	3	1	0	0	0	0	0	0	1	0	0
16	3	2	0	0	0	0	0	0	0	1	0
17	3	2	0	0	0	0	0	0	0	1	0

Denote the population mean of the response under FLC $\boldsymbol{\ell}$ by $\eta_{\boldsymbol{\ell}}$. It is called the ***cell mean*** under FLC $\boldsymbol{\ell}$. The a_*-vector of cell means is $\boldsymbol{\eta} = (\eta_{\boldsymbol{\ell}})$, listed in lexicographic order on $\boldsymbol{\ell}$. The expected value of the s-th response is $\eta_{\boldsymbol{\ell}_s}$, the $\boldsymbol{\ell}_s$-th entry of $\boldsymbol{\eta}$.

The population means of $y_1, \ldots, y_n$ are entries μ_i of the n-vector $\boldsymbol{\mu}$. For each $s \in \{1, \ldots, n\}$, define the s-th row of the $n \times a_*$ matrix $\mathbb{K}$ to have 1 in the $\boldsymbol{\ell}_s$-th column and zeros in all the other columns. Then $\boldsymbol{\mu} = \mathbb{K}\boldsymbol{\eta}$. The $\boldsymbol{\ell}$-th column of $\mathbb{K}$ has $n_{\boldsymbol{\ell}}$ ones, the number of subjects observed under FLC $\boldsymbol{\ell}$. If the $\boldsymbol{\ell}$-th cell is empty, then $n_{\boldsymbol{\ell}} = 0$ and the $\boldsymbol{\ell}$-th column of $\mathbb{K}$ is $\mathbf{0}_n$. See Table 14.1 for an example.

In general, the responses and their FLCs can appear in any order in $\boldsymbol{y}$ and $\mathbb{K}$. Occasionally, it will be convenient to consider them listed in lexicographic order on $\boldsymbol{\ell}, s$. That might entail permuting the rows of $\boldsymbol{y}$ and $\mathbb{K}$, which is the same as premultiplying by a permutation matrix. In any case, though, note that the rows of $\mathbb{K}'\boldsymbol{y}$ are in lexicographic order on $\boldsymbol{\ell}$.

The following chapters deal with models for $\boldsymbol{\mu}$ through models for $\boldsymbol{\eta}$. A model $M\boldsymbol{\beta}$ for $\boldsymbol{\eta}$ becomes the model $\mathbb{K}M\boldsymbol{\beta}$ for $\boldsymbol{\mu}$.

A dot in a subscript of a symbol indicates the sum of the values of the symbol over the range of that subscript, and an overbar indicates the average over the range of that subscript. Thus, for example, for $i = 1, \ldots, a$, $j = 1, \ldots, b$, and $s = 1, \ldots, n_{ij}$, $n_{i\cdot} = n_{i1} + \cdots + n_{ib}$, and $n_{\cdot\cdot} = \sum_{i,j} n_{ij} = n$. If $n_{ij} > 0$, then $y_{ij\cdot} = y_{ij1} + \cdots + y_{ijn_{ij}}$, $\bar{y}_{ij\cdot} = y_{ij\cdot}/n_{ij}$, and $\bar{y}_{i\cdot\cdot} = (1/n_{i\cdot}) \sum_{j=1}^{b} \sum_{s=1}^{n_{ij}} y_{ijs}$.

In this context, replace a sum or average over an empty range by 0. Although this seems arbitrary, because models take the form $\mathbb{K}M\boldsymbol{\beta}$, it can be shown that this convention causes no problems. For example, the a_*-vector $\mathbb{K}'\boldsymbol{y} = (y_{\boldsymbol{\ell},\cdot})$ has 0s in rows for which $n_{\boldsymbol{\ell}} = 0$.

As a simplification, a subscript entirely of dots can be replaced by a single dot, or by none, if it is unambiguous to do so. With that understood, in context n or $n.$ can be used instead of $n_{\ldots}$, and $\bar{y}.$ or $\bar{y}$ can be used instead of $\bar{y}_{\ldots}$. Finally, unnecessary final dots can be omitted, so that $\bar{y}_{i\cdot\cdot}$ can be written simply as $\bar{y}_{i\cdot}$ and $\bar{y}_{ij\cdot}$ as $\bar{y}_{ij}$.

It will be convenient at times to use "(*indexed expression*)" to mean a vector whose entries are *expression* evaluated sequentially over the range of the index or indexes in lexicographic order. The result depends on the context. If the range of the indexes is $i = 1, \ldots, a$, $j = 1, \ldots, b$, $s = 1, \ldots, n_{ij}$, and $n = \sum_{i,j} n_{ij}$, then (y_{ijs}), $(\bar{y}_{i\cdot})$, $(\bar{y}_{ij})$, and $(\bar{y})$ are all n-vectors. The expression $(\bar{y}_{i\cdot})$, for example, repeats each $\bar{y}_{i\cdot}$ $n_{i\cdot}$ times sequentially for $i = 1, \ldots, a$.

For general notation, strings of factor names, like A or AB or ACD, will be designated by f-tuples that concatenate strings of 0s and 1s. With $f = 4$, for example, the tuple 1000 designates the string A; 1100, AB; and 1011 designates ACD. An empty string, conventionally denoted (1), is denoted 0000; denote it by 0_f instead of $0 \cdots 0$ to symbolize it when the value of f is not specified.

Symbolize such f-tuples as $\boldsymbol{j} = j_1 \ldots j_f$. Let $\mathcal{B}^f$ denote the set of all such f-tuples. There are 2^f members of $\mathcal{B}^f$. They are customarily listed as if they are counting from 0 to $2^f - 1$ in binary, but reversing the order of the bits. Listed this way for $f = 3$, they are $000, 100, 010, 110, 001, 101, 011, 111$, corresponding to (1), A, B, AB, C, AC, BC, ABC. In balanced, multi-factor settings, this traditional ordering organizes computations of sums of squares conveniently.

We will occasionally need to extend dot notation as follows. Given an f-tuple $\boldsymbol{j}$ and a FLC $\boldsymbol{\ell}$, let $\boldsymbol{\ell}_{\boldsymbol{j}}$ denote the symbol obtained by replacing each ℓ_k in $\boldsymbol{\ell} = (\ell_1, \ldots, \ell_f)$ by a dot if $j_k = 0$. Thus, with $\boldsymbol{j} = 0110$, $\boldsymbol{\ell}_{\boldsymbol{j}} = (\cdot, \ell_2, \ell_3, \cdot)$.

15

ANOVA Effects Defined

This chapter defines ANOVA effects for factorial treatment arrangements. This provides definitions and notation for a general setting in which subjects are observed under multiple factor-level combinations and for models that permit any given combination of ANOVA effects.

15.1 Introduction

The canonization of ANOVA is due to R. A. Fisher. While the notion of additional variation due to factor effects is a natural feature of least squares, Fisher's formalization of the ANOVA table and corresponding terminology became ubiquitous. His description for analyzing effects of two factors is straightforward:

> It is often necessary to divide the total variance into more than two portions; it sometimes happens both in experimental and in observational data that the observations may be grouped into classes in more than one way; each observation belongs to one class of type A and to a different class of type B. In such a case we can find separately the variance between classes of type A and between classes of type B; the balance of the total variance may represent only the variance within each subclass, or there may be in addition an interaction of causes, so that a change in class of type A does not have the same effect in all B classes. If the observations do not occur singly in the subclasses, the variance within the subclasses may be determined independently, and the presence or absence of interaction verified. (R. A. Fisher, 1934, p. 222.)

In this setting there are two factors, designated A and B, with levels $i = 1, \ldots, a$ and $j = 1, \ldots, b$, respectively. At each combination i, j of levels, responses y_{ijs} of n_{ij} subjects are observed, $s = 1, \ldots, n_{ij}$, sampled independently from a population with mean η_{ij} and variance σ^2 of the response variable. Section 15.4 shows how Fisher's description defines factor effects in this two-factor setting.

The factor-level combination (FLC) i, j is also called a *cell*, and η_{ij} is called a *cell mean.* Row (corresponding to factor A) and column (B) *marginal means* are $\bar{\eta}_{i\cdot}$ and $\bar{\eta}_{\cdot j}$; the corresponding *sample* cell means and marginal means are $\bar{y}_{ij\cdot}$, $\bar{y}_{i\cdot\cdot}$, and $\bar{y}_{\cdot j\cdot}$.

Fisher called levels of A or B *classes* and FLCs of A and B *subclasses.* The cell sample sizes n_{ij} are traditionally called *subclass numbers.* For this discussion, assume that $n_{i\cdot} > 0$ and $n_{\cdot j} > 0$, but keep open the possibility that some cells might be *empty,* with $n_{ij} = 0$. A model or setting is *balanced* if all subclass numbers n_{ij} are equal, and otherwise it is *unbalanced.*

Let us reserve the acronym SOS for a sum of squared deviations from the average. That is, for $\boldsymbol{z} \in \Re^n$, $SOS(\boldsymbol{z}) = \sum_i (z_i - \bar{z})^2 = \boldsymbol{z}' S_n \boldsymbol{z}$.

15.2 Effects: Differences and Contrasts

Under the sampling model that $\boldsymbol{y} \sim (X\boldsymbol{\beta}, \sigma^2 \mathrm{I})$, effects of factors on populations of subjects are differences among population means corresponding to different FLCs. Throughout this discussion and the ensuing treatment of analysis of variance, it is important to keep in mind some simple, central facts about differences. To see them, consider a vector $\boldsymbol{m} = (m_1, \ldots, m_q)'$. That all its elements are the same, $m_1 = m_2 = \cdots = m_q$, is equivalent to $m_i = (1/m)\sum_i m_i = \bar{m}$, $i = 1, \ldots, q$; and that is equivalent to $SOS(\boldsymbol{m}) = \sum_{i=1}^q (m_i - \bar{m})^2 = 0$. That is, there are differences among $m_1, \ldots, m_q$ iff $SOS(\boldsymbol{m}) > 0$.

This fact is readily apparent in terms of vectors and orthogonal projections. That all entries in $\boldsymbol{m}$ are the same is equivalent to $\boldsymbol{m} \in \mathrm{sp}(\mathbf{1}_q)$. The orthogonal projection matrix onto $\mathrm{sp}(\mathbf{1}_q)$ is $U_q = (1/q)\mathbf{1}_q\mathbf{1}_q'$. That $\boldsymbol{m}$ is in $\mathrm{sp}(\mathbf{1}_q)$ is equivalent to $\boldsymbol{m} = U_q\boldsymbol{m}$, or to $S_q\boldsymbol{m} = \mathbf{0}$, where $S_q = \mathrm{I}_q - U_q$, which is the orthogonal projection matrix onto $\mathrm{sp}(\mathbf{1}_q)^\perp$.

Note that $S_q\boldsymbol{m} = (m_1 - \bar{m}, \ldots, m_q - \bar{m})'$. That there are no differences among $m_1, \ldots, m_q$ is equivalent to $S_q\boldsymbol{m} = \mathbf{0}$, which in turn is equivalent to $\boldsymbol{m}'S_q\boldsymbol{m} = \sum_{i=1}^q (m_i - \bar{m})^2 = 0$. Differences among $m_1, \ldots, m_q$ show up as non-zero entries in $S_q\boldsymbol{m}$, which cause $SOS(\boldsymbol{m}) = \boldsymbol{m}'S_q\boldsymbol{m}$ to be positive.

Linear functions of $S_q\boldsymbol{m}$ are called ***contrasts*** on $\boldsymbol{m}$. Then $\boldsymbol{c}'\boldsymbol{m}$ is a contrast on $\boldsymbol{m}$ iff $\boldsymbol{c} = S_q\boldsymbol{\ell}$ for some vector $\boldsymbol{\ell}$, and hence $\boldsymbol{c} \in \mathrm{sp}(S_q) = \mathrm{sp}(\mathbf{1}_q)^\perp$: $\boldsymbol{c}'\boldsymbol{m}$ is a contrast on $\boldsymbol{m}$ iff $\mathbf{1}_q'\boldsymbol{c} = 0$. All members of $\boldsymbol{m}$ are equal iff all contrasts on $\boldsymbol{m}$ are equal to 0. Differences among members of $\boldsymbol{m}$ are contrasts on $\boldsymbol{m}$.

This definition and use of the word "contrast" has been standard terminology in statistics since Fisher's time. Regrettably, that clarity has been lost by its use in some R packages to mean any old linear combination of cell means. It is used here only as it is defined above.

15.3 Fisher's ANOVA SSs in the Balanced Two-Factor Setting

Fisher's definitions of ANOVA effects, p. 141, implicitly assume that the model is balanced, with all the subclass numbers equal, $n_{ij} = m$. These definitions extend readily to balanced models for any number of factors. Then appropriate numerator sums of squares for any effects (main effects, interaction effects) can be expressed in terms of SOSs, sums of squared deviations from the average. See Exercise 2, p. 176. Thus practically all basic computations require only this one algorithm.

Following Fisher's prescriptions, "the variance between classes of type A" is

$$SS_A = \sum_{i=1}^{a} \sum_{j=1}^{b} \sum_{s=1}^{m} (\bar{y}_{i\cdot\cdot} - \bar{y}_{\cdot\cdot\cdot})^2 = mb \sum_{i=1}^{a} (\bar{y}_{i\cdot\cdot} - \bar{y}_{\cdot\cdot\cdot})^2, \tag{15.1}$$

and "between classes of type B" is

$$SS_B = \sum_{i=1}^{a} \sum_{j=1}^{b} \sum_{s=1}^{m} (\bar{y}_{\cdot j\cdot} - \bar{y}_{\cdot\cdot\cdot})^2 = ma \sum_{j=1}^{b} (\bar{y}_{\cdot j\cdot} - \bar{y}_{\cdot\cdot\cdot})^2. \tag{15.2}$$

Treating the ab combinations of levels as a single factor, "total variance" (which came to be called "subtotal sum of squares") is

$$SS_{A,B} = \sum_{i=1}^{a} \sum_{j=1}^{b} \sum_{s=1}^{m} (\bar{y}_{ij\cdot} - \bar{y}_{\cdot\cdot\cdot})^2 = m \sum_{i=1}^{a} \sum_{j=1}^{b} (\bar{y}_{ij\cdot} - \bar{y}_{\cdot\cdot\cdot})^2. \tag{15.3}$$

Then "the balance of total variance" due to "an interaction of causes" is

$$\begin{aligned} SS_{AB} &= SS_{A,B} - SS_A - SS_B \\ &= m \sum_{i=1}^{a} \sum_{j=1}^{b} (\bar{y}_{ij\cdot} - \bar{y}_{i\cdot\cdot} - \bar{y}_{\cdot j\cdot} + \bar{y}_{\cdot\cdot\cdot})^2. \end{aligned} \tag{15.4}$$

These computations were easy to codify, teach, and master. It was obvious, though, that they didn't work for unbalanced models. However, models with ***proportional subclass numbers*** (psn), defined as $n_{ij} = n_{i\cdot} n_{\cdot j}/n_{\cdot\cdot}$ for all i, j, were thought to have all the conveniences of balanced models. See Chapter 20. Snedecor and Cox (1935) noted that, "There is only a slight increase in the complications of reduction if the subclass numbers are proportional but unequal."

Because they have the form of SOS, SS_A, SS_B, and $SS_{A,B}$ can be adapted readily to unbalanced models simply by replacing the range of s to 1 to n_{ij}. However, Snedecor and Cox (1935) give an example, not psn, in which the difference $SS_{A,B} - SS_A - SS_B$ is negative, and hence SS_{AB} is not a proper SS. Such examples fostered the widespread attitude that ANOVA was fine for balanced models or models with psn, but that it was complicated to adapt to unbalanced models.

15.4 ANOVA Effects in the Two-Factor Setting

Let us apply Fisher's definitions to the cell means (η_{ij}) as if all the subclass numbers are $n_{ij} = 1$. As he used the terms, "total variance" is

$$\sum_{i,j}(\eta_{ij} - \bar{\eta})^2. \tag{15.5}$$

The ab population means are all equal iff this sum of squares is 0.

The objective of analysis is to assess differences among cell means η_{ij} and how they correspond to levels of A and B. The "variance between classes of type A" and "variance between classes of type B" are, respectively,

$$\sum_{i=1}^{a}\sum_{j=1}^{b}(\bar{\eta}_{i\cdot} - \bar{\eta})^2 \text{ and } \sum_{i=1}^{a}\sum_{j=1}^{b}(\bar{\eta}_{\cdot j} - \bar{\eta})^2. \tag{15.6}$$

The "balance of the total variance" is

$$\begin{aligned}\sum_{i=1}^{a}\sum_{j=1}^{b}(\eta_{ij} - \bar{\eta})^2 - \left[\sum_{i=1}^{a}\sum_{j=1}^{b}(\bar{\eta}_{i\cdot} - \bar{\eta})^2 + \sum_{i=1}^{a}\sum_{j=1}^{b}(\bar{\eta}_{\cdot j} - \bar{\eta})^2\right]& \\ = \sum_{i=1}^{a}\sum_{j=1}^{b}(\eta_{ij} - \bar{\eta}_{i\cdot} - \bar{\eta}_{\cdot j} + \bar{\eta})^2,&\end{aligned} \tag{15.7}$$

due to "the interaction of causes," more commonly called now "AB interaction effects." It is 0 iff all the differences $\eta_{ij} - \bar{\eta}_{i\cdot} - \bar{\eta}_{\cdot j}$ are the same, that is, iff $\eta_{ij} = \bar{\eta}_{i\cdot} + \bar{\eta}_{\cdot j} - \bar{\eta}$ for all factor-level combinations i, j. In that case, the A and B effects on the response are said to be ***additive***.

ANOVA effects can be defined succinctly by expressions involving Kronecker products. Before going on, review some of their properties as shown in Exercise 30, p. 54.

Effects are differences among the ab cell means, which are linear functions of $S_{ab}\boldsymbol{\eta} = (\eta_{ij} - \bar{\eta})$. "Total variance" is $\boldsymbol{\eta}' S_{ab}\boldsymbol{\eta}$. Fisher's "variance between classes of type A" came to identify factor A ***main effects***. The sum of squared differences among the η_{ij}s attributable to A main effects can be expressed as

$$\begin{aligned}
\sum_{i,j}(\bar{\eta}_{i\cdot} - \bar{\eta})^2 &= (\bar{\eta}_{i\cdot} - \bar{\eta})'(\bar{\eta}_{i\cdot} - \bar{\eta}) \\
&= [(\mathrm{I}_a \otimes U_b - U_a \otimes U_b)\boldsymbol{\eta}]'[(\mathrm{I}_a \otimes U_b - U_a \otimes U_b)\boldsymbol{\eta}] \\
&= [(S_a \otimes U_b)\boldsymbol{\eta}]'[(S_a \otimes U_b)\boldsymbol{\eta}] \\
&= \boldsymbol{\eta}'(S_a \otimes U_b)\boldsymbol{\eta}, \qquad (15.8)
\end{aligned}$$

since $S_a \otimes U_b$ is symmetric and idempotent. Thus there are no A main effects iff $\boldsymbol{\eta}'(S_a \otimes U_b)\boldsymbol{\eta} = 0$, which in turn is equivalent to $(S_a \otimes U_b)\boldsymbol{\eta} = \mathbf{0}$, which in turn is equivalent to all A marginal means being equal, $\bar{\eta}_{1\cdot} = \cdots = \bar{\eta}_{a\cdot}$. A main effects are linear functions of $(S_a \otimes U_b)\boldsymbol{\eta} = (\bar{\eta}_{i\cdot} - \bar{\eta})$.

Show similarly that the "variance between classes of type B" can be expressed as $\boldsymbol{\eta}'(U_a \otimes S_b)\boldsymbol{\eta}$, implicitly defining B main effects as differences among the B marginal means, which are linear functions of $(U_a \otimes S_b)\boldsymbol{\eta} = (\bar{\eta}_{\cdot j} - \bar{\eta})$.

At this point, note the following identity:

$$\begin{aligned}
\mathrm{I}_{ab} &= \mathrm{I}_a \otimes \mathrm{I}_b \\
&= (U_a + S_a) \otimes (U_b + S_b) \\
&= U_a \otimes U_b + S_a \otimes U_b + U_a \otimes S_b + S_a \otimes S_b. \qquad (15.9)
\end{aligned}$$

Noting that $U_a \otimes U_b = U_{ab}$, this means that

$$S_{ab} = \mathrm{I}_{ab} - U_{ab} = S_a \otimes U_b + U_a \otimes S_b + S_a \otimes S_b. \qquad (15.10)$$

"Total variance" is $\boldsymbol{\eta}'S_{ab}\boldsymbol{\eta}$. Then the "balance of the total variance" is

$$\boldsymbol{\eta}'(S_{ab} - S_a \otimes U_b - U_a \otimes S_b)\boldsymbol{\eta} = \boldsymbol{\eta}'(S_a \otimes S_b)\boldsymbol{\eta}.$$

The identity (15.10) establishes that differences $S_{ab}\boldsymbol{\eta}$ among the cell means can be resolved into three orthogonal components: differences due to A main effects, $(S_a \otimes U_b)\boldsymbol{\eta} = (\bar{\eta}_{i\cdot} - \bar{\eta})$; differences due to B main effects, $(U_a \otimes S_b)\boldsymbol{\eta} = (\bar{\eta}_{\cdot j} - \bar{\eta})$; and differences due to AB interaction effects, $(S_a \otimes S_b)\boldsymbol{\eta} = (\eta_{ij} - \bar{\eta}_{i\cdot} - \bar{\eta}_{\cdot j} + \bar{\eta})$.

We will take these as the definitions of ANOVA effects in the two-factor setting. A main effects are linear functions of $(S_a \otimes U_b)\boldsymbol{\eta}$. There are no A main effects iff $(S_a \otimes U_b)\boldsymbol{\eta} = \mathbf{0}$, which is the same as equality of all the A marginal means $\bar{\eta}_{i\cdot}$. Corresponding usage applies for B main effects and AB interaction effects.

15.5 ANOVA Effects for Multiple Factors

The identity (15.10) can be extended readily to any number of factors. With three, named A, B, and C, at a, b, and c levels, respectively,

$$\mathrm{I}_{abc} = (U_a + S_a) \otimes (U_b + S_b) \otimes (U_c + S_c),$$

and hence

$$\begin{aligned} S_{abc} &= S_a \otimes U_b \otimes U_c + U_a \otimes S_b \otimes U_c \\ &\quad + S_a \otimes S_b \otimes U_c + U_a \otimes U_b \otimes S_c \\ &\quad + U_a \otimes S_a \otimes S_c + S_a \otimes S_b \otimes S_c. \end{aligned} \tag{15.11}$$

The terms on the right are identified with, and define, A main effects, B main effects, AB interaction effects, C main effects, AC interaction effects, BC interaction effects, and ABC interaction effects, respectively. AB interaction effects, for example, comprise all linear functions of $(S_a \otimes S_b \otimes U_c)\boldsymbol{\eta}$.

Recall the notation with f-tuples $\boldsymbol{j}$ of 0s and 1s described in Chapter 14 and the set $\mathcal{B}^f$ of all such f-tuples. A partial ordering, ***containment***, is defined on $\mathcal{B}^f$. For two such f-tuples, $\boldsymbol{j}_2$ *contains* $\boldsymbol{j}_1$ ($\boldsymbol{j}_1$ *is contained in* $\boldsymbol{j}_2$) iff $j_{2k} \geqslant j_{1k}$, $k = 1, \ldots, f$. Equivalently, $j_{1k} = 1$ only if $j_{2k} = 1$. Denote this by $\boldsymbol{j}_2 \succeq \boldsymbol{j}_1$, or by $\boldsymbol{j}_2 \succ \boldsymbol{j}_1$ to exclude $\boldsymbol{j}_2 = \boldsymbol{j}_1$. Define $\boldsymbol{j}_1 \preceq \boldsymbol{j}_2$ and $\boldsymbol{j}_1 \prec \boldsymbol{j}_2$ correspondingly. With three factors A, B, and C, for example, 100 and 010, designating A and B, are both contained in 110, which indicates the string AB, and C (001) is not. In the strings A, B, C, AB, etc., the string AB contains A and B but it does not contain C.

For each $\boldsymbol{j} \in \mathcal{B}^f$ define

$$H_{\boldsymbol{j}} = \bigotimes_{k=1}^{f} \begin{cases} U_{a_k} & \text{if } j_k = 0, \\ S_{a_k} & \text{if } j_k = 1. \end{cases} \tag{15.12}$$

For example, with $f = 4$, $H_{0101} = U_a \otimes S_b \otimes U_c \otimes S_d$. Each of these 2^f matrices is symmetric and idempotent, and the product of any two different ones is 0. The sum of any set of distinct such matrices is also symmetric and idempotent.

We will define ANOVA factor effects in terms of these matrices: the $\boldsymbol{j}$ ANOVA factor effects are linear functions of $H_{\boldsymbol{j}}\boldsymbol{\eta}$. With $f = 4$ factors, A main effects are linear functions of $H_{1000}\boldsymbol{\eta}$, and ACD interaction effects are linear functions of $H_{1011}\boldsymbol{\eta}$. There are no $\boldsymbol{j}$ ANOVA factor effects iff $H_{\boldsymbol{j}}\boldsymbol{\eta} = \boldsymbol{0}$, which is equivalent to $\boldsymbol{\eta}' H_{\boldsymbol{j}} \boldsymbol{\eta} = 0$.

With f factors, note that

$$\begin{aligned} \mathrm{I}_{a*} &= \mathrm{I}_{a_1} \otimes \cdots \otimes \mathrm{I}_{a_f} \\ &= (U_{a_1} + S_{a_1}) \otimes \cdots \otimes (U_{a_f} + S_{a_f}) \\ &= \sum \{H_{\boldsymbol{j}} : \boldsymbol{j} \in \mathcal{B}^f\}. \end{aligned} \tag{15.13}$$

It follows that

$$\begin{aligned} S_{a*} &= \mathrm{I}_{a*} - U_{a*} = \mathrm{I}_{a*} - H_{0_f} \\ &= \sum \{H_{\boldsymbol{j}} : \boldsymbol{j} \in \mathcal{B}^f, \boldsymbol{j} \succ 0_f\}. \end{aligned} \tag{15.14}$$

Call this the ***ANOVA Identity***. For an a_*-vector $\boldsymbol{\eta}$ of cell means, this means that differences $S_{a_*}\boldsymbol{\eta}$ can be resolved into the sum of $2^f - 1$ orthogonal components $H_{\boldsymbol{j}}\boldsymbol{\eta}$. The $\boldsymbol{j}$ ANOVA effects are defined to be the set of all linear combinations of $H_{\boldsymbol{j}}\boldsymbol{\eta}$.

The following result will be useful. Let $\boldsymbol{j}_0 = \boldsymbol{j}_L 0 \boldsymbol{j}_R$ and $\boldsymbol{j}_1 = \boldsymbol{j}_L 1 \boldsymbol{j}_R$ be two f-tuples of 0s and 1s, the same except that $\boldsymbol{j}_0$ has 0 in the k-th place and $\boldsymbol{j}_1$ has 1 in the k-th place. For example, let $\boldsymbol{j}_L = 010$ and $\boldsymbol{j}_R = 10$, so that $\boldsymbol{j}_0 = 010010$ and $\boldsymbol{j}_1 = 010110$. Then

$$H_{\boldsymbol{j}_0} + H_{\boldsymbol{j}_1} = H_{\boldsymbol{j}_L} \otimes (U_{a_k} + S_{a_k}) \otimes H_{\boldsymbol{j}_R} = H_{\boldsymbol{j}_L} \otimes \mathrm{I}_{a_k} \otimes H_{\boldsymbol{j}_R}. \tag{15.15}$$

Given an f-tuple $\boldsymbol{j}_*$, it follows that

$$\sum\{H_{\boldsymbol{j}} : \boldsymbol{j} \in \mathcal{B}^f \text{ and } \boldsymbol{j} \preceq \boldsymbol{j}_*\} = \bigotimes_{k=1}^{f} \begin{cases} U_{a_k} & \text{if } j_{*k} = 0, \\ \mathrm{I}_{a_k} & \text{if } j_{*k} = 1. \end{cases} \tag{15.16}$$

As a particular example, with $\boldsymbol{j}_* = 1\ldots1$, this gives the identity (15.13) above. As another example, with $f = 3$ factors let $\boldsymbol{j}_* = 101$. The 3-tuples contained in $\boldsymbol{j}_*$ are $000, 100, 001$, and 101. Then

$$\begin{aligned}
&H_{000} + H_{100} + H_{001} + H_{101} \\
&\quad = U_a \otimes U_b \otimes U_c + S_a \otimes U_b \otimes U_c + U_a \otimes U_b \otimes S_c + S_a \otimes U_b \otimes S_c \\
&\quad = (U_a + S_a) \otimes U_b \otimes U_c + (U_a + S_a) \otimes U_b \otimes S_c \\
&\quad = \mathrm{I}_a \otimes U_b \otimes U_c + \mathrm{I}_a \otimes U_b \otimes S_c \\
&\quad = \mathrm{I}_a \otimes U_b \otimes (U_c + S_c) = \mathrm{I}_a \otimes U_b \otimes \mathrm{I}_c.
\end{aligned}$$

The definitions of the matrices $H_{\boldsymbol{j}}$ resemble closely and serve the same purpose as Hocking's (2013, p. 413) H_t. The difference is that H_t is formulated so that its rows are linearly independent, and so the corresponding orthogonal projection becomes $H_t'(H_t H_t')^{-1} H_t$ in place of what is denoted here $H_{\boldsymbol{j}}$.

16

Cell Means Models with ANOVA Effects

This chapter formulates models for cell means in multi-factor settings. Models that permit any specified set of factor effects (main effects, interaction effects) can be formulated in terms of the contrasts that define ANOVA effects in balanced models. The default formulation has been in terms of indicator variables, but it has containment relations that limit the subsets of ANOVA effects that it can accommodate. It and other formulations can be expressed in terms of balanced-model ANOVA effects. It is shown that restricted models to test ANOVA effects can be had by deleting their predictor variables from the full model.

16.1 Introduction

This chapter focuses on models for $\boldsymbol{\eta}$, the a_*-vector of cell means, in terms of ANOVA effects, in the general form $\boldsymbol{\eta} = M\boldsymbol{\beta} \in \text{sp}(M)$. The matrix M is fixed and known in each specific setting. This induces a model for $\boldsymbol{\mu}$, the mean of the n-vector response $\boldsymbol{y}$, as $\mathbb{K}M\boldsymbol{\beta}$, in the form of $X\boldsymbol{\beta}$ with $X = \mathbb{K}M$. This setting and notation are described in Section 14.2.

There has been heated discussion of what effects may be included in a model and what effects may be tested. See Nelder (1977) for its thorough treatment and disputatious colloquy. Much of the discussion following the main paper would not seem out of place today. The unfortunate features of the traditional models built on dummy variables and the habits and ways of thinking they have engendered are the source of much of this conflict.

The overarching objective here is to explore ways to formulate models for the mean vector that include any desired set of factor effects, as well as covariates and combinations of covariate and factor effects. With that capability, correct restricted models can be formulated for testing hypotheses about any effects in the model. Among other things, it will be shown here that, in one form, restricted models to test factor effects can be formed simply by deleting a corresponding set of predictor variables from the model matrix.

The earliest formulation of multiple regression models for ANOVA effects seems to have been in terms of dummy variables. That has been the standard, and its features, good and bad, have had great influence on practice, teaching, statistical computing packages, and folklore about models for ANOVA effects. Other approaches use different schemes to code FLCs. These approaches will be described in this chapter and their relations to ANOVA effects will be shown. First, though, it is shown in the next section that models for the mean vector $\boldsymbol{\mu}$ can be formulated that include exclusively any specified subset of the possible ANOVA effects.

16.2 Models for Sets of ANOVA Effects

Recall that $\mathcal{B}^f$ denotes the set of all possible binary f-tuples. Each member $\boldsymbol{j}$ of $\mathcal{B}^f$ corresponds to the ANOVA effect defined by $H_{\boldsymbol{j}}$. Let $\mathcal{J}$ denote a subset of $\mathcal{B}^f$, and let $H_{\mathcal{J}} = \sum\{H_{\boldsymbol{j}} : \boldsymbol{j} \in \mathcal{J}\}$. The model $\text{sp}(H_{\mathcal{J}})$ for the a_*-vector $\boldsymbol{\eta}$ of cell means includes the ANOVA effects in $\mathcal{J}$ and it excludes those not in $\mathcal{J}$. That is, for any $\boldsymbol{j} \in \mathcal{J}$, $\text{sp}(H_{\boldsymbol{j}}) \subset \text{sp}(H_{\mathcal{J}})$. And for any $\boldsymbol{j}$ not in $\mathcal{J}$ and any $\boldsymbol{\eta} \in \text{sp}(H_{\mathcal{J}})$, $H_{\boldsymbol{j}}\boldsymbol{\eta} = \mathbf{0}$. In summary, a model for $\boldsymbol{\eta}$ that permits exclusively the ANOVA effects in $\mathcal{J}$ can be formulated as $\text{sp}(H_{\mathcal{J}})$, which becomes the model $\text{sp}(\mathbb{K}H_{\mathcal{J}})$ for the mean vector $\boldsymbol{\mu}$. (At times "the model $\mathcal{J}$" will be used to mean the set $\mathcal{J}$ when the meaning is clear in context.)

For an effect $\boldsymbol{j}_*$ in this model, the model for $\boldsymbol{\eta}$ restricted by the condition $H_{\boldsymbol{j}*}\boldsymbol{\eta} = \mathbf{0}$ is

$$\begin{aligned} \text{sp}(H_{\mathcal{J}}) \cap \text{sp}(H_{\boldsymbol{j}*})^{\perp} &= \text{sp}(H_{\mathcal{J}} - H_{\boldsymbol{j}*}) \\ &= \text{sp}\left(\sum\{H_{\boldsymbol{j}},\ \boldsymbol{j} \in \mathcal{J} \text{ and } \boldsymbol{j} \neq \boldsymbol{j}_*\}\right) \\ &= \text{sp}(H_{\mathcal{J}\setminus \boldsymbol{j}*}), \end{aligned} \tag{16.1}$$

by excluding $H_{\boldsymbol{j}_*}$ from the sum $H_{\mathcal{J}}$. (The backslash denotes set difference.) The restricted model for $\boldsymbol{\mu}$ is then

$$\begin{aligned} \{\mathbb{K}\boldsymbol{\eta} : \boldsymbol{\eta} \in \text{sp}(H_{\mathcal{J}}) \text{ and } H_{\boldsymbol{j}*}\boldsymbol{\eta} = \mathbf{0}\} &= \text{sp}[\mathbb{K}(H_{\mathcal{J}} - H_{\boldsymbol{j}*})] \\ &= \text{sp}(\mathbb{K}H_{\mathcal{J}\setminus \boldsymbol{j}*}). \end{aligned} \tag{16.2}$$

This way of formulating models for the vector of cell means directly in terms of ANOVA effects is not common. Hocking's formulation (2013) seems to be the exception, and he uses it in a slightly different form.

The notation $H_{\mathcal{J}}$ is reserved for sums of $H_{\boldsymbol{j}}$ matrices that define ANOVA effects. Other methods to create models for factor effects are built by concatenating matrices column-wise instead of summing them. For a subset $\mathcal{J} = \{\boldsymbol{j}_1, \ldots, \boldsymbol{j}_t\}$ of $\mathcal{B}^f$, and with matrices $M_{\boldsymbol{j}}$ defined over $\mathcal{B}^f$, $M_{\mathcal{J}}$ is defined as the matrix $(M_{\boldsymbol{j}_1}, \ldots, M_{\boldsymbol{j}_t})$. Note, though, that the model formed by the sum $H_{\mathcal{J}}$ is the same as the model formed by concatenation as $(H_{\boldsymbol{j}_1}, \ldots, H_{\boldsymbol{j}_t})$.

16.3 Dummy-Variable Models for Factor Effects

This section describes models for ANOVA effects formulated with dummy variables. This approach can be seen as arising from the notion of *additive effects.*

For two factors, A and B effects are additive if the effect of level i of A is to add a constant α_i to the response, and the effect of level j of B is to add β_j to the response. If the population mean is η_0 to begin with, then under the FLC i, j it becomes

$$\eta_{ij} = \eta_0 + \alpha_i + \beta_j, i = 1, \ldots, a, j = 1, \ldots, b. \tag{16.3}$$

Under this model, the effect of each level of A is the same, irrespective of the level of B. This corresponds to Fisher's description in the quotation on p. 141 to "find separately" effects of A and B. To illustrate, with $a = 2$ and $b = 3$, putting these together in the ab vector $\boldsymbol{\eta}$, the model for $\boldsymbol{\eta}$ appears as

$$\boldsymbol{\eta} = \begin{pmatrix} \eta_{11} \\ \eta_{12} \\ \eta_{13} \\ \eta_{21} \\ \eta_{22} \\ \eta_{23} \end{pmatrix} = \left(\begin{array}{c|cc|ccc} 1 & 1 & 0 & 1 & 0 & 0 \\ 1 & 1 & 0 & 0 & 1 & 0 \\ 1 & 1 & 0 & 0 & 0 & 1 \\ 1 & 0 & 1 & 1 & 0 & 0 \\ 1 & 0 & 1 & 0 & 1 & 0 \\ 1 & 0 & 1 & 0 & 0 & 1 \end{array}\right) \begin{pmatrix} \eta_0 \\ \boldsymbol{\alpha} \\ \boldsymbol{\beta} \end{pmatrix} \tag{16.4}$$

$$= (\mathbf{1}_a \otimes \mathbf{1}_b, \mathrm{I}_a \otimes \mathbf{1}_b, \mathbf{1}_a \otimes \mathrm{I}_b) \begin{pmatrix} \eta_0 \\ \boldsymbol{\alpha} \\ \boldsymbol{\beta} \end{pmatrix}. \tag{16.5}$$

It is convenient and useful to recognize how (16.4) can be expressed in (16.5) in terms of Kronecker products. The first column, $\mathbf{1}_{ab}$, is the same as $\mathbf{1}_a \otimes \mathbf{1}_b$. The columns of $\mathrm{I}_a \otimes \mathbf{1}_b$ are indicator variables of levels of A. Each entry in the first column of $\mathrm{I}_a \otimes \mathbf{1}_b$ is 1 if A is at level 1 in its row, and 0 otherwise; and each entry in the second column is 1 if A is at level 2 in its row, and 0 otherwise. Indicator variables like this are often called *dummy variables.* The columns of $\mathbf{1}_a \otimes \mathrm{I}_b$ are dummy variables for levels of factor B. This is a dummy-variable formulation of an additive-effects model for the vector $\boldsymbol{\eta}$ of cell means.

It should be clear that not all possible ab-vectors can be represented as (16.3). For example, with $a = 2$ and $b = 3$, arranging η_{ij} in rows and columns corresponding to i and j, respectively, in the array

$$\begin{pmatrix} 87 & 101 & 95 \\ 90 & 104 & 98 \end{pmatrix},$$

it can be seen that A (row) and B (column) effects are additive. Identify values of η_0, α_1, α_2, β_1, β_2, and β_3 such that (16.3) fits these six values. On the other hand, the set of values η_{ij} in

$$\begin{pmatrix} 87 & 105 & 95 \\ 90 & 104 & 98 \end{pmatrix}$$

is not additive in A and B effects.

To include non-additive effects due to "interaction of causes," the model is customarily formulated as

$$\eta_{ij} = \eta_0 + \alpha_i + \beta_j + (\alpha\beta)_{ij} \tag{16.6}$$

for the two-factor setting. The notation $(\alpha\beta)_{ij}$ is used here to denote a single scalar, one for each combination of levels of A and B; it is a two-letter name, not a product. For $a = 2$ levels of factor A and $b = 3$ levels of factor B, for example,

$$\boldsymbol{\eta} = \begin{pmatrix} \eta_{11} \\ \eta_{12} \\ \eta_{13} \\ \eta_{21} \\ \eta_{22} \\ \eta_{23} \end{pmatrix} = \left(\begin{array}{c|cc|ccc|cccccc} 1 & 1 & 0 & 1 & 0 & 0 & 1 & 0 & 0 & 0 & 0 & 0 \\ 1 & 1 & 0 & 0 & 1 & 0 & 0 & 1 & 0 & 0 & 0 & 0 \\ 1 & 1 & 0 & 0 & 0 & 1 & 0 & 0 & 1 & 0 & 0 & 0 \\ 1 & 0 & 1 & 1 & 0 & 0 & 0 & 0 & 0 & 1 & 0 & 0 \\ 1 & 0 & 1 & 0 & 1 & 0 & 0 & 0 & 0 & 0 & 1 & 0 \\ 1 & 0 & 1 & 0 & 0 & 1 & 0 & 0 & 0 & 0 & 0 & 1 \end{array}\right) \begin{pmatrix} \eta_0 \\ \boldsymbol{\alpha} \\ \boldsymbol{\beta} \\ (\boldsymbol{\alpha\beta}) \end{pmatrix}$$

$$= (\mathbf{1}_a \otimes \mathbf{1}_b, \mathrm{I}_a \otimes \mathbf{1}_b, \mathbf{1}_a \otimes \mathrm{I}_b, \mathrm{I}_a \otimes \mathrm{I}_b) \begin{pmatrix} \eta_0 \\ \boldsymbol{\alpha} \\ \boldsymbol{\beta} \\ (\boldsymbol{\alpha\beta}) \end{pmatrix}. \tag{16.7}$$

The last term alone, $\mathrm{I}_a \otimes \mathrm{I}_b = \mathrm{I}_{ab}$, enables this model to include any ab-vector $\boldsymbol{\eta}$ because $\mathrm{sp}(\mathrm{I}_{ab}) = \Re^{ab}$. Such a model is said to be ***saturated***.

Let $E_{00} = \mathbf{1}_a \otimes \mathbf{1}_b$, $E_{10} = \mathrm{I}_a \otimes \mathbf{1}_b$, $E_{01} = \mathbf{1}_a \otimes \mathrm{I}_b$, and $E_{11} = \mathrm{I}_a \otimes \mathrm{I}_b$. Concatenate these matrices as $E = (E_{00}, E_{10}, E_{01}, E_{11})$. Then the model for $\boldsymbol{\eta}$ is $\mathrm{sp}(E)$.

This scheme can be extended to f factors in terms of matrices defined by

$$E_{\boldsymbol{j}} = \bigotimes_{k=1}^{f} \begin{cases} \mathbf{1}_{a_k} & \text{if } j_k = 0, \\ \mathrm{I}_{a_k} & \text{if } j_k = 1, \end{cases} . \tag{16.8}$$

With three factors, for example, $E_{000} = \mathbf{1}_a \otimes \mathbf{1}_b \otimes \mathbf{1}_c$, $E_{100} = \mathrm{I}_a \otimes \mathbf{1}_b \otimes \mathbf{1}_c$, and so on, up to $E_{111} = \mathrm{I}_a \otimes \mathrm{I}_b \otimes \mathrm{I}_c$. Subsets of these matrices can be concatenated column-wise to build models for $\boldsymbol{\eta}$. Thus, for example, the model specified by the subset $\mathcal{J} = \{000, 100, 010, 001\}$ is $\mathrm{sp}(E_{000}, E_{100}, E_{010}, E_{001})$, which includes an intercept (1) and additive effects of A, B, and C.

A central, and troubling, feature of dummy-variable models is their redundancy. In particular, if $\boldsymbol{j}_1 \prec \boldsymbol{j}_2$, then $\mathrm{sp}(E_{\boldsymbol{j}_1}) \subset \mathrm{sp}(E_{\boldsymbol{j}_2})$. For example, $\mathrm{sp}(E_{100})$ is contained in each of $\mathrm{sp}(E_{110})$, $\mathrm{sp}(E_{101})$ and $\mathrm{sp}(E_{111})$, while $\mathrm{sp}(E_{000})$, $\mathrm{sp}(E_{010})$, $\mathrm{sp}(E_{001})$, and $\mathrm{sp}(E_{011})$ do not contain $\mathrm{sp}(E_{100})$. Indeed, this fact is the sole need to define containment, as used in this context.

Note that, based on (15.16), for any $\boldsymbol{j}_*$ in $\mathcal{B}^f$,

$$\mathrm{sp}(E_{\boldsymbol{j}*}) = \sum\{\mathrm{sp}(H_{\boldsymbol{j}}) : \boldsymbol{j} \preceq \boldsymbol{j}_*\}, \tag{16.9}$$

due to the fact that, for any $\boldsymbol{j} \in \mathcal{B}^f$,

$$\mathbf{P}_{E_{\boldsymbol{j}}} = \bigotimes_{k=1}^{f} \begin{cases} U_{a_k} & \text{if } j_k = 0, \\ \mathrm{I}_{a_k} & \text{if } j_k = 1. \end{cases} \tag{16.10}$$

When $E_{\boldsymbol{j}}$ is included in a model for $\boldsymbol{\eta}$, the model then includes the ANOVA effect $\boldsymbol{j}$, and it also includes all the ANOVA effects that $\boldsymbol{j}$ contains. This means, for example, that it is impossible to express a model that includes (1), B, and AB effects, and excludes A main effects, with dummy variables, and so it is impossible to test A main effects by additional SSE due to deleting E_{10} from the full model $\mathrm{sp}(E_{00}, E_{10}, E_{01}, E_{11})$.

Let $\mathcal{J} = \{\boldsymbol{j}_1, \ldots, \boldsymbol{j}_t\}$ be a subset of $\mathcal{B}^f$, and let $E_{\mathcal{J}} = (E_{\boldsymbol{j}_1}, \ldots, E_{\boldsymbol{j}_t})$ be the matrix formed by concatenating the matrices $E_{\boldsymbol{j}_i}$, $i = 1, \ldots, t$. Let $\bar{\mathcal{J}}$ denote the set of members of $\mathcal{B}^f$ that are contained in some member of $\mathcal{J}$. (For example, if $\mathcal{J} = \{00, 10, 01\}$, then $\bar{\mathcal{J}} = \{00, 10, 01\}$. If $\mathcal{J} = \{101\}$, then $\bar{\mathcal{J}} = \{000, 100, 001, 101\}$.) Then

$$\mathrm{sp}(E_{\mathcal{J}}) = \mathrm{sp}(H_{\bar{\mathcal{J}}}). \tag{16.11}$$

The dummy-variable model for $\boldsymbol{\eta}$ specified by $\mathcal{J}$ includes exactly those ANOVA effects $\boldsymbol{j}$ such that $\boldsymbol{j} \preceq \boldsymbol{j}_i$ for some i in $1, \ldots, t$. This means, among other things, that the dummy-variable model for $\mathcal{J} = \{011, 101\}$ includes the ANOVA effects 000, 010, 001, 011, 100, and 101.

In order to extend notation to models based on any subset $\mathcal{J}$ of $\mathcal{B}^f$, denote the coefficient vector of $E_{\boldsymbol{j}}$ by $\boldsymbol{\beta}_{\boldsymbol{j}}$, and let $\boldsymbol{\beta}' = (\boldsymbol{\beta}'_{\boldsymbol{j}_1}, \ldots, \boldsymbol{\beta}'_{\boldsymbol{j}_t})$. In the saturated two-factor model, then, $\boldsymbol{\beta}' = (\beta_{00}, \boldsymbol{\beta}'_{10}, \boldsymbol{\beta}'_{01}, \boldsymbol{\beta}'_{11})$ replaces $(\eta_0, \boldsymbol{\alpha}', \boldsymbol{\beta}', (\boldsymbol{\alpha\beta})')$. The model (16.7) for $\boldsymbol{\eta}$ can be expressed as $E\boldsymbol{\beta}$, $\boldsymbol{\beta} \in \Re^t$, with $t = 1 + a + b + ab$. In general, the dimension of $\boldsymbol{\beta}_{\boldsymbol{j}}$ is the column dimension of $E_{\boldsymbol{j}}$, which is

$$p_{\boldsymbol{j}} = \Pi_{k=1}^f \begin{cases} 1 & \text{if } j_k = 0, \\ a_k & \text{if } j_k = 1. \end{cases}$$

The corresponding model for the mean vector is $\boldsymbol{\mu} = \mathbb{K} E_{\mathcal{J}} \boldsymbol{\beta}$.

16.4 Other Coding Schemes

Other coding schemes for factor effects have appeared. In each one, models are formed by concatenating matrices $M_{\boldsymbol{j}}$ defined for each f-tuple $\boldsymbol{j}$ by

$$M_{\boldsymbol{j}} = \bigotimes_{k=1}^{f} \begin{cases} \mathbf{1}_{a_k} & \text{if } j_k = 0, \\ M_{a_k} & \text{if } j_k = 1. \end{cases} \tag{16.12}$$

ℓ	Dummy				Effect			Poly.			Poly. Contr.		
1	1	0	0	0	1	0	0	1	1^2	1^3	-3	1	-1
2	0	1	0	0	0	1	0	2	2^2	2^3	-1	-1	3
3	0	0	1	0	0	0	1	3	2^3	3^3	1	-1	-3
4	0	0	0	1	-1	-1	-1	4	2^4	4^3	3	1	1

TABLE 16.1: Some coding schemes. Matrices are M_a matrices for $a = 4$ factor levels.

Note that $M_{0_f} = \mathbf{1}_{a*}$. A model for the vector $\boldsymbol{\eta}$ of cell means built from these takes the form $\text{sp}(M_{\mathcal{J}})$, where $\mathcal{J} = \{\boldsymbol{j}_1, \ldots, \boldsymbol{j}_t\}$ is a subset of $\mathcal{B}^f$ and $M_{\mathcal{J}} = (M_{\boldsymbol{j}_1}, \ldots, M_{\boldsymbol{j}_t})$. As above, denote the model also as $M_{\mathcal{J}}\boldsymbol{\beta}$ where the coefficient vector $\boldsymbol{\beta}$ concatenates the coefficients $\boldsymbol{\beta}_{\boldsymbol{j}_i}$. The model induced for the mean vector is $\boldsymbol{\mu} = \mathbb{K}\boldsymbol{\eta} = \mathbb{K}M_{\mathcal{J}}\boldsymbol{\beta}$.

The essential property that each matrix M_{a_k} must have is that $\text{sp}(\mathbf{1}_{a_k}, M_{a_k}) = \Re^{a_k}$. Then it may be seen that $\sum\{\text{sp}(M_{\boldsymbol{j}}) : \boldsymbol{j} \in \mathcal{B}^f\} = \Re^{a*}$, thus including all possible vectors of cell means when all 2^f such matrices are concatenated. For example, for two factors, A and B, at a and b levels, suppose $\text{sp}(\mathbf{1}_a, M_a) = \Re^a$ and $\text{sp}(\mathbf{1}_b, M_b) = \Re^b$. Recall that, for matrices A and B, $\mathbf{P}_{A \otimes B} = \mathbf{P}_A \otimes \mathbf{P}_B$, and thus $\text{sp}(A \otimes B) = \text{sp}(\mathbf{P}_A \otimes \mathbf{P}_B)$. Show also that $(A, B) \otimes C = (A \otimes C, B \otimes C)$ and $\text{sp}(A \otimes B, A \otimes C) = \text{sp}(A \otimes (B, C))$. Then

$$\begin{aligned} &\text{sp}(\mathbf{1}_a \otimes \mathbf{1}_b, M_a \otimes \mathbf{1}_b, \mathbf{1}_a \otimes M_b, M_a \otimes M_b) \\ &\quad = \text{sp}((\mathbf{1}_a, M_a) \otimes \mathbf{1}_b, (\mathbf{1}_a, M_a) \otimes M_b) \\ &\quad = \text{sp}((\mathbf{1}_a, M_a) \otimes (\mathbf{1}_b, M_b)) \\ &\quad = \text{sp}(\mathrm{I}_a \otimes \mathrm{I}_b) = \Re^{ab}. \end{aligned}$$

M matrices for some coding schemes are shown in Table 16.1. Recall that $\boldsymbol{e}_i$ is used to denote the i-th column of an identity matrix. For dummy-variable coding, $M_a = \mathrm{I}_a$. For effect coding, $M_a = \mathrm{I}_a - \boldsymbol{e}_{i*}\mathbf{1}_a'$ and the column of zeros may be omitted. In reference-level coding, which is not shown, a column of I_a is omitted, so that $M_a = \mathrm{I}_a - \boldsymbol{e}_{i*}\boldsymbol{e}_{i*}'$, with the column of zeros omitted; the level $i*$ is the reference level. Polynomial coding is based on the fact that $\mathbf{1}_a$, $(1, \ldots, a)'$, $(1^2, \ldots, a^2)'$, …, $(1^{a-1}, \ldots, a^{a-1})'$ span $\Re^a$. The polynomial contrasts are an orthogonal spanning set built from $\mathbf{1}_a$ and the columns of the polynomial coding M_a.

In dummy-variable coding, the columns of $(\mathbf{1}_a, \mathrm{I}_a)$ are linearly dependent. Note that the columns of $(\mathbf{1}_a, M_a)$ are linearly independent for reference-level coding and the other three schemes shown in Table 16.1.

If the columns of M_a are contrasts (that is, $\text{sp}(M_a) \subset \text{sp}(S_a)$), then $\text{sp}(\mathbf{1}_a, M_a) = \Re^a$ implies that $\text{sp}(M_a) = \text{sp}(S_a)$. It follows that, for any binary f-tuple $\boldsymbol{j}$, $\text{sp}(M_{\boldsymbol{j}}) = \text{sp}(H_{\boldsymbol{j}})$ and hence that $\mathbf{P}_{M_{\boldsymbol{j}}} = H_{\boldsymbol{j}}$. Furthermore, for distinct f-tuples $\boldsymbol{j}_1$ and $\boldsymbol{j}_2$, the columns of $M_{\boldsymbol{j}_1}$ are orthogonal to the columns of $M_{\boldsymbol{j}_2}$.

Models for $\boldsymbol{\eta}$ that include exclusively the ANOVA effects corresponding to any subset $\mathcal{J}$ of $\mathcal{B}^f$ can be built from $H_{\mathcal{J}}$ or, with matrices $M_{\boldsymbol{j}}$ of contrasts, from $M_{\mathcal{J}}$. While this is a simple observation, it is important enough for creating restricted models for ANOVA effects that it deserves to be stated as a proposition.

Proposition 16.1. *Suppose that, for each $k = 1, \ldots, f$, M_{a_k} is a matrix such that $\text{sp}(M_{a_k}) = \text{sp}(S_{a_k})$. For each $\boldsymbol{j} \in \mathcal{B}^f$ define $M_{\boldsymbol{j}}$ by (16.12). Let $\mathcal{J}$ be a subset of $\mathcal{B}^f$, and define $M_{\mathcal{J}}$ as it is in the paragraph that includes (16.12). Then*

$$\text{sp}(M_{\mathcal{J}}) = \text{sp}(H_{\mathcal{J}}).$$

16.5 Restricted Models for Testing ANOVA Effects

Models for the mean vector $\boldsymbol{\mu}$ that include exclusively any given set $\mathcal{J}$ of ANOVA effects can be constructed as $\text{sp}(\mathbb{K}H_{\mathcal{J}})$, in terms of the matrices that define ANOVA effects, or as $\text{sp}(\mathbb{K}M_{\mathcal{J}})$ built on matrices of contrasts. Other coding schemes, like dummy-variable or reference-level coding or polynomial coding, cannot capture some configurations of effects.

To test an effect $\boldsymbol{j}_*$, the restricted model can be formed with $\mathbb{K}\sum\{H_{\boldsymbol{j}} : \boldsymbol{j} \in \mathcal{J}, \boldsymbol{j} \neq \boldsymbol{j}_*\} = \mathbb{K}(H_{\mathcal{J}} - H_{\boldsymbol{j}*})$. In $\mathbb{K}M_{\mathcal{J}}$, based on contrasts, the same restricted model can be formed by omitting the columns of $M_{\boldsymbol{j}*}$ from $M_{\mathcal{J}}$. Both require additional computation. While omitting columns seems simpler, it is not clear which is more efficient.

This is worth emphasizing, because it does not seem to be widely recognized. In a model comprising the ANOVA effects listed in $\mathcal{J}$, the correct restricted model for testing $\text{H}_0 : H_{\boldsymbol{j}*}\boldsymbol{\eta} = \mathbf{0}$ is $\text{sp}(\mathbb{K}H_{\mathcal{J}\setminus\boldsymbol{j}*})$, which is the same as $\text{sp}(\mathbb{K}M_{\mathcal{J}\setminus\boldsymbol{j}*})$ (based on contrasts) formed by deleting the columns of $M_{\boldsymbol{j}*}$ from $M_{\mathcal{J}}$. Either way, the RMFM SS is the same, and, by Proposition 10.3, it tests exclusively the estimable part of $H_{\boldsymbol{j}*}\boldsymbol{\eta}$. See LaMotte (2023).

16.6 Testing Main Effects in the Saturated Two-Factor Model.

The two-factor model has been written about and discussed widely in this context. Particularly, the question how to test for A main effects in an unbalanced model that also permits B main effects and AB interaction effects, has been controversial, going back at least as far as Yates's (1934) seminal paper. Some have argued vehemently that A main effects should not be examined at all when the model permits interaction effects. It was the perceived difficulty of this question that, in part, prompted SAS to invent Type III tests.

The question, how, and even whether, to test A main effects when the model permits AB interaction effects, is surprisingly controversial. It is clear that the cell means can have both main effects and interaction effects, and so whether there are main effects is a legitimate question, not obviated by the existence of interaction effects. However, Nelder (1977) abjures such tests because they violate a principle he calls *marginality*. Kempthorne (1975, p. 483) (who famously disagreed with Nelder's 1977 presentation) concurs colorfully, that "testing of main effects in the presence of interaction, without additional input, is an exercise in fatuity."

In terms of contrast matrices, the full model matrix is

$$\mathbb{K}M = \mathbb{K}(M_{00}, M_{10}, M_{01}, M_{11}). \tag{16.13}$$

For testing $\mathrm{H}_0 : H_{10}\boldsymbol{\eta} = \mathbf{0}$, the restricted model matrix is $\mathbb{K}M_0 = \mathbb{K}(M_{00}, M_{01}, M_{11})$. The RMFM SS is the difference between SSEs for these two models,

$$SS_{10} = \boldsymbol{y}'(\mathbf{P}_{\mathbb{K}M} - \mathbf{P}_{\mathbb{K}M_0})\boldsymbol{y}. \tag{16.14}$$

By (10.3), its ncp is 0 iff the estimable part of $H_{10}\boldsymbol{\eta} = M_{10}\boldsymbol{\beta}_{10}$ is $\mathbf{0}$. If there are no empty cells, then $\mathrm{sp}(M'\mathbb{K}') = \mathrm{sp}(M')$. Note that

$$\begin{aligned} H_{10}M &= (S_a \otimes U_b)[(\mathbf{1}_a \otimes \mathbf{1}_b), (M_a \otimes \mathbf{1}_b), (\mathbf{1}_a \otimes M_b), (M_a \otimes M_b)] \\ &= (\mathbf{0}, M_a \otimes \mathbf{1}_b, 0, 0), \end{aligned} \tag{16.15}$$

because $(S_a \otimes U_b)(M_a \otimes \mathbf{1}_b) = M_a \otimes \mathbf{1}_b$, and the other products are 0. Then $(\mathbf{0}, M_{10}, 0, 0)' = M'H_{10}$ is in $\mathrm{sp}(M') = \mathrm{sp}(M'\mathbb{K}')$, and hence all of $H_{10}\boldsymbol{\eta} = M_{10}\boldsymbol{\beta}_{10}$ is estimable. In this case, with no empty cells, SS_{10} tests exclusively that $H_{10}\boldsymbol{\eta} = \mathbf{0}$, i.e., that there are no A main effects.

If there are empty cells, then, depending on the model, it can happen that not all of $M_{10}\boldsymbol{\beta}_{10}$ is estimable. One empty cell in a saturated model, for example, reduces A degrees of freedom from $a-1$ to $a-2$. Still, the RMFM SS, SS_{10}, tests its estimable part and nothing else.

16.7 The 'Usual' Conditions

The containment relations among components of dummy-variable models cause the relation $E_{\mathcal{J}}\boldsymbol{\beta} = \boldsymbol{\eta}$, given $\boldsymbol{\eta}$, to have many solutions for $\boldsymbol{\beta}$. A long-standing practice is to impose conditions on $\boldsymbol{\beta}$ to identify a single solution. In some textbook accounts, these conditions are included as part of the definition of the coefficients.

In the two-factor dummy-variable model with $\mathcal{J} = \{00, 10, 01, 11\}$, the usual conditions are $\sum_\ell \alpha_\ell = \sum_\ell \beta_\ell = \sum_\ell (\alpha\beta)_{\ell j} = \sum_\ell (\alpha\beta)_{i\ell} = 0$ for all i, j. In terms of $\boldsymbol{\beta}$, these are $\mathbf{1}_a'\boldsymbol{\beta}_{10} = 0$, $\mathbf{1}_b'\boldsymbol{\beta}_{01} = 0$, $(\mathbf{1}_a \otimes \mathrm{I}_b)'\boldsymbol{\beta}_{11} = \mathbf{0}$, and $(\mathrm{I}_a \otimes \mathbf{1}_b)'\boldsymbol{\beta}_{11} = \mathbf{0}$. Let M_a and M_b be matrices of contrasts such that $\mathrm{sp}(M_a) = \mathrm{sp}(\mathbf{1}_a)^\perp$ and $\mathrm{sp}(M_b) = \mathrm{sp}(\mathbf{1}_b)^\perp$. These conditions imply, respectively, that $\boldsymbol{\beta}_{10} = M_a\boldsymbol{\beta}_{10*}$, $\boldsymbol{\beta}_{01} = M_b\boldsymbol{\beta}_{01*}$, and $\boldsymbol{\beta}_{11} = (M_a \otimes M_b)\boldsymbol{\beta}_{11*}$. Then

$$E_{\mathcal{J}}\boldsymbol{\beta} = (\mathbf{1}_a \otimes \mathbf{1}_b, M_a \otimes \mathbf{1}_b, \mathbf{1}_a \otimes M_b, M_a \otimes M_b)\boldsymbol{\beta}_* = M_{\mathcal{J}}\boldsymbol{\beta}_*, \tag{16.16}$$

effectively reparameterizing the dummy-variable model for $\boldsymbol{\eta}$ in terms of matrices of contrasts.

It is particularly noteworthy that this transformation managed to transform the respective components from $E_{\boldsymbol{j}}$ to $M_{\boldsymbol{j}}$, and, in the process, transformed their meanings from sets of ANOVA effects to individual ANOVA effects. At the extreme, $\mathrm{sp}(E_{11}) = \Re^{ab}$ is transformed to $\mathrm{sp}(S_a \otimes S_b)$, corresponding exactly to $\boldsymbol{j} = 11$, the AB interaction effect.

These conditions are so well known among ANOVA aficionados that they are called the "usual" or "customary" side conditions. It is not altogether clear, though, what form they would take in other settings and models. For example, with three factors and a model that includes A, B, C, AC, and BC effects ($\mathcal{J} = \{000, 100, 010, 001, 101, 011\}$), what conditions can be imposed to transform the model from $E_{\mathcal{J}}\boldsymbol{\beta}$ to $M_{\mathcal{J}}\boldsymbol{\beta}_*$? One might guess, just by similarity, that the conditions might be $\mathbf{1}_a'\boldsymbol{\beta}_{100} = \mathbf{1}_b'\boldsymbol{\beta}_{010} = \mathbf{1}_c'\boldsymbol{\beta}_{001} = 0$, $(\mathrm{I}_a \otimes \mathbf{1}_c)'\boldsymbol{\beta}_{101} = \mathbf{0}$, $(\mathbf{1}_a \otimes \mathrm{I}_c)'\boldsymbol{\beta}_{101} = \mathbf{0}$, $(\mathrm{I}_b \otimes \mathbf{1}_c)'\boldsymbol{\beta}_{011} = \mathbf{0}$, and $(\mathbf{1}_b \otimes \mathrm{I}_c)'\boldsymbol{\beta}_{011} = \mathbf{0}$.

Such a guess is less apparent if the model comprises (1), A, AC, and BC effects, for $\mathcal{J} = \{000, 100, 011, 101\}$. This model includes AC and BC while it does not include B or C. More generally, $\mathcal{J} \neq \bar{\mathcal{J}}$. Some authors have asserted that such models are improper and should never be considered. If formulated in terms of the eponymous dummy-variable terms, the model includes the listed effects, but it also includes all those contained in them, those in $\bar{\mathcal{J}}$. If the intent is to include only effects in $\mathcal{J}$, then the model can be formulated directly as $\mathrm{sp}(H_{\mathcal{J}})$ or, in terms of contrast matrices, as $\mathrm{sp}(M_{\mathcal{J}})$. As an (unnecessary) exercise, one could formulate conditions on the coefficients of the dummy-variable model that transform it into the intended model. Or, if $\bar{\mathcal{J}}$ is in fact the intended model, then conditions can be found on the coefficients in the dummy-variable model $E_{\mathcal{J}}\boldsymbol{\beta}$ that convert it to the equivalent model $\mathrm{sp}(M_{\bar{\mathcal{J}}})$.

As an exercise, it can be shown that, for f factors and any $\boldsymbol{j} \in \mathcal{B}^f$, a matrix $N_{\boldsymbol{j}}$ can be found such that $\mathrm{sp}(E_{\boldsymbol{j}}N_{\boldsymbol{j}}) = \mathrm{sp}(H_{\boldsymbol{j}})$. With $C_{\boldsymbol{j}}$ such that $\mathrm{sp}(C_{\boldsymbol{j}}) = \mathrm{sp}(N_{\boldsymbol{j}})^\perp$, then, $\{E_{\boldsymbol{j}}\boldsymbol{\beta}_{\boldsymbol{j}} : C_{\boldsymbol{j}}'\boldsymbol{\beta}_{\boldsymbol{j}} = \mathbf{0}\} = \mathrm{sp}(E_{\boldsymbol{j}}N_{\boldsymbol{j}}) = \mathrm{sp}(H_{\boldsymbol{j}})$. There always are conditions that transform terms in the dummy-variable model into ANOVA effects. Such transformations generally have some estimable part, and so they reduce the model. The model can be left unreduced, though, by imposing conditions separately on $\boldsymbol{\beta}_{\boldsymbol{j}}$ for each ANOVA effect contained in $\boldsymbol{j}$.

Formulating conditions that transform dummy-variable terms into ANOVA effects is always possible, if that is, for some reason, a useful goal.

If the objective is to express the model for $\boldsymbol{\eta}$ in terms of contrasts, as $\boldsymbol{\eta} = E_{\mathcal{J}}\boldsymbol{\beta} = M_{\mathcal{J}}\boldsymbol{\beta}_*$, there is, of course, no need to go through the exercise of imposing conditions on $\boldsymbol{\beta}$. Instead, just formulate the model directly.

16.8 Factor-Effects Models

Define a ***factor-effects model*** for $\boldsymbol{\eta}$ as a linear subspace that is spanned by a sum of distinct $H_{\boldsymbol{j}}$ matrices. Any coding scheme in which columns of M_a are contrasts has the property that for any list $\boldsymbol{j}_1, \ldots, \boldsymbol{j}_t$ of distinct effect designators, $M = (M_{\boldsymbol{j}_1}, \ldots, M_{\boldsymbol{j}_t})$ gives a factor-effects model that includes exactly the ANOVA effects corresponding to $\boldsymbol{j}_1, \ldots, \boldsymbol{j}_t$.

When all 2^f $M_{\boldsymbol{j}}$ matrices are concatenated to form the matrix M for the model sp(M) for the vector of cell means $\boldsymbol{\eta}$, the model is $\Re^{a*}$, and it includes all possible factor effects. Depending on the coding scheme, including some, but not all, $M_{\boldsymbol{j}}$s in the model may not produce a factor effects model at all. This may be seen, for example, with reference-level coding.

For coding schemes defined by $M_{\boldsymbol{j}}$ based on matrices like those in Table 16.1, it can be seen that the model sp$(M_{\mathcal{J}})$ is a factor effects model if $\mathcal{J} = \bar{\mathcal{J}}$, that is, if $\mathcal{J}$ includes all effects contained in members of $\mathcal{J}$.

16.9 What Does SS_P Test?

In an orderly world, if we are interested in inference about, say, $G'\boldsymbol{\beta}$ in the model $X\boldsymbol{\beta}$ for the mean vector of the response $\boldsymbol{Y}$, we could get a numerator SS, $SS_P = \boldsymbol{y}'P\boldsymbol{y}$, for an F-statistic via an RMFM SS, and we would know that its ncp is 0 iff the estimable part of $G'\boldsymbol{\beta}$ is $\mathbf{0}$. However, if $G'\boldsymbol{\beta}$ is not estimable, more work still would be required to deduce what its estimable part is.

Using statistical computing packages, most often what we see in output are sums of squares, their names, and the effects that they are associated with, like those in Tables 2.6–2.8, pp. 15–16. There, in the model with one empty cell, both Type III and RMFM SSs nominally test A main effects. But they are different, and apparently they test different sets of contrasts. In fact, there is only one estimable A main effects contrast, the RMFM SS tests it, and so it is clear that the Type III SS does not test exclusively A main effects.

The purpose of this section is to illustrate steps that can be taken to deduce the set of linear functions, $G'\boldsymbol{\beta}$, that a given sum of squares, $SS_P = \boldsymbol{y}'P\boldsymbol{y}$, tests in a given model $X\boldsymbol{\beta}$. If the symmetric idempotent matrix P is known, then the ncp of SS_P is 0 iff $PX\boldsymbol{\beta} = \mathbf{0}$. That is, SS_P tests exclusively $\mathrm{H}_0 : G'\boldsymbol{\beta} = \mathbf{0}$ for any matrix G such that $\mathrm{sp}(G) = \mathrm{sp}(X'P)$.

The subclass numbers in the example with one empty cell in Table 2.6, p. 15, are

$$(n_{ij}) = \begin{pmatrix} 1 & 2 & 3 \\ 3 & 1 & 2 \\ 3 & 2 & 0 \end{pmatrix}.$$

Factor A main effects contrasts in this design take the form $\boldsymbol{c}'\boldsymbol{\eta} = (c_1\mathbf{1}_3', c_2\mathbf{1}_3', c_3\mathbf{1}_3')\boldsymbol{\eta}$ with $c_1 + c_2 + c_3 = 0$.

The dummy-variable model that includes an intercept and A, B, and AB effects is $\mathbb{K}M\boldsymbol{\beta}$, with $M = (E_{00}, E_{10}, E_{01}, E_{11})$. The Type III SS for A main effects takes the form $SS_{3A} = \boldsymbol{y}'P_{3A}\boldsymbol{y}$. Based on the formulations in Chapter 17, the matrix Q_{3A} can be computed such that $P_{3A} = Q_{3A}Q_{3A}'$. Then SS_{3A} tests exclusively $G_{3A}'\boldsymbol{\beta} = (M'\mathbb{K}'Q_{3A})'\boldsymbol{\beta} = (\mathbf{P}_M\mathbb{K}'Q_{3A})'\boldsymbol{\eta}$, where $\boldsymbol{\eta} = M\boldsymbol{\beta}$. In this case $\mathbf{P}_M = \mathrm{I}$. See Table 16.2 for the numerical results. Columns of G_{3A} are rescaled to integers in $G_{3A*} = (\boldsymbol{g}_{1*}, \boldsymbol{g}_{2*})$. The (rescaled) column $\boldsymbol{c}_1$ spans $\mathrm{sp}(H_{10}) \cap \mathrm{sp}(G_{3A})$: it is the one A main effects contrast tested by the Type III A SS in this design. To reexpress $\mathrm{sp}(G_{3A})$, GS on $(\boldsymbol{c}_1, G_{3A})$ produces $\boldsymbol{q}_1$ and $\boldsymbol{q}_2$, which become $\boldsymbol{c}_1$ and $\boldsymbol{c}_2$ after rescaling. Then $\mathrm{sp}(G_{3A}) = \mathrm{sp}(\boldsymbol{c}_1, \boldsymbol{c}_2)$.

The Type III A main effects sum of squares in this design tests exclusively $C'\boldsymbol{\eta} = \mathbf{0}$, with $C = (\boldsymbol{c}_1, \boldsymbol{c}_2)$. Note that $\boldsymbol{c}_1'\boldsymbol{\eta} = 3(\bar{\eta}_{1\cdot} - \bar{\eta}_{2\cdot})$ and $\boldsymbol{c}_2'\boldsymbol{\eta} = (\eta_{11} + \eta_{12}) + (\eta_{21} + \eta_{22}) - 2(\eta_{31} + \eta_{32})$. The second is not an A main effects contrast, but it would be if factor B had only two levels.

It can happen that a statistical computing package outputs results (sum of squares, degrees of freedom, F-statistic, p-value) for a specified factor effect, like A main effects, without a clear description of the symmetric, idempotent matrix P by which the SS was computed. SAS's Type III SSs, as a prominent example, are described in SAS's documentation by a set of instructions and examples. Deducing a mathematical formula for P in terms of the design is not altogether simple.

Even in such cases, though, it is always possible to reproduce P by a process called "synthesis," as described in Hartley et al. (1978). This entails successively feeding columns $\boldsymbol{e}_i$ and $\boldsymbol{e}_i + \boldsymbol{e}_j$ from I_n as response vectors through the procedure that produces SS_P. Then P_{ii} is the value of the SS for input $\boldsymbol{e}_i$, and $P_{ii} + P_{jj} + 2P_{ij}$ is the value when $\boldsymbol{e}_i + \boldsymbol{e}_j$ $(i \neq j)$ is input as the response. In this way P can be determined numerically, and from it the steps outlined above can be followed to elucidate the contrasts on the cell means that SS_P tests. In any case, if SS_P can be computed, it is always possible to deduce what it tests. Once G is determined, then column transformations on it might be found that reexpress it in simpler, perhaps even recognizable, form.

TABLE 16.2: Type III contrasts for A main effects for the design with one empty cell in Table 2.6.

$G_{3A} = \mathbb{K}'Q_{3A}$		Rescaled		$\cap H_{10}$	From GS($\boldsymbol{c}_1, G_{3A}$)		Rescaled	
$\boldsymbol{g}_1$	$\boldsymbol{g}_2$	$\boldsymbol{g}_{1*}$	$\boldsymbol{g}_{2*}$	$\boldsymbol{c}_1$	$\boldsymbol{q}_1$	$\boldsymbol{q}_2$	$\boldsymbol{c}_1$	$\boldsymbol{c}_2$
0.6371768	0.1211195	150	1	1	0.4082483	0.2886751	1	1
0.6371768	0.1211195	150	1	1	0.4082483	0.2886751	1	1
0.4630151	-0.2422389	109	-2	1	0.4082483	0	1	0
-0.2888535	0.6055974	-68	5	-1	-0.408248	0.2886751	-1	1
-0.2888535	0.6055974	-68	5	-1	-0.408248	0.2886751	-1	1
-0.4630151	0.2422389	-109	2	-1	-0.408248	0	-1	0
-0.3483233	-0.7267168	-82	-6	0	0	-0.57735	0	-2
-0.3483233	-0.7267168	-82	-6	0	0	-0.57735	0	-2
0.0000000	0.0000000	0	0	0	0	0	0	0

16.10 Covariates

Models for the mean vector $\boldsymbol{\mu}$ may include possible effects of covariates, such as an n-vector $\boldsymbol{x}_1$. It may appear as a linear effect with a term like $\boldsymbol{x}_1\beta_1$. Or its coefficient may be allowed to differ with FLCs of categorical factors in the model. Let $\mathbb{K}_1 = \text{diag}(\boldsymbol{x}_1)\mathbb{K}$, and let $\boldsymbol{\eta}_1$ denote the a_*-vector of coefficients of its columns. In general, the coefficient may follow a factor-effects model in the FLCs, in the same way that the cell means in $\boldsymbol{\eta}$ do, which can be defined by a list of the form $\mathcal{J}_1 = \{\boldsymbol{j}_{11}, \ldots, \boldsymbol{j}_{1t_1}\}$. For example, if $\mathcal{J}_1 = \{0\}$, then $\boldsymbol{\eta}_1 \in \text{sp}(H_0) = \text{sp}(\mathbf{1}_{a*})$, and so $\boldsymbol{\eta}_1 = \mathbf{1}_{a*}\beta_1$. Then, because $\mathbb{K}_1\mathbf{1}_{a*} = \boldsymbol{x}_1$, in the model for $\boldsymbol{\mu}$, $\mathbb{K}_1\boldsymbol{\eta}_1 = \boldsymbol{x}_1\beta_1$, thus representing a linear term in $\boldsymbol{x}_1$. Any other model in the FLCs is possible. If $\mathcal{J}_1 = \{00, 10, 01\}$, with two factors, for example, then the model for $\boldsymbol{\eta}_1$ is $\text{sp}(H_{00} + H_{10} + H_{01})$, comprising a constant term and additive effects of A and B. The model could be written $\boldsymbol{\eta}_1 = (H_{00} + H_{10} + H_{01})\boldsymbol{\beta}$ or, with dummy variables, as $E_{00}\beta_{00} + E_{10}\boldsymbol{\beta}_{10} + E_{01}\boldsymbol{\beta}_{01}$. It permits the coefficients of $\boldsymbol{x}_1$ to differ with levels of A and B, but additively, so that the effect on the coefficient of a level of A is the same for all levels of B.

Extending this notation and structure to c covariates, $\boldsymbol{x}_1, \ldots, \boldsymbol{x}_c$, and renaming the basic model without covariates with a subscript 0, lists $\mathcal{J}_0, \ldots, \mathcal{J}_c$ specify models for the intercept (without any covariates) and the coefficients of the respective covariates. $\mathbb{K}$ becomes $\mathbb{K}_0 = \mathbb{K} = \text{diag}(\mathbf{1}_n)\mathbb{K}$, and $\mathbb{K}_i = \text{diag}(\boldsymbol{x}_i)\mathbb{K}_0$, $i = 1, \ldots, c$. For example, with two factors and two covariates, $\mathcal{J}_0 = \{00, 10, 01, 11\}$, $\mathcal{J}_1 = \{00, 10\}$, and $\mathcal{J}_2 = \{00\}$ specifies a basic model for the intercept (with respect to $\boldsymbol{x}_1$ and $\boldsymbol{x}_2$) with (1), A, B, and AB effects; a constant coefficient (with respect to FLCs) for $\boldsymbol{x}_2$; and a model so that the coefficient of $\boldsymbol{x}_1$ can differ with levels of A but not B.

A model for the mean vector $\boldsymbol{\mu}$ of $\boldsymbol{y}$ with f factors and c covariates takes the general form

$$\boldsymbol{\mu} = \mathbb{K}_0 M_{\mathcal{J}_0}\boldsymbol{\beta}_0 + \mathbb{K}_1 M_{\mathcal{J}_1}\boldsymbol{\beta}_1 + \cdots + \mathbb{K}_c M_{\mathcal{J}_c}\boldsymbol{\beta}_c. \tag{16.17}$$

Each set $\mathcal{J}_i$ is a subset of $\mathcal{B}^f$. Substituting $H_{\mathcal{J}_i}$ matrices for $M_{\mathcal{J}_i}$ matrices produces the same model (linear subspace) if all the M matrices are based on contrasts.

I think the prevailing attitude about covariates is that it is unusual for them to be linearly dependent. While that clearly is not universally so, usually there is the expectation that the linear subspaces corresponding to the models $0, 1, \ldots, c$ meet only at $\mathbf{0}$. In that case, effects within each of the $c + 1$ models can be addressed separately, because estimability of, say, $G'\boldsymbol{\beta}_i$ is the same in the full model for $\boldsymbol{\mu}$ as it is in the i-th submodel, $\mathbb{K}_i M_{\mathcal{J}_i}\boldsymbol{\beta}_i$ for the i-th covariate.

17

Type III Sums of Squares and Tests

Type III sums of squares were devised to test ANOVA effects in unbalanced designs. They were presented as step-by-step instructions to construct a matrix of estimable functions, which is then used to compute a GLH-style numerator SS. Introducing them, SAS asserted that they were invariant to distributions of subclass numbers and that they produced Yates's MWSM sum of squares in two-factor models with no empty cells. Until recently, none of those properties had been established mathematically. This chapter refers to those results to describe the properties of Type III sums of squares.

17.1 Introduction

Type III estimable functions, hypotheses, and sums of squares came to light in SAS publications in the 1970s, mainly Goodnight (1976) and SAS (1978). The recipes given there and in SAS documentation are detailed, but it is difficult to discern a general algorithm or the rationale behind the construction.

Type III estimable functions are defined in exclusive reference to dummy-variable formulations of multiple linear regression models for factor effects. They address fundamental problems encountered in analysis of variance and known almost since R. A. Fisher first expounded it. In balanced settings, there is practically no disagreement about how main effects and interaction effects should be defined and tested. In unbalanced settings, though, some crucial properties no longer hold. ANOVA sums of squares either are not distributed as proportional to chi-squared random variables or they do not test the same hypotheses as in balanced settings. See Herr (1986) for a historical perspective.

The Type III construction provided answers in situations where before there had been no consensus and certainly no single, ideal, proven answer. Procedures in use at the time had some well-known undesirable properties, one being that the effects they tested differed with different configurations of subclass numbers, and that they were not the same as ANOVA effects in balanced models. Introducing Type III, Goodnight (1976) asserted that Type

III estimable functions "have one major advantage in that they are invariant with respect to the cell frequencies" and that "[w]hen no missing cells exist in a factorial model, Type III SS will coincide with Yate[s]'s weighted squares of means technique."

Type III methods have been criticized in strong words. Milliken and Johnson (1984, p. 185) say that, when there are empty cells, "... we think that the Type III hypotheses are the worst hypotheses to consider ... because there seems to be no reasonable way to interpret them." Venables (2000, p. 12) says, "I was profoundly disappointed when I saw that S-PLUS 4.5 now provides 'Type III' sums of squares as a routine option ... ." The debate on the merits of Type III methodology continues: see Macnaughton (1998), Langsrud (2003), Hector et al. (2010), Venables (2000), and Smith and Cribbie (2014).

The objective in this chapter is to provide a concise mathematical description of Type III SSs and some of their properties. These are detailed in LaMotte (2020) and LaMotte (2024).

17.2 Type III Properties

The setting here is the general GM Model, where the n-variate response $\boldsymbol{Y}$ (realized value $\boldsymbol{y}$) has a mean vector in $\{X\boldsymbol{\beta} : \boldsymbol{\beta} \in \Re^{k+1}\} = \text{sp}(X)$, where X is a fixed, known $n \times (k+1)$ matrix. Assume further that $\text{Var}(\boldsymbol{Y}) = \sigma^2 \text{I}_n$ for some positive scalar parameter σ^2.

An accurate description of Type III SSs is the following from IBM Corporation (2021):

> The Type III sum of squares for an effect F can best be described as the sum of squares for F adjusted for effects that do not contain it, and orthogonal to effects (if any) that contain it.

Consider columns of the model matrix X to be partitioned as $X = (X_0, X_1, X_2)$, corresponding roughly to the target effect (X_1), "effects that do not contain it" (X_0), and "effects (if any) that contain it" (X_2). In the developments in this section, "containment" relations play no role. They are crucially relevant, though, in the dummy-variable models, Section 17.3, for which SAS developed Type III methods (Goodnight, 1976). A formulation of Type III SSs based on this description is developed in LaMotte (2020). An alternative development is shown here.

Define

$$X_{1|0} = (\mathbf{P}_X - \mathbf{P}_{X_0})X_1,$$

which often is called "X_1 adjusted for X_0." In the general developments in this section, consider that $X_{1|0}\boldsymbol{\beta}_1$ is the target effect. That is, the goal is to formulate a SS that tests exclusively that $X_{1|0}\boldsymbol{\beta}_1 = \mathbf{0}$.

Denote (X_0, X_1) jointly by X_{01}, so that $X = (X_{01}, X_2)$. A conventional approach to fitting $\boldsymbol{y}$ to X is to fit $\boldsymbol{y}$ to X_{01} and then to "X_2 adjusted for X_{01}," defined by $X_{2|01} = (\mathbf{P}_X - \mathbf{P}_{X_{01}})X_2$ (as would result from applying GS to X). The model becomes

$$\begin{aligned}\mathrm{sp}(X) &= \mathrm{sp}(X_{01}) + \mathrm{sp}(X_{2|01}) \\ &= \mathrm{sp}(X_{01}) + \mathrm{sp}(X_{2|01}X'_{2|01}) \\ &= \mathrm{sp}(X_{01}) + \mathrm{sp}[(\mathbf{P}_X - \mathbf{P}_{X_{01}})X_2X'_{2|01}] \\ &\subset \mathrm{sp}(X_{01}) + \mathrm{sp}(-\mathbf{P}_{X_{01}}X_2X'_{2|01}) + \mathrm{sp}(X_2X'_{2|01}) \\ &= \mathrm{sp}(X_{01}) + \mathrm{sp}(X_2X'_{2|01}). \qquad (17.1)\end{aligned}$$

Each step in this development can be justified by results already proven, mostly in Chapter 7.

With N_{01} such that $\mathrm{sp}(N_{01}) = \mathrm{sp}(X_{01})^{\perp} \cap \mathrm{sp}(X) = \mathrm{sp}(\mathbf{P}_X - \mathbf{P}_{X_{01}})$, it follows that $\mathrm{sp}(X_2X'_{2|01}) = \mathrm{sp}(X_2X'_2N_{01})$. Define $X_{2*} = X_2X'_2N_{01}$ and $X_* = (X_{01}, X_{2*})$. Then $\mathrm{sp}(X_*) = \mathrm{sp}(X_{01}) + \mathrm{sp}(X_{2*}) \subset \mathrm{sp}(X)$, and it follows that $\mathrm{sp}(X_*) = \mathrm{sp}(X)$.

To show that the sum $\mathrm{sp}(X_{01}) + \mathrm{sp}(X_{2*})$ is direct, let $\boldsymbol{z}$ be a vector in $\mathrm{sp}(X_{01}) \cap \mathrm{sp}(X_{2*})$. Then there exist $\boldsymbol{b}_1$ and $\boldsymbol{b}_2$ such that

$$\boldsymbol{z} = X_{01}\boldsymbol{b}_1 = X_2X'_2N_{01}\boldsymbol{b}_2.$$

This implies that $\boldsymbol{b}'_2N'_{01}X_{01}\boldsymbol{b}_1 = 0 = \boldsymbol{b}'_2N'_{01}X_2X'_2N_{01}\boldsymbol{b}_2$, which in turn implies that $X_2X'_2N_{01}\boldsymbol{b}_2 = \boldsymbol{z} = \mathbf{0}$. Taken together, these arguments establish Proposition 17.1.

Proposition 17.1. $\mathrm{sp}(X) = \mathrm{sp}(X_*) = \mathrm{sp}(X_0, X_1) \oplus \mathrm{sp}(X_{2*})$.

In addition, because $\mathrm{sp}(X_0, X_1) \cap \mathrm{sp}(X_{2*}) = \{\mathbf{0}\}$, estimability of linear functions of $\boldsymbol{\beta}_0$ and $\boldsymbol{\beta}_1$ is determined entirely by (X_0, X_1), by Exercise 14, p. 74. Furthermore, because $\mathrm{sp}(X_0, X_1) = \mathrm{sp}(X_0) \oplus \mathrm{sp}(X_{1|0})$, $X_{1|0}\boldsymbol{\beta}_1$ is estimable in the model X_*.

The conventional way to test $X_{1|0}\boldsymbol{\beta}_1$ in this model is as additional SSE due to deleting X_1 from the model $X_* = (X_0, X_1, X_{2*})$. That is, the numerator SS would be $SS_3 = \boldsymbol{y}'P_3\boldsymbol{y}$, with

$$P_3 = \mathbf{P}_X - \mathbf{P}_{(X_0, X_{2*})}. \qquad (17.2)$$

These are general definitions of Type III* SS and df = $\mathrm{tr}(P_3)$. The asterisk indicates that it cannot be confirmed that SS_3 is mathematically identical to the Type III SS that SAS produces in its `proc glm` in all possible models and settings. I have tried many examples designed to show a difference, in 2- and 3-factor unbalanced settings with and without empty cells, with and without covariates, and the results have all agreed numerically to at least ten digits. There was a difference with a nested design in which the number of nested levels differed within nesting levels.

The properties shown here are for Type III* SSs. Detailed derivations and proofs are given in LaMotte (2020, 2024).

P_3 can be computed in several ways. One way is in two steps, with the Gram-Schmidt (GS) construction. From GS on (X_0, X_1, X), take N_{01} as the columns in the orthonormal spanning set contributed by X after (X_0, X_1). Compute X_{2*}, then compute Q_3 as the columns in the orthonormal spanning set from GS on (X_0, X_{2*}, X) contributed by X after (X_0, X_{2*}). Then $\mathbf{P}_3 = Q_3 Q_3'$, and its df is the number of columns in Q_3.

Proposition 17.2 establishes what the Type III* SS tests in this general setting, the model defined by X_*. As it is an RMFM SS, Proposition 17.2 follows directly from Proposition 10.3, p. 85, and the fact established above that $X_{1|0}\boldsymbol{\beta}_1$ is estimable in the model X_*. However, see also the proof on p. 237.

Proposition 17.2. *The Type III* SS tests exclusively* $H_{03} : X_{1|0}\boldsymbol{\beta}_1 = \mathbf{0}$ *in the model* $\text{sp}(X_0, X_1, X_{2*})$.

In this general setting, SAS's Type II SS for $X_1\boldsymbol{\beta}_1$ is $\boldsymbol{y}'\mathbf{P}_{X_{1|0}}\boldsymbol{y}$. $X_{1|0}\boldsymbol{\beta}_1$ is estimable in the model $\text{sp}(X_0, X_1)$, corresponding to Type II, and it is estimable in the model $\text{sp}(X_0, X_1, X_{2*})$, corresponding to Type III. Consequently, the dfs for the Type II SS and the Type III SS are the same, namely $\text{tr}(\mathbf{P}_{X_{1|0}})$.

17.3 Type III in Dummy-Variable Models

The Type III construction is specific to ANOVA models built on dummy variables, in the general form $\text{sp}(X_0, X_1, X_2) = \text{sp}(\mathbb{K}E)$, where E is formed by concatenating matrices $E_{\jmath}$ column-wise over some set $\mathcal{J}$ of f-tuples, as described in Section 16.3.

The original exposition of Type III (SAS (1978), for example) is in terms of estimable functions of $\boldsymbol{\eta} = E\boldsymbol{\beta}$ in the model $\boldsymbol{\mu} = \mathbb{K}(E\boldsymbol{\beta})$. These are functions $\boldsymbol{q}'E\boldsymbol{\beta}$ with $E'\boldsymbol{q} \in \text{sp}(E'\mathbb{K}')$. It is clear that $\text{sp}(E'\mathbb{K}') = \text{sp}(E'\mathbb{K}_0')$, where $\mathbb{K}_0$ is defined as $\mathbb{K}$ would be if each positive cell sample size $n_{\boldsymbol{\ell}}$ were replaced by 1. Then $\mathbb{K}_0$ has exactly one 1 in each row and at most one 1 in each column. This construction establishes a basic property of Type III, that its set of estimable functions depends only on the pattern of empty cells, and it does not depend otherwise on the distribution of cell sample sizes.

The set of Type III estimable functions for an effect depends also on the list of effects included in the model, of course; depending on it, it can happen that the estimability of some effects is not affected by empty cells. So-called "connected" designs with additive-effects models provide one well-known example.

The pivotal role of the target effect is central and clear when numerator SSs are constructed as RMFM (Section 10.3) or GLH SSs (Section 10.5). That role is not apparent in the Type III construction, and so the connection between the SS and the target effect is obscured. Section 17.2 establishes that the Type III SS tests exclusively $X_{1|0}\boldsymbol{\beta}_1 = \mathbf{0}$ in the model $\text{sp}(X_0)\oplus\text{sp}(X_{1|0})\oplus\text{sp}(X_{2*})$. The purpose of this section is to explicate how $X_{1|0}\boldsymbol{\beta}_1$ in the general formulation translates into ANOVA effects in dummy-variables, factor-effects models.

The model matrix is $X = \mathbb{K}E$. The Type III partition of it as $X = (X_0, X_1, X_2)$ is dictated by the target effect, say $\boldsymbol{j}_* \in \mathcal{J}$, and containment. Thus $\mathcal{J}_0 = \{\boldsymbol{j} \in \mathcal{J} : \boldsymbol{j} \not\geq \boldsymbol{j}_*\}$, $\mathcal{J}_1 = \{\boldsymbol{j}_*\}$, and $\mathcal{J}_2 = \{\boldsymbol{j} \in \mathcal{J} : \boldsymbol{j} > \boldsymbol{j}_*\}$. Then E_k is formed by concatenating the columns of $\{E_{\boldsymbol{j}} : \boldsymbol{j} \in \mathcal{J}_k\}$, and then $X_k = \mathbb{K}E_k$, $k = 0, 1, 2$.

Given $X = (X_0, X_1, X_2)$, consider $\boldsymbol{\beta}$ partitioned correspondingly as $\boldsymbol{\beta} = (\boldsymbol{\beta}_0', \boldsymbol{\beta}_1', \boldsymbol{\beta}_2')'$. Let $E_{1|0} = (\text{I} - \mathbf{P}_{E_0})E_1$ and $E_{2*} = E_2E_2'\mathbb{K}'N_{01}$, where as above $\text{sp}(N_{01}) = \text{sp}(X_0, X_1)^{\perp}$. Note that $\text{sp}(E_0, E_1) = \text{sp}(E_0, E_{1|0})$. Let $E_* = (E_0, E_{1|0}, E_{2*})$. Reexpress the model $\text{sp}(X)$ as $\text{sp}(\mathbb{K}E_*)$. Think of this as (X_0, X_1, X_{2*}), but keep in mind that $X_1 = \mathbb{K}E_{1|0}$ is not the same as $X_{1|0} = (\text{I} - \mathbf{P}_{X_0})X_1$.

The Type III SS for effect $\boldsymbol{j}_*$ is defined by (17.2). The effects that it tests exclusively are the linear functions of $\boldsymbol{\delta}_3$, which is

$$\begin{aligned}
\boldsymbol{\delta}_3 &= \mathbf{P}_3X\boldsymbol{\beta} && (17.3)\\
&= (\mathbf{P}_{\mathbb{K}(E_0,E_{1|0},E_{2*})} - \mathbf{P}_{\mathbb{K}(E_0,E_{2*})})(\mathbb{K}E_0\boldsymbol{\beta}_0 + \mathbb{K}E_{1|0}\boldsymbol{\beta}_1 + \mathbb{K}E_{2*}\boldsymbol{\beta}_{2*})\\
&= \mathbb{K}E_{1|0}\boldsymbol{\beta}_1 - \mathbf{P}_{\mathbb{K}(E_0,E_{2*})}\mathbb{K}E_{1|0}\boldsymbol{\beta}_1. && (17.4)
\end{aligned}$$

Proposition 17.3. *Let R be a matrix such that $R'E_{1|0}\boldsymbol{\beta}_1$ is estimable in the model $\mathbb{K}E_*$. Then there exists a matrix L such that $R'E_{1|0}\boldsymbol{\beta}_1 = L'\boldsymbol{\delta}_3$. That is, in the model $\mathbb{K}E_*$, estimable linear functions of $E_{1|0}\boldsymbol{\beta}_1$ are linear functions of $\boldsymbol{\delta}_3$.*

By Proposition 17.2, SS_3 tests exclusively that $X_{1|0}\boldsymbol{\beta}_1 = (\mathbb{K}E_1|\mathbb{K}E_0)\boldsymbol{\beta}_1 = \mathbf{0}$. By Proposition 17.3, SS_3 tests all the estimable functions of $E_{1|0}\boldsymbol{\beta}_1$. If not all linear functions of $E_{1|0}\boldsymbol{\beta}_1$ are estimable, then there may be other linear functions of $\boldsymbol{\beta}$ that SS_3 tests too. If $E_{1|0}\boldsymbol{\beta}_1$ is estimable ($R = \text{I}$), then $\boldsymbol{\delta}_3 = \mathbf{0}$ implies that $E_{1|0}\boldsymbol{\beta}_1 = \mathbf{0}$; and, by (17.4), $E_{1|0}\boldsymbol{\beta}_1 = \mathbf{0}$ implies that $\boldsymbol{\delta}_3 = \mathbf{0}$. In that case, SS_3 tests exclusively that $E_{1|0}\boldsymbol{\beta}_1 = \mathbf{0}$.

The model for the cell means is $\boldsymbol{\eta} = (E_0, E_{1|0}, E_{2*})\boldsymbol{\beta}$. To see the connection between $E_{1|0}\boldsymbol{\beta}_1$ and ANOVA effects, note that

$$\begin{aligned}
\mathbf{P}_{E_{1|0}} &= \mathbf{P}_{(E_0,E_1)} - \mathbf{P}_{E_0}\\
&= \sum\{H_{\boldsymbol{j}} : \boldsymbol{j} \in \bar{\mathcal{J}}_0 \cup \bar{\mathcal{J}}_1\} - \sum\{H_{\boldsymbol{j}} : \boldsymbol{j} \in \bar{\mathcal{J}}_0\}\\
&= \sum\{H_{\boldsymbol{j}} : \boldsymbol{j} \in \mathcal{J}_*\}, && (17.5)
\end{aligned}$$

where $\mathcal{J}_* = \bar{\mathcal{J}}_1\backslash\bar{\mathcal{J}}_0$ is the set of $\boldsymbol{j}$s contained in at least one member of $\mathcal{J}_1$ and not contained in any member of $\mathcal{J}_0$. Let $H_* = \mathbf{P}_{E_{1|0}}$. $H_*\boldsymbol{\eta}$ is a sum of ANOVA effects.

Proposition 17.4. *$H_*E_0 = 0$, $H_*E_{1|0} = E_{1|0}$, and $H_*E_{2*} = 0$.*

Now it follows that $E_{1|0}\boldsymbol{\beta}_1 = H_*(E_0, E_{1|0}, E_{2*})\boldsymbol{\beta} = H_*\boldsymbol{\eta}$ in the model for $\boldsymbol{\eta}$. If $H_*\boldsymbol{\eta}$ is estimable, then the Type III test, with $\boldsymbol{j}_*$ effects as the nominal target, tests exclusively that $H_*\boldsymbol{\eta} = \mathbf{0}$. Otherwise, it tests the estimable part of $H_*\boldsymbol{\eta}$ plus some other effects, up to Type II degrees of freedom, as detailed in Section 17.4.

Consider two-factor models, for example. The model specified by $\mathcal{J} = \{00, 10, 01, 11\}$ is, in terms of dummy variables, $\text{sp}(E_{00}, E_{10}, E_{01}, E_{11})$. With $\boldsymbol{j}_* = 10$ for A main effects as the target, $\mathcal{J}_0 = \{00, 01\}$, $\mathcal{J}_1 = \{10\}$, and $\mathcal{J}_2 = \{11\}$. Then $\mathcal{J}_* = \bar{\mathcal{J}}_1 \backslash \bar{\mathcal{J}}_0 = \{00, 10\} \backslash \{00, 01\} = \{10\}$. In this case, $H_* = H_{10}$, and $H_*\boldsymbol{\eta}$ comprises exactly all A main effects contrasts. Consider another model with $\mathcal{J} = \{00, 10, 11\}$ and $\boldsymbol{j}_* = 11$. For it, $\mathcal{J}_* = \{00, 10, 01, 11\} \backslash \{00, 10\} = \{01, 11\}$, and $H_* = H_{01} + H_{11} = \text{I}_a \otimes S_b$. The effects $H_*\boldsymbol{\eta}$ are sums of B effects and AB effects, which sometimes are called collectively "B within A" effects.

A conjecture is that the only ANOVA effects included among Type III $\boldsymbol{j}_*$ effects are the estimable part of $H_*\boldsymbol{\eta}$. If so, then the other included effects (contrasts), if any, are not ANOVA effects.

In general, with $\boldsymbol{j}$ as the target effect, Type III tests, say, $G'_{\boldsymbol{j}}\boldsymbol{\eta}$. Another conjecture is that, if $\boldsymbol{j}_1 \neq \boldsymbol{j}_2$, then $\text{sp}(G_{\boldsymbol{j}_1}) \cap \text{sp}(G_{\boldsymbol{j}_2}) = \{\mathbf{0}\}$. This is certainly true if each $G_{\boldsymbol{j}} = \text{sp}(H_{\boldsymbol{j}})$ for all $\boldsymbol{j} \succ \mathbf{0}_f$, which would hold if all effects were estimable. In that case the $G_{\boldsymbol{j}}$s would partition all possible contrasts on $\boldsymbol{\eta}$. The question is whether, or under what conditions, the functions tested by Type III do the same.

Consider a two-factor setting with both factors at five levels, but with positive n_{ij}s only for $i = 1, 2, 3$ and $j = 1, 2, 3$ and for $i = 4, 5$ and $j = 4, 5$. Together these comprise disconnected 3×3 and 2×2 designs. It can be seen that the only estimable ANOVA effects are five AB interaction contrasts. RMFM SSs would test only the five interaction contrasts. The Type III bonus contrasts for main effects pool the SSs for the two sub-designs, so that they test $2 + 1 = 3$ df for each main effect plus $2 \times 2 + 1 \times 1 = 5$ df for interaction effects. In this case, it can be demonstrated that $\text{sp}(G_{10})$, $\text{sp}(G_{01})$, and $\text{sp}(G_{11})$ meet only at $\mathbf{0}$. Taken together, the Type III A, B, and AB SSs test 11 of the possible 12 degrees of freedom for differences among the means of the 13 non-empty cells.

17.4 Type II and Type III Degrees of Freedom

It was noted above that Type II and Type III degrees of freedom are the same, say $\nu_2 = \nu_3$. The Type II full model is $\text{sp}[\mathbb{K}(E_0, E_1)]$. It can be represented in two ways, as

$$\begin{aligned} \text{sp}[\mathbb{K}(E_0, E_1)] &= \text{sp}(\mathbb{K}E_0) \oplus \text{sp}(\mathbb{K}E_1|\mathbb{K}E_0), \text{ and as} \\ &= \text{sp}(\mathbb{K}E_0) + \text{sp}(\mathbb{K}E_{1|0}), \end{aligned}$$

the second because $\text{sp}(E_0, E_1) = \text{sp}(E_0) \oplus \text{sp}(E_{1|0})$, but the direct sum may not carry over.

Dimensions of these linear subspaces are dfs, which are the same as the ranks of the matrices that generate them. Let $\nu_{01} = \dim \text{sp}[\mathbb{K}(E_0, E_1)]$, $\nu_0 = \dim \text{sp}(\mathbb{K}E_0)$, $\nu_2 = \nu_3 = \dim \text{sp}(\mathbb{K}E_1|\mathbb{K}E_0)$, and $\nu_{1|0} = \dim \text{sp}(\mathbb{K}E_{1|0})$. Let $\nu_* = \dim \text{sp}(E_{1|0})$: it is the *innate* df of the effect. Note that $\nu_* \geqslant \nu_{1|0}$. The effect's df is ν_* if it is entirely estimable. For main effects of factor A at a levels, for example, $\nu_* = a - 1$.

Dimensions of direct sums of linear subspaces are the sums of their respective dimensions; and dimensions of sums of linear subspaces are not greater than the sums of their respective dimensions. It follows that

$$\begin{aligned} \nu_{01} &= \nu_0 + \nu_3 \\ &\leqslant \nu_0 + \nu_{1|0}, \end{aligned}$$

and hence that $\nu_3 \leqslant \nu_{1|0} \leqslant \nu_*$.

Let ν_{*0} denote the dimension of the estimable part of $E_{1|0}\boldsymbol{\beta}_1$. Because the Type III SS tests this estimable part, it follows that $\nu_{*0} \leqslant \nu_3$. If all of $E_{1|0}\boldsymbol{\beta}_1$ is estimable, then $\nu_{*0} = \nu_*$, which implies also that $\nu_{*0} = \nu_3 = \nu_{1|0} = \nu_*$. We showed above a stronger result, that in this case Type III SS tests exclusively that $H_*\boldsymbol{\eta} = E_{1|0}\boldsymbol{\beta}_1 = \mathbf{0}$, which implies that $\nu_3 = \nu_*$.

When testing the same effect in a given setting (characterized by $\mathbb{K}$) and model (characterized by E), Type II and Type III degrees of freedom are equal. Within the inequalities just shown, practically any relation is possible. It is not unusual to see, for example, that none of the effect is estimable ($\nu_{*0} = 0$) and that $\nu_3 = \nu_*$. Often, too, not all of the effect is estimable ($\nu_{*0} < \nu_*$) but $\nu_3 = \nu_*$, that is, the Type III degrees of freedom is the same as if all of the effect were estimable.

17.5 What About SAS Type IV SS?

Conventional wisdom about SAS's four types of SSs seems to be that Type II works for models with proportional subclass numbers, Type III for unbalanced models with no empty cells, and that SAS intended Type IV for models with empty cells. SAS asserts, without proof, that the Type III and IV SSs are the same when there are no empty cells.

It has been established here that Type III SSs always test the estimable part of the target effect, and that they implicitly identify contrasts to make up the difference, up to Type II df. Those extra contrasts seem to be quite reasonable, even appealing, in specific examples. Is Type IV really needed?

Consider a 3×3 ANOVA setting with only the $(3, 3)$ cell empty. For A main effects, Type III tests the one estimable A contrast, $\boldsymbol{c}_A = (1, 1, 1, -1, -1, -1, 0, 0, 0)'$, along with $\boldsymbol{c}_{A32} = (1, 1, 0, 1, 1, 0, -2, -2, 0)'$, which is not an A contrast. Type IV tests $\boldsymbol{c}_{A41} = (1, 1, 0, 0, 0, 0, -1, -1, 0)'$ and $\boldsymbol{c}_{A42} = (0, 0, 0, 1, 1, 0, -1, -1, 0)'$. Clearly $\boldsymbol{c}_A$ is not in the column space of $\boldsymbol{c}_{A41}$ and $\boldsymbol{c}_{A42}$. However, $\boldsymbol{c}_{A32} = \boldsymbol{c}_{A41} + \boldsymbol{c}_{A42}$, so Type III and Type IV share one df in common, but it is not the one estimable A contrast.

What the Type IV SS tests has nothing to do with A main effects in this setting. It doesn't satisfy the basic requirement, that it should at least test the estimable part of the target effect. There can be A main effects and the Type IV SS ncp equal to zero, and the ncp can be positive while there are no A main effects. Its results should not be attributed to any effect, and particularly not to the target effect. For these reasons, in my opinion, Type IV should be deprecated.

17.6 Descriptions and Implementations of Type III

Since Type III methods appeared, and as they became widely used, textbooks, journal articles, and online publications undertook to describe them. Several sources describe Type III SSs as deleted-variables extra SSEs when models are formulated by imposing the "usual non-estimable conditions" or (equivalently) with FLCs coded with contrast matrices. See Section 16.7, p. 156. Venables (2000, pp. 15–17) gives an ironic version of this:

> The irony is, of course, that Type III sums of squares were available all along if only people understood what they really were and how to get them. … *Provided* you have used a contrast matrix … they will be unique, and they are none other than the notorious 'Type III sums of squares.' … Well don't look at me like that – you didn't really expect me to tell them how to do it the easy way, did you?

Taken in its full context, this seems to dismiss Type III as being no more than RMFM SSs with contrast coding. Neither that author, nor any other such sources I have seen, gave any basis for this assertion. It seems, therefore, to have been based on observations of examples in which Type III SSs and RMFM SSs from contrasts were numerically the same. That would not have been observed if the target ANOVA effects had not been estimable. So the empirical evidence might have been limited to designs with no empty cells, and perhaps only saturated models. This is not a broad base for such a sweeping assertion – which is in fact false unless the target effect is estimable.

A cursory internet search comes up with many attempts to describe Type III SSs and methods. The great majority make the same assertion as just quoted above, that Type III SSs are computed as extra SSEs due to deleting an effect's terms in the model when effects are coded by contrasts. Several give instructions to obtain Type III SSs in the R `lm` package, by such statements as `options(contrasts = c("contr.sum","contr.poly"))` or with similar statements in the **Anova** function in the `car` package. By the fact that both Type III and RMFM SSs based on contrasts test exclusively the target ANOVA effects when they are estimable, it is clear that they will coincide in these package functions in the same way. However, when the target effect is not estimable, they will not.

The R package **emmeans** (Lenth, 2025, p. 53) includes a function `joint_tests` that produces results for tests that "correspond to 'type III' tests a la **SAS** [*sic*]." Apparently, it uses RMFM SSs with contrast coding and then creates contrasts that fill in the rest of the available df for differences among not-empty-cell means. Results are the same when the target effect is estimable, but they can differ considerably from what SAS Type III tests otherwise. Examples show that the extra contrasts tested with this function change with subclass numbers. This is a comprehensive and versatile package, but it does not produce Type III SSs when the target effect is not estimable.

17.7 Discussion and Summary

When SAS's Type III methods were introduced, some core results established here were not widely known. It was not known that the RMFM SS tests exclusively the estimable part of the hypothesized effect (Proposition 10.3), and it was not clear that every given set of ANOVA effects could be modeled exactly with contrast coding (Proposition 16.1). The GLH form of numerator SS was more widely used than the RMFM form, because the RMFM form required formulating and fitting a restricted model. Definitions of ANOVA effects were not universally agreed-upon, and so it was not clear how to formulate the GLH form. Had these been known and standard, then a natural way to test a given ANOVA effect would have been as an RMFM SS after deleting that effect's terms from the contrast-coded model or as a GLH SS starting with $H_{\mathfrak{J}} M_{\mathcal{J}} \hat{\boldsymbol{\beta}}$.

In the state of knowledge and practice at that time, SAS's (unproven) assertion that the Type III SS for A main effects agreed with Yates's MWSM SS in a two-factor model with no empty cells was enough to convince practitioners (and funding sources and regulatory agencies) that using Type III was acceptable. As such, it provided a unified solution to what had been problems with no consensus solutions.

No other properties of Type III SSs were established mathematically, in large part because there was no explicit mathematical formula for them. What became attributed to them was based only on anecdotes and experience. Based on the formulation shown here, it has been established that they test the estimable part of the target ANOVA effect plus other effects up to Type II degrees of freedom.

From today's perspective, testing the estimable part of the target effect was possible with the state of knowledge in 1976, if definitions of effects had been widely agreed-upon. So, that leaves the lagniappe contrasts that Type III tests as its only novel contribution.

It is appropriate to question whether Type III SSs should be considered a general form along with the RMFM and GLH forms of numerator SSs. Like RMFM and GLH, Type III SSs test the estimable part of the target effect. RMFM SSs test exclusively the estimable part of the target effect. Type III and GLH SSs both test additional, unasked-for effects if the target effect is not estimable. All three test exclusively the target effect if it is estimable, and in that case all three SSs are identical, by Exercise 17, p. 99.

My limited experience with multiple configurations, like those in Table 18.1, p. 179, does not suggest whether Type III or GLH identifies better lagniappe contrasts when the target effect is only partly estimable. For configuration N_{24}, for example, with an empty cell in the first row, only two degrees of freedom for A main effects are estimable. Arranging the coefficients in an $a \times b$ matrix, GLH and Type III test those plus one additional contrast each,

$$\begin{pmatrix} 0 & 9 & 0 \\ -4 & 5 & -4 \\ 8 & -19 & 8 \\ -4 & 5 & -4 \end{pmatrix} \text{(GLH) and} \begin{pmatrix} 0 & 3 & 3 \\ 0 & -1 & -1 \\ 0 & -1 & -1 \\ 0 & -1 & -1 \end{pmatrix} \text{(III)}, \tag{17.6}$$

respectively. In my opinion, the Type III contrast is more appealing because it is like an A effect contrast ignoring the first level of B.

Configuration N_{14} in Table 18.1 permits no estimable A main effects, and so both GLH and Type III substitute two pairs of contrasts,

$$\begin{pmatrix} 0 & 1 & 2 \\ -1 & 0 & -2 \\ 1 & -1 & 0 \end{pmatrix}, \begin{pmatrix} 0 & 1 & 2 \\ 5 & 0 & -2 \\ -5 & -1 & 0 \end{pmatrix} \quad \text{(GLH), and}$$

$$\begin{pmatrix} 0 & 1 & 2 \\ -1 & 0 & -2 \\ 1 & -1 & 0 \end{pmatrix}, \begin{pmatrix} 0 & 1 & 0 \\ 1 & 0 & 0 \\ -1 & -1 & 0 \end{pmatrix} \quad \text{(III)}. \tag{17.7}$$

They share one df. In my opinion, there is little basis to prefer one to the other. Taken together with tests substituting for B main effects and AB interaction effects, both forms parse the available 5 df for treatment effects into a panel of 2, 2, and 1 degrees of freedom. RMFM, on the other hand, produces only one contrast, the one estimable degree of freedom for interaction effects.

If the research objective is to assess a particular set of effects, then only those effects need to be tested. There is no need for tests of an organized panel of effects, like those produced by a complete ANOVA-effect partition. Indeed, even considering other kinds of contrasts than the prespecified research hypothesis can give the appearance of significance-shopping. If the target effects are estimable, then all three schemes – RMFM, GLH, and Type III – give identical results. If not, then only the RMFM SS is guaranteed to test exclusively the estimable part.

Often, though, the objective of analysis is not focused on just one or a few pre-specified hypotheses. Rather it is in part, even mostly, exploratory, seeking a broad view of how the cell means might differ. Choosing effects to be included – model-building – often is also an objective. For that purpose a good panel of effects can provide an organized, parsimonious, and familiar structure in trying to assess and categorize differences. The partition of ANOVA effects has traditionally served that purpose. Again, if all ANOVA effects are estimable, then the three methods are equivalent. If some are not, then all three capture the estimable parts, but GLH and Type III provide a broader perspective than RMFM.

The Type III method for sums of squares and tests has been controversial since it was introduced. I think some adverse opinions have been due in part to the fact that it was not clearly and explicitly defined; that none of its properties were known except anecdotally, and none were established mathematically; and that it was not introduced first in a high-visibility, peer-reviewed statistical journal. In addition, though, I think that much of the skepticism about the method has been due to the perception and resentment of SAS's hegemony in statistical practice.

In the process of establishing Type III's properties, my long-standing skepticism changed to admiration. Whether it was sharp insight or a little bit of good luck, what SAS created not only provided a good general tool, it also happened to provide a method that always tests the estimable part of the target effect, and it fills in the remaining degrees of freedom with contrasts that seem to be relevant and reasonable.

Still, it would be misleading to include these lagniappe contrasts under the name of the target effect: those that are produced for A main effects are not in fact A main effect contrasts. It would be more accurate to call them *Type III* effects, as in "Type III A main effects." Even then, though, the set of contrasts tested depends on the pattern of empty cells and its effect on the set of estimable functions. The problem remains that "Type III A main effects" means different things in different settings. One resolution, in uses where one is desirable, would be to describe explicitly the set of contrasts tested and to distinguish those that are part of the target effect and those that are not.

17.8 Exercises

The setting for these exercises is the dummy-variable formulation $\boldsymbol{\mu} = \mathbb{K}\boldsymbol{\eta} = \mathbb{K}E_{\mathcal{J}}\boldsymbol{\beta}$ – the setting for which Type III methods were devised.

See Section 14.2 for definitions of notation, mainly $\mathcal{L}$, $\boldsymbol{\ell} \in \mathcal{L}$, $n_{\boldsymbol{\ell}}$, and $\boldsymbol{j} \in \mathcal{B}^f$ for f factors. The model for the cell means $\boldsymbol{\eta}$ comprises the set $\mathcal{J}$ of effects in $\mathcal{B}^f$.

Denote the Type III SS for an effect $\boldsymbol{j}_* \in \mathcal{J}$ by $SS_{III\boldsymbol{j}*} = \boldsymbol{y}'P_3\boldsymbol{y}$.

1. The ncp of $SS_{III\boldsymbol{j}*}$ is proportional to $\delta_3^2 = \boldsymbol{\delta}_3'\boldsymbol{\delta}_3$ formulated in Equation 17.4. It tests exclusively that $\boldsymbol{\beta}$ is in $\{\boldsymbol{\beta} \in \Re^{k+1} : \delta_3^2(\boldsymbol{\beta}) = 0\}$. Introducing it, Goodnight (1976) asserted that what it tested did not depend on the subclass numbers, which we may interpret as meaning that what it tests is the same if all positive $n_{\boldsymbol{\ell}}$s are replaced by 1. Try to formulate an argument that proves this assertion.

2. Goodnight (1976) asserted also that, when used to compute SS_A for A main effects in a saturated two-factor model with no empty cells, the Type III SS is the same as Yates's MWSM SS. Using the formulation in Chapter 19, prove this assertion.

3. Compute the SSs for A main effects shown in Tables 2.7 (excluding Type IV) and 2.8, p. 16. The two GLH SSs are based on two least-squares solutions, $TQ'\boldsymbol{y}$ from GS on X and $(X'X)^+X'\boldsymbol{y}$. $(X'X)^+$ is the Moore-Penrose pseudoinverse of $X'X$: it is what SAS proc iml returns as ginv$(X'X)$.

18

ANOVA Exercises and Projects

This chapter provides exercises that establish further properties of ANOVA SSs and procedures. In addition, it provides two major projects that involve specific examples and considerable computations.

18.1 Exercises

For each $\boldsymbol{j} \in \mathcal{B}^f$, define

$$ST_{\boldsymbol{j}} = \mathbf{P}_{\mathbb{K}E_{\boldsymbol{j}}} - \mathbf{P}_{\mathbf{1}_n} \tag{18.1}$$

and

$$Q_{\boldsymbol{j}} = ST_{\boldsymbol{j}} - \sum\{Q_{\boldsymbol{i}} : 0 < \boldsymbol{i} < \boldsymbol{j}\}. \tag{18.2}$$

("ST" is intended to suggest "subtotal.")

For cells $\boldsymbol{\ell} \in \mathcal{L}$, define cell and marginal totals by

$$\begin{aligned} T_{\boldsymbol{\ell}} &= \begin{cases} \sum_{s=1}^{n_{\boldsymbol{\ell}}} y_{\boldsymbol{\ell},s} \text{ if } n_{\boldsymbol{\ell}} > 0, \\ 0 \text{ if } n_{\boldsymbol{\ell}} = 0, \text{ and} \end{cases} \\ T_{\boldsymbol{\ell}_{\boldsymbol{j}}} &= \sum\{T_{\boldsymbol{\ell}} : \ell_k = 1, \ldots, a_k \forall k \text{ such that } j_k = 0\} \\ &= n_{\boldsymbol{\ell}_{\boldsymbol{j}}} \bar{y}_{\boldsymbol{\ell}_{\boldsymbol{j}}}. \end{aligned} \tag{18.3}$$

(The notation $\boldsymbol{\ell}_{\boldsymbol{j}}$ is defined on p. 139.) Define the grand total $T = T.$ to be the sum of all responses $y_{\boldsymbol{\ell},s}$. Define the *correction factor* to be $CF = T^2/n$.

Assume that rows of $\mathbb{K}$ and $\boldsymbol{y}$ are in lexicographic order on the subscripts $\boldsymbol{\ell}, s$, where $\boldsymbol{\ell}$ ranges over $\mathcal{L}$ and, within each cell $\boldsymbol{\ell}$, $s = 1, \ldots, n_{\boldsymbol{\ell}}$. There are no row entries in either $\boldsymbol{y}$ or $\mathbb{K}$ for empty cells. However, if $n_{\boldsymbol{\ell}} = 0$, the $\boldsymbol{\ell}$-th column of $\mathbb{K}$ is $\mathbf{0}_n$: the mean vector does not depend on $\eta_{\boldsymbol{\ell}}$ at all, but a column for it is still included in $\mathbb{K}$.

1. The purpose of this exercise is to illustrate that, for balanced two-factor models, computations of ANOVA SSs are simple, that each boils down to a straightforward sum of squared deviations from the average.

   For a balanced model with $f = 2$ factors A and B at a and b levels and m observations per FLC, verify the following expressions.

(a) "Variance between classes of type A" is

$$\begin{aligned} SS_A &:= \sum_{i,j,s=1}^{a,b,m} (\bar{y}_{i\cdot} - \bar{y})^2 \\ &= (bm) \sum_{i=1}^{a} (\bar{y}_{i\cdot} - \bar{y})^2 \\ &= \frac{1}{bm} \sum_{i=1}^{a} T_{i\cdot}^2 - CF, \end{aligned} \tag{18.4}$$

(b) "variance between classes of type B" is

$$\begin{aligned} SS_B &:= \sum_{i,j,s=1}^{a,b,m} (\bar{y}_{\cdot j} - \bar{y})^2 \\ &= (am) \sum_{j=1}^{b} (\bar{y}_{\cdot j} - \bar{y})^2 \\ &= \frac{1}{am} \sum_{j=1}^{b} T_{\cdot j}^2 - CF, \end{aligned} \tag{18.5}$$

(c) and "total variance" between subclasses is

$$\begin{aligned} SS_{A,B} &:= \sum_{i,j,s=1}^{a,b,m} (\bar{y}_{ij} - \bar{y})^2 \\ &= m \sum_{i=1}^{a} \sum_{j=1}^{b} (\bar{y}_{ij} - \bar{y})^2 \\ &= \frac{1}{m} \sum_{i=1}^{a} \sum_{j=1}^{b} T_{ij}^2 - CF. \end{aligned} \tag{18.6}$$

(d) Show that "the balance of total variance" is

$$\begin{aligned} SS_{AB} &:= SS_{A,B} - SS_A - SS_B \\ &= m \sum_{i=1}^{a} \sum_{j=1}^{b} (\bar{y}_{ij} - \bar{y}_{i\cdot} - \bar{y}_{\cdot j} + \bar{y})^2. \end{aligned} \tag{18.7}$$

2. The purpose of this exercise is to show that all the ANOVA computations required for balanced models reduce to repeated applications of SOSs, sums of squared deviations from the average. With that one basic computation, repeated many times, any SS to test any ANOVA effect could be had. This made ANOVA accessible, even by hand calculations. Some measures were used, too, to make that easier, such as centering and scaling to reduce data to mostly small positive integers.

   For a balanced model with f factors and m observations per FLC, verify the following items.

(a) $\mathbb{K} = \mathrm{I}_{a*} \otimes \mathbf{1}_m$ (perhaps after permuting rows), and $\mathbf{P}_{\mathbb{K}} = \mathrm{I}_{a*} \otimes U_m$.
(b) $\mathbf{P}_{\mathbb{K}}\boldsymbol{y} = \bar{\boldsymbol{y}} \otimes \mathbf{1}_m$, where $\bar{\boldsymbol{y}} = (1/m)\mathbb{K}'\boldsymbol{y}$ is the a_*-vector of cell means.
(c) $\mathbb{K}E_{\boldsymbol{j}} = E_{\boldsymbol{j}} \otimes \mathbf{1}_m$.
(d) $\mathbf{P}_{\mathbb{K}E_{\boldsymbol{j}}} = \mathbf{P}_{E_{\boldsymbol{j}}} \otimes U_m = \sum\{H_{\boldsymbol{i}} \otimes U_m : \boldsymbol{i} \in \mathcal{B}^f \text{ and } \boldsymbol{i} \preceq \boldsymbol{j}\}$.
(e) $ST_{\boldsymbol{j}} = \sum\{H_{\boldsymbol{i}} \otimes U_m : 0 < \boldsymbol{i} \preceq \boldsymbol{j}\}$.
(f) $Q_{\boldsymbol{j}} = H_{\boldsymbol{j}} \otimes U_m$.
(g) $\boldsymbol{y}'Q_{\boldsymbol{j}}\boldsymbol{y} = m\bar{\boldsymbol{y}}'H_{\boldsymbol{j}}\bar{\boldsymbol{y}}$.
(h) For any full model specified by ANOVA effects in $\mathcal{J}$, if $\boldsymbol{j} \in \mathcal{J}$, then the restricted model for $\mathrm{H}_0 : H_{\boldsymbol{j}}\boldsymbol{\eta} = \mathbf{0}$ is $\mathcal{J}\backslash\boldsymbol{j}$, and the RMFM numerator SS is $\boldsymbol{y}'(H_{\boldsymbol{j}} \otimes U_m)\boldsymbol{y} = m\bar{\boldsymbol{y}}'H_{\boldsymbol{j}}\bar{\boldsymbol{y}}$ with $\mathrm{tr}(H_{\boldsymbol{j}})$ degrees of freedom. The ncp of this quadratic form is $m\boldsymbol{\eta}'H_{\boldsymbol{j}}\boldsymbol{\eta}/\sigma^2$, and it is 0 iff $H_{\boldsymbol{j}}\boldsymbol{\eta} = \mathbf{0}$, that is, iff there are no $\boldsymbol{j}$ effects.

The basic statistics required are $\bar{\boldsymbol{y}}$ and $SSE = SSW = \boldsymbol{y}'(\mathrm{I}_{a*} \otimes S_m)\boldsymbol{y}$. From these, the numerator SS for ANOVA effect $\boldsymbol{j}$ is $SS_{\boldsymbol{j}} = m\bar{\boldsymbol{y}}'H_{\boldsymbol{j}}\bar{\boldsymbol{y}}$. The same SS can be had as $m\bar{\boldsymbol{y}}'Q_{\boldsymbol{j}}\bar{\boldsymbol{y}}$, where $Q_{\boldsymbol{j}}$, (18.2), is computed from successive subtotal sums of squares, $\boldsymbol{y}'ST_{\boldsymbol{j}}\boldsymbol{y}$, all of which are SOSs.

3. Compute ANOVA SSs ($SS_A, SS_B, SS_{A,B}, SS_{AB}, SS_W$) for the balanced data in Table 2.6, p. 15. In order to work only with positive integers, start by multiplying each datum by 10 and then subtracting 200. I suggest that you use a spreadsheet, which makes it easy to look at alternative formulations.

4. For a general setting with subclass numbers $(n_{\boldsymbol{\ell}})$, show that

$$\mathbf{P}_{\mathbb{K}}\boldsymbol{y} = \mathbb{K}\bar{\boldsymbol{y}},$$

where $\bar{\boldsymbol{y}}$ is the a_*-vector of *sample cell means* with entries $\bar{y}_{\boldsymbol{\ell}}$, and for each FLC $\boldsymbol{\ell}$ in $\mathcal{L}$, $\bar{y}_{\boldsymbol{\ell}} = 0$ if $n_{\boldsymbol{\ell}} = 0$, and $\bar{y}_{\boldsymbol{\ell}} = (1/n_{\boldsymbol{\ell}})\sum_{s=1}^{n_{\boldsymbol{\ell}}} y_{\boldsymbol{\ell},s}$ if $n_{\boldsymbol{\ell}} > 0$. This is equivalent to defining $\bar{\boldsymbol{y}}$ to be $N^+\mathbb{K}'\boldsymbol{y}$, where $N = \mathbb{K}'\mathbb{K} = \mathrm{diag}(n_{\boldsymbol{\ell}})$ and $N^+ = \mathrm{diag}\begin{pmatrix} 1/n_{\boldsymbol{\ell}} \text{ if } n_{\boldsymbol{\ell}} > 0 \\ 0 \text{ if } n_{\boldsymbol{\ell}} = 0 \end{pmatrix}$.

5. Show that, for $\boldsymbol{j} \in \mathcal{B}^f$,

$$\begin{aligned}
\mathbf{P}_{\mathbb{K}E_{\boldsymbol{j}}}\mathbf{1}_n &= \mathbf{1}_n, \\
\mathbf{P}_{\mathbb{K}E_{\boldsymbol{j}}}\boldsymbol{y} &= \{\bar{y}_{\boldsymbol{\ell}_{\boldsymbol{j}}} : \boldsymbol{\ell} \in \mathcal{L}, s = 1, \ldots, n_{\boldsymbol{\ell}}\}, \text{ or} \\
&= \{\bar{y}_{\boldsymbol{\ell}_{\boldsymbol{j}}}\mathbf{1}_{n_{\boldsymbol{\ell}}} : \boldsymbol{\ell} \in \mathcal{L} \text{ and } n_{\boldsymbol{\ell}} > 0\} \\
&\quad \text{when rows are in lexicographic order on } \boldsymbol{\ell}, s, \\
U_n\mathbf{P}_{\mathbb{K}E_{\boldsymbol{j}}}\boldsymbol{y} &= \bar{y}\mathbf{1}_n, \\
ST_{\boldsymbol{j}} &= \mathbf{P}_{\mathbb{K}E_{\boldsymbol{j}}}S_n\mathbf{P}_{\mathbb{K}E_{\boldsymbol{j}}}, \text{ and} \\
\boldsymbol{y}'ST_{\boldsymbol{j}}\boldsymbol{y} &= \sum\{n_{\boldsymbol{\ell}}(\bar{y}_{\boldsymbol{\ell}_{\boldsymbol{j}}} - \bar{y})^2 : \boldsymbol{\ell} \in \mathcal{L}\} \\
&= \sum_{\boldsymbol{\ell}_{\boldsymbol{j}}}\left\{\frac{T_{\boldsymbol{\ell}_{\boldsymbol{j}}}^2}{n_{\boldsymbol{\ell}_{\boldsymbol{j}}}} : n_{\boldsymbol{\ell}_{\boldsymbol{j}}} > 0\right\} - \frac{T^2}{n}. \qquad (18.8)
\end{aligned}$$

6. Prove (16.9).
7. Prove: In the model $\mathbb{K}H_{\mathcal{J}}\boldsymbol{\beta}$, for any $\boldsymbol{j} \in \mathcal{J}$ the estimable part of $H_{\boldsymbol{j}}\boldsymbol{\beta}$ is $\text{sp}(H_{\boldsymbol{j}}\mathbb{K}')$. That is,

$$\text{sp}(H_{\mathcal{J}}\mathbb{K}') \cap \text{sp}(H_{\boldsymbol{j}}) = \text{sp}(H_{\boldsymbol{j}}\mathbb{K}'). \tag{18.9}$$

8. Show an example of subclass numbers (n_{ij}) for two factors such that Q_{11} (18.2) is idempotent, and show another such that it is not. Verify both numerically.
9. Show that, for each $\boldsymbol{j} \in \mathcal{B}^f$, $\mathbf{P}_{\mathbb{K}E_j}\boldsymbol{y} = (\bar{y}_{\boldsymbol{\ell}_j})$.
10. Show that $\mathbf{P}_{\mathbb{K}E_j} - \mathbf{P}_{\mathbf{1}} = \mathbf{P}_{\mathbb{K}E_j} S_n \mathbf{P}_{\mathbb{K}E_j}$.
11. If the design is balanced, so that $\mathbb{K}$ may be considered to be $\mathrm{I}_{a*} \otimes \mathbf{1}_m$, show that

$$\mathbf{P}_{\mathbb{K}E_j} = \mathbf{P}_{E_j \otimes U_m} = \sum\{H_{\boldsymbol{i}} \otimes U_m : \boldsymbol{i} \preceq \boldsymbol{j}\}. \tag{18.10}$$

12. For the same balanced design, show also that, for every $\boldsymbol{y} \in \Re^n$,

$$SS_{\boldsymbol{j}} = \boldsymbol{y}'Q_{\boldsymbol{j}}\boldsymbol{y} = m\bar{\boldsymbol{y}}'H_{\boldsymbol{j}}\bar{\boldsymbol{y}}. \tag{18.11}$$

18.2 What SSs Test

The purpose of these exercises is to investigate and compare the properties of three types of numerator SSs to test A main effects in four two-factor models and eight sets of subclass numbers. They can be repeated for B and AB effects and then to compare properties of whole ANOVA tables across types of SSs, models, and subclass numbers.

Table 18.1 shows the eight sets of subclass numbers. The models are:

1. (1), A.
2. (1), A, B.
3. (1), A, B, AB.
4. (1), A, B, AB, x.

The models can all be represented in the form $\text{sp}(\mathbb{K}M_m) = \{\mathbb{K}M_m\boldsymbol{\beta} : \boldsymbol{\beta} \in \Re^{c_m}\}$ ($m = 1, 2, 3$, and c_m is the number of columns in M_m) or $\text{sp}(\mathbb{K}M_4, \boldsymbol{x})$ for model $m = 4$. In each, the model for the ab-vector of cell means is $\boldsymbol{\eta} = M_m\boldsymbol{\beta} \in \text{sp}(M_m)$. Except for Type III SSs, M_m can be defined in any way that produces the correct linear subspace of effects.

TABLE 18.1: Eight sets of subclass numbers, $N_{1,1}$–$N_{2,4}$, and values of a covariate x. In each set, entries are n_{ij}, $i = 1, \ldots, a$, $j = 1, \ldots, b$, arranged in rows (i) and columns (j). Use as many column-wise consecutive values of the covariate x as needed. $N_{1,1}$ is balanced and $N_{2,1}$ has proportional subclass numbers (psn, p. 143). All others are not psn. $N_{1,3}$, $N_{1,4}$, and $N_{2,4}$ have empty cells. $N_{2,3}$ has three balanced rows.

Eight sets of subclass numbers, n_{ij}												Use these for x				
2	2	2	1	2	3	0	2	3	0	2	3	23	94	82	15	45
2	2	2	3	1	2	3	1	2	3	0	2	49	97	63	70	97
2	2	2	3	2	1	3	2	1	3	2	0	43	46	31	18	67
6	4	4	1	2	3	2	2	2	0	2	3	41	28	59	77	77
3	2	2	3	1	2	3	3	3	3	1	2	23	33	85	42	79
3	2	2	3	1	2	1	1	1	3	1	2	72	22	1	68	32
			3	2	1	3	2	1	3	2	1	77	44	7	20	33

The three types of numerator SSs are RMFM ($t = 1$), the GLH form ($t = 2$), and Type III ($t = 3$), which is defined below. For consistency, agree that the least-squares solution on which GLH is based is $TQ'\boldsymbol{y}$ from GS on X, that is, $A'\boldsymbol{y}$ with $A' = TQ'$. Type t SS for model m has the form $SS_{tm} = \boldsymbol{y}'P_{tm}\boldsymbol{y}$. P_{tm} is symmetric and idempotent. This defines $3 \times 4 = 12$ numerator SSs. Define Q_{tm} such that $\text{sp}(Q_{tm}) = \text{sp}(P_{tm})$ and $P_{tm} = Q_{tm}Q'_{tm}$.

Chapter 17 presents Type III SSs. For these exercises, it is important to realize that the Type III construction is based on the dummy-variable formulation for factor effects. In the models here, A main effects are the target effects, which defines $X_1 = \mathbb{K}E_{10}$. The containment relations for dummy variables define the Type III partition of X (either $\mathbb{K}M$ or $(\mathbb{K}M, \boldsymbol{x})$) as $X = (X_0, X_1, X_2)$. See Table 18.2.

For Type III, let N_{01} be such that $\text{sp}(N_{01}) = \text{sp}(X_0, X_1)^{\perp} \cap \text{sp}(X)$, and let $X_{2*} = X_2X_2'N_{01}$. Then $P_{3m} = \mathbf{P}_X - \mathbf{P}_{(X_0, X_{2*})}$. Note that $X_2 = 0$ in models 1 and 2, and so in those models Type III SS is the same as the RMFM SS.

We have established that the RMFM SS tests exclusively the estimable part of $H_{10}\boldsymbol{\eta}$, and that the GLH SS tests the same estimable part plus other contrasts up to Type II degrees of freedom (at most $a - 1$). It is established in Chapter 17 that the Type III SS has this same property. If all of $H_{10}\boldsymbol{\eta}$

TABLE 18.2: Partition of X for Type III, $X = (X_0, X_1, X_2)$. For all models, $X_1 = \mathbb{K}E_{10}$.

Model	X_0	X_2
1	$\mathbb{K}E_{00}$	
2	$\mathbb{K}(E_{00}, E_{01})$	
3	$\mathbb{K}(E_{00}, E_{01})$	$\mathbb{K}E_{11}$
4	$\mathbb{K}(E_{00}, E_{01}), \boldsymbol{x}$	$\mathbb{K}E_{11}$

is estimable in model m, then the three types of SSs for model m all test exclusively that $H_{10}\boldsymbol{\eta} = \mathbf{0}$. In fact, it can be shown that in that case the three SSs for model m are identical.

Proposition 10.4 established that whatever SS_{tm} tests in model m, it also tests in models that model m contains. For example, for subclass numbers N_{12}, it can be verified that SS_{13} (RMFM SS for model 3) tests the two degrees of freedom for A main effects in model 3, and it also tests them in model 2. In model 2, then, both SS_{12} and SS_{13} test exclusively all A main effect contrasts. Proposition 10.2 established that in that case the ncp for SS_{12} is everywhere $\geqslant$ the ncp of SS_{13}. If the true model were model 2, then SS_{12} would be better than SS_{13} even though both test the same contrasts. (The proof of Proposition 10.2 shows a simple way to verify this dominance.)

In some cases, a SS for a lesser model works as well in a greater model. For example, for N_{21} (which is psn) SS_{11} (which is simply SS_A, (15.1)) tests all A main effects both in model 1 and in model 2. It is widely believed that, because the subclass numbers are psn, SS_{11} also tests A main effects in model 3. Verify that in fact it does not test any part of A main effects in model 3.

Properties of these 12 SSs in four models for each of the eight sets of subclass numbers can be addressed systematically. For example, for each N_{ij}, for each type t of SS, tabulate $G_{tm_1m_2}$ for the contrasts $G'_{tm_1m_2}\boldsymbol{\eta}$ that SS_{tm_1} tests in model m_2. See Table 18.3 as an exemplar.

Tabulate degrees of freedom (dimensions) shared by pairs of the 12 SSs (Table 18.4). Note that, for the proportional subclass numbers N_{21}, $P_{11} = P_{12} = P_{21} = P_{22} = P_{31} = P_{32}$, because they all are projection matrices onto the same 2-dimensional linear subspace.

Tabulate the degrees of freedom of the A contrasts tested by each SS in each model (Table 18.5 for subclass numbers N_{21}). For example, SS_{12} (RMFM for model 2) tests both A degrees of freedom in models 1 and 2 but none in models 3 and 4.

The contrasts shown in Table 18.3 were computed as $\mathbf{P}_{M_{m_2}}\mathbb{K}'Q_{tm_1}$ and then rescaled, column by column, to produce integers. Hocking (2013), citing Hocking, Hackney, and Speed (1978), gives formulas for these contrasts for RMFM SSs for models 1 and 2 in model 3, that is, G_{113} and G_{123}. These formulas are rational functions of the subclass numbers, and it is a simple matter to re-scale the coefficients to be integers. They do not necessarily give the same results as $\mathbf{P}_{M_{m_2}}\mathbb{K}'Q_{tm_1}$, but both formulas produce the same linear subspaces, namely $\text{sp}(G_{tm_1m_2})$.

Displayed as in Table 18.3, with each column of G re-arranged as an $a \times b$ array, A contrasts must have equal entries in each row, and each column must sum to 0. It is clear, then, that RMQG12 comprises two A contrasts, while neither contrast in RMQG13 is an A contrast. Keep in mind, though, that it is possible that a linear combination of the two contrasts in RMQG13 could be an A contrast. However, as noted in the caption to Table 18.3, A contrasts in sp(RMQGm_1m_2), if any, were extracted first, which means that if no A contrast is shown, then there is none.

TABLE 18.3: Subclass numbers are N_{21}, which has proportional subclass numbers. $Q_{\text{RMi}}Q'_{\text{RMi}}$ is the matrix of the RMFM SS to test $H_{10}\boldsymbol{\eta}$ in model i. The table shows G_{RMij}, heading RMQGij, such that Q_{RMi} tests $G'_{\text{RMij}}\boldsymbol{\eta}$ in model j. $\text{sp}(G_{\text{RMij}}) = \text{sp}(\mathbf{P}_{M_j}\mathbb{K}'Q_{\text{RMi}})$. Each G is ab ($ab+1$ in model 4) by 2. As displayed, each column of G is arranged in an $a \times b$ array, corresponding to levels of A (rows) and B (columns). Contrasts are separated by horizontal lines. Such Gs are not unique: any G_* with $\text{sp}(G) = \text{sp}(G_*)$ works as well. Contrasts for A main effects were extracted first for these, so contrasts like RMQG13 that are not A contrasts signify that no A contrasts are in sp(G). See also Table 18.5.

RMQG11			RMQG12			RMQG13			RMQG14			
2	2	2	2	2	2	18	12	12	18	12	12	.
-1	-1	-1	-1	-1	-1	-3	-2	-2	-3	-2	-2	.
-1	-1	-1	-1	-1	-1	-15	-10	-10	-15	-10	-10	-402
0	0	0	0	0	0	-3	-2	-2	-3	-2	-2	.
1	1	1	1	1	1	6	4	4	6	4	4	.
-1	-1	-1	-1	-1	-1	-3	-2	-2	-3	-2	-2	-120
RMQG21			RMQG22			RMQG23			RMQG24			
2	2	2	2	2	2	18	12	12	18	12	12	.
-1	-1	-1	-1	-1	-1	-3	-2	-2	-3	-2	-2	.
-1	-1	-1	-1	-1	-1	-15	-10	-10	-15	-10	-10	-402
0	0	0	0	0	0	-3	-2	-2	-3	-2	-2	.
1	1	1	1	1	1	6	4	4	6	4	4	.
-1	-1	-1	-1	-1	-1	-3	-2	-2	-3	-2	-2	-120
RMQG31			RMQG32			RMQG33			RMQG34			
2	2	2	2	2	2	2	2	2	2	2	2	.
-1	-1	-1	-1	-1	-1	-1	-1	-1	-1	-1	-1	.
-1	-1	-1	-1	-1	-1	-1	-1	-1	-1	-1	-1	-68
0	0	0	0	0	0	0	0	0	0	0	0	.
1	1	1	1	1	1	1	1	1	1	1	1	.
-1	-1	-1	-1	-1	-1	-1	-1	-1	-1	-1	-1	-57
RMQG41			RMQG42			RMQG43			RMQG44			
2	2	2	2	2	2	2	2	2	2	2	2	.
-1	-1	-1	-1	-1	-1	-1	-1	-1	-1	-1	-1	.
-1	-1	-1	-1	-1	-1	-1	-1	-1	-1	-1	-1	0.
0	0	0	0	0	0	0	0	0	0	0	0	
1	1	1	1	1	1	1	1	1	1	1	1	.
-1	-1	-1	-1	-1	-1	-1	-1	-1	-1	-1	-1	0.

TABLE 18.4: Subclass numbers are N_{21}. Entries are $\dim[\text{sp}(P_{t_1 m_1}) \cap \text{sp}(P_{t_2 m_2})]$ for the matrices P_{tm} of numerator SSs of type t and model m.

	1	1	1	1	2	2	2	2	3	3	3	3
tm	1	2	3	4	1	2	3	4	1	2	3	4
11	2	2	0	0	2	2	0	0	2	2	0	0
12	2	2	0	0	2	2	0	0	2	2	0	0
13	0	0	2	1	0	0	2	1	0	0	2	1
14	0	0	1	2	0	0	1	2	0	0	1	2
21	2	2	0	0	2	2	0	0	2	2	0	0
22	2	2	0	0	2	2	0	0	2	2	0	0
23	0	0	2	1	0	0	2	1	0	0	2	1
24	0	0	1	2	0	0	1	2	0	0	1	2
31	2	2	0	0	2	2	0	0	2	2	0	0
32	2	2	0	0	2	2	0	0	2	2	0	0
33	0	0	2	1	0	0	2	1	0	0	2	1
34	0	0	1	2	0	0	1	2	0	0	1	2

TABLE 18.5: Subclass numbers are N_{21}. Entries are $\dim[\text{sp}(G_{tm_1 m_2}) \cap \text{sp}(H_{10})]$, the A main effects degrees of freedom that $P_{tm_1} = Q_{tm_1} Q'_{tm_1}$ tests in model m_2. $G_{tm_1 m_2} = \mathbf{P}_{M_{m_2}} \mathbb{K}' Q_{tm_1}$, so $G_{tm_1, m_2 - 1} = \mathbf{P}_{M_{m_2 - 1}} G_{tm_1 m_2}$. Row "0." lists estimable A df by model.

	m_2			
tm_1	1	2	3	4
0.	2	2	2	2
11	2	2	0	0
12	2	2	0	0
13	2	2	2	1
14	2	2	2	2
21	2	2	0	0
22	2	2	0	0
23	2	2	2	1
24	2	2	2	2
31	2	2	0	0
32	2	2	0	0
33	2	2	2	1
34	2	2	2	2

18.3 ANOVA Synthesis

Refer to Tables 2.6–2.8, pp. 15–16. Values of the Type IV SS were computed in SAS's proc glm; they are included here only for illustration and comparison.

All the SSs shown there purport to test A main effects. The differences show that they test different things in different data sets and under different codings of the model matrix. Obvious differences like these have provoked much discussion, particularly about how to define ANOVA effects and how to test them in unbalanced designs, not to mention with empty cells.

Each of the four data sets specifies a different matrix $\mathbb{K}$. Each coding scheme (DV, RL, EC: see Table 16.1, p. 154) is built with a different model matrix M for the vector $\boldsymbol{\eta}$ (which is a $3 \times 3 = 9$-vector). These matrices take the form $M = (M_{00}, M_{10}, M_{01}, M_{11})$, as described on p. 153. The model for the mean vector $\boldsymbol{\mu}$ is then $\mathbb{K}\boldsymbol{\eta} = \mathbb{K}M\boldsymbol{\beta}$, or $X\boldsymbol{\beta}$ with $X = \mathbb{K}M$. For each such M, $\text{sp}(M) = \Re^9$. All three span the same saturated model for $\boldsymbol{\eta}$, which includes an intercept term, A and B main effects, and AB interaction effects.

The target of inference here is A main effects. That there are none is equivalent to the hypothesis $\text{H}_{A0} : C_{10}'\boldsymbol{\eta} = \mathbf{0}$, where

$$C_{10} = \begin{pmatrix} 1 & 0 \\ -1 & 1 \\ 0 & -1 \end{pmatrix} \otimes \mathbf{1}_3.$$

Then $C_{10}'\boldsymbol{\eta} = 3\binom{\bar{\eta}_{1\cdot} - \bar{\eta}_{2\cdot}}{\bar{\eta}_{2\cdot} - \bar{\eta}_{3\cdot}}$, and H_{A0} is equivalent to equality of the three A marginal means. In the parameterization $\boldsymbol{\eta} = M\boldsymbol{\beta}$, this becomes $G'\boldsymbol{\beta}$ with $G = M'C_{10}$. This G was used to compute the results shown here for the two GLH SSs and the RMFM SSs.

SAS's Types I-IV SSs are based on DV models for $\boldsymbol{\eta}$. Type I SS is extra SSE due to deleting $\mathbb{K}E_{10}$ from the model $\mathbb{K}(E_{00}, E_{10})$. Type II SS is extra SSE due to deleting $\mathbb{K}E_{10}$ from the model $\mathbb{K}(E_{00}, E_{10}, E_{01})$. SAS's Type III SSs for A main effects are computed as described in Section 17.3.

GLH SSs take the form $\boldsymbol{y}'\mathbf{P}_{AG}\boldsymbol{y}$, where A is a matrix such that $XA' = \mathbf{P}_X$, as described on p. 86. The two shown here, GLH_1 and GLH_2, use respectively: $A_1' = TQ'\mathbf{P}_X$ with Q and T from GS on X; and $A_2' = (X'X)^+X'$. $(X'X)^+$ is the Moore-Penrose pseudoinverse of $X'X$: for example, it is what SAS proc iml returns for `ginv(`$X'X$`)`.

In each coding scheme, the RMFM SS for A main effects in each data set is computed as described in Section 10.4 for the conditions $G'\boldsymbol{\beta} = \mathbf{0}$.

The DelVars SS is extra SSE due to deleting columns of X that correspond to A main effects. In the model $X\boldsymbol{\beta} = \mathbb{K}(M_{00}, M_{10}, M_{01}, M_{11})\boldsymbol{\beta}$, it is extra SSE due to deleting the columns $\mathbb{K}M_{10}$.

Questions here relate to SAS Types I-III, GLH, RMFM, and DelVars SSs. We know already that Type III, GLH, and RMFM SSs each test the estimable part of $G'\boldsymbol{\beta}$; and if $G'\boldsymbol{\beta}$ is estimable, they test exclusively $G'\boldsymbol{\beta}$. The RMFM

SS tests exclusively the estimable part of $G'\boldsymbol{\beta}$. Type III and GLH SSs test the estimable part and, if the effect is not estimable, they test those together with other contrasts up to at most $a - 1 = 2$ degrees of freedom. About DelVars, we know only that, under contrast coding, it tests what RMFM tests.

The items listed below are intended to provoke discussion and further questions.

1. With these definitions, you should be able to compute all the dfs and SSs shown in Tables 2.7 and 2.8 except for SAS's Type IV.

   The remaining questions address properties that are dependent only on the subclass numbers n_{ij}. They are not specific to the particular responses in Table 2.6.

2. Verify that, for each formulation M, $\text{sp}(M) = \Re^9$ and $\text{sp}(H_{10}) \subset \text{sp}(M)$, that is, each formulation M is saturated and each includes A main effects.

   Show that the estimable part of A main effects is $\text{sp}(\mathbb{K}') \cap \text{sp}(H_{10})$. Within each of the four data sets, identify a minimal spanning set for this subspace.

3. Show that each of Type III, GLH, RMFM, and DelVars SSs takes the form $\boldsymbol{y}'P\boldsymbol{y}$ with $\text{sp}(P) \subset \text{sp}(X)$ in each data set and each coding scheme. Is the same true for Types I and II?

   Verify that this implies that, if two such SSs are different for the same $\boldsymbol{y}$, then they test different functions of $\boldsymbol{\eta}$.

4. Which dfs and SSs would we know to be equal before doing the computations?

5. In the unbalanced data set with one empty cell, under RL coding, verify that the DelVars SS for A main effects tests exclusively $\boldsymbol{c}'\boldsymbol{\eta} = \eta_{13} - \eta_{23}$. Is this an A main effect? That is, is $\boldsymbol{c} \in \text{sp}(H_{10})$?

6. Verify that, if (all) A main effects are estimable, then Type III SS is the same as the DelVars SS when factor-level combinations are coded in terms of contrasts. Verify and substantiate that this is true in general, beyond the specific examples shown here. The examples here illustrate that this assertion is not true if A effects are not estimable.

7. We have shown that GLH, RMFM, and Type III SSs always test the estimable part of the target effect, and that GLH and Type III test that plus more contrasts up to $\text{tr}(H_{10})$ df, which is 2 here. Is that evident here?

8. The RMFM SSs and the DelVars SSs under EC are the same in all the designs shown here. How general is that relation?

9. Verify that, in the unbalanced design with three empty cells, no A main effect contrast is estimable. The Type III SS and the two GLH SSs test something there, but it can't be A main effects contrasts. What do they test?

10. Try to find an explanation why, in the two data sets with empty cells, the GLH SSs are the same under some codings and different under others.

11. For each coding scheme, $\text{sp}(M) = \Re^9$, and so the model for the mean vector could be expressed simply as $\text{sp}(\mathbb{K})$. This could be considered another coding scheme, that is, with $M = \text{I}_3 \otimes \text{I}_3$. The GLH form and the RMFM form would work under it, but DelVars would not make sense. How would the GLH and RMFM SSs and dfs for A main effects compare to the others in the four data sets?

19

Yates's Method of Weighted Squares of Means

The method that F. Yates described to compute a sum of squares for main effects in unbalanced two-factor models that permit interaction effects set the gold standard for all the efforts that followed. However, its form was never extended to more general settings, and so it is not widely used in today's statistical computing packages. The purpose of this chapter is to present it, mainly for its historical significance, but also to formulate it in clear matrix-algebraic terms, and to describe the heuristic behind Yates's approach.

19.1 Introduction

In a seminal and influential paper, F. Yates (1934) described the "method of weighted squares of means" (MWSM) to obtain a numerator sum of squares for testing main effects of factor A in unbalanced models for main effects of factors A and B and their interaction effects. This solved a problem inherent in conventional ANOVA computations and thereby defined main effects SSs unambiguously.

He reasoned that, if $\boldsymbol{U} \sim \mathbf{N}(\mu \mathbf{1}_p, \sigma^2 D)$, with $D = \text{Diag}(1/w_i)$, all $w_i > 0$, then, quoting his equation (A),

$$\begin{aligned} Q &= (p-1)s^2 = w_1(u_1 - \bar{u})^2 + w_2(u_2 - \bar{u})^2 + \cdots \\ &= w_1 u_1^2 + w_2 u_2^2 + \cdots - (w_1 + w_2 + \cdots)\bar{u}^2 \\ &\qquad \text{where } \bar{u} = \frac{w_1 u_1 + w_2 u_2 + \cdots}{w_1 + w_2 + \cdots} \end{aligned} \tag{19.1}$$

"provides an efficient estimate" of $(p-1)\sigma^2$ from the realized value $\boldsymbol{u} = (u_1, \ldots, u_p)'$ of $\boldsymbol{U}$. In matrix terms, Q can be expressed as

$$Q = \boldsymbol{u}'(D^{-1} - D^{-1}\mathbf{1}_p(\mathbf{1}_p' D^{-1}\mathbf{1}_p)^{-1}\mathbf{1}_p' D^{-1})\boldsymbol{u}. \tag{19.2}$$

The MWSM SS for A main effects comes from this expression upon substituting the "marginal means of the subclass means" for u_i, with corresponding substitutions for the diagonal entries of D.

Yates proffered no further rationale. He did not invoke a general approach or set of criteria. For that reason, it is not clear how to develop Q from basics, or how (or whether) it is related to alternatively-developed sums of squares, like RMFM or GLH, or how to extend the MWSM to other settings with multiple factors and covariates.

General methods to compute SSs for factor effects (main effects, interaction effects) were developed after Yates's paper, but they did not extend the MWSM directly. Like the MWSM, they were developed to be "suitable for desk calculation" and to avoid "solving different sets of simultaneous equations for each sum of squares." (Federer and Zelen, 1966). One of the methods described by Paik and Federer (1974) was, like the MWSM, based on the vector of cell sample means, and so it might be regarded as an extension of the MWSM to multiple factors. It provided SSs for non-saturated models (where some effects are excluded) and settings in which some factor-level combinations contain no observations (empty cells). Its scope did not extend to general linear models that might include in addition covariates and factor-by-covariate interaction effects.

The advent and widespread use of statistical computing packages required general-purpose procedures. The MWSM SS became a touchstone for attempts at more general approaches to assess factor effects in general linear models. Statistical packages did not provide it directly, though, because it was not clear how it could be extended beyond two-factor analysis of variance settings. When SAS introduced Type III estimable functions and SSs four decades later, for example, Goodnight (1976) alluded to the possibility of other good approaches, reflecting the thinking at that time that different definitions of effects and SSs could be equally satisfactory:

> Perhaps (and just perhaps) we may someday be able to agree on the estimable functions we want to use in any given situation. If this day ever comes, we can then consolidate the different types of estimable functions (and live happily ever after).

Herr (1986) notes that in an earlier paper Yates (1933) "indicated that [the MWSM] is a least squares procedure when he said that the variance for treatment in [the MWSM] is 'identical with the residual variance when constants representing [B main effects] and [AB interaction effects] are fitted' (p. 118)." In Yates's usage, "residual variance" meant the RMFM SS, the increase in error SS (SSE) upon imposing a set of conditions on a full model. Relating the MWSM SS to least-squares "residual variance" lent it some credibility in 1933. Now, by Proposition 10.3 and Proposition 10.2 and its corollaries, proving this connection would establish that Yates's MWSM SS is optimal in the same ways that the RMFM SS is.

The assertion that the MWSM SS "is a least squares procedure" (an RMFM SS) was not derived or proven in either of the Yates (1933, 1934) papers (Yates (1933, p. 118) says "It can be shown ..."), nor did Herr (1986) give any justification for the assertion. It is widely held, apparently, that this is true. Perhaps this belief was based on direct experience, but direct proofs are hard to find. Anderson and Bancroft (1952, p. 279) say that "[the MWSM] provides exact tests of the main effects when interaction is present," but they give scant basis for that assertion. They cite, among others, Snedecor and Cox (1935, p. 246), who suggest that the MWSM

> ... is especially appropriate if the postulated population has equal subclass numbers. ... [I]f the method is applied to a sample with equal subclass numbers it yields exactly the same results as the standard method for such numbers; but if it is applied to a sample with proportional (but not equal) subclass numbers the results do not coincide with those obtained from the standard method for proportional numbers.

This last statement is puzzling, as it asserts that "exact tests of the main effects when interaction is present" when subclass numbers are proportional but not equal, on the one hand, and "the standard method for proportional numbers," on the other, are inconsistent. Chapter 20 resolves this issue by showing that in fact the numerator SS produced by "the standard method" does not provide "exact tests of the main effects" unless the subclass numbers satisfy more restrictive conditions. It is established here that the MWSM does indeed provide "exact tests of the main effects when interaction is present."

It appears that these sources relied mainly on examples and experience rather than mathematical constructions. However, some relations were established. Searle (1971, p. 371) showed that the MWSM SS tests equality of the A marginal means by showing that its non-centrality parameter is 0 if and only if the marginal means are all equal. That can be deduced from (19.2) upon substituting the population marginal means for $\boldsymbol{u}$. Searle et al. (1981, appendix B) related it directly to least squares by showing "after some tedious algebra" that it could be derived from the GLH numerator SS for testing factor main effects. Searle (1987, p. 90) quoted the MWSM SS directly as shown in Yates (1934) and then justified that the resulting F-statistic "is a test statistic for" the hypothesis of equal A marginal means because, if the marginal means are equal, then the MWSM SS is distributed as proportional to a central chi-squared random variable.

In 1934, a very positive feature of the MWSM SS was that it was an explicit formula that humans could manage. Today, as we have already established in the foregoing chapters, an appropriate numerator SS can be had for any linear hypothesis in any linear model for the mean vector. In models that involve effects of combinations of levels of multiple factors, it is widely thought that the SAS Type III SS (SAS, 1978) is a correct numerator SS for testable hypotheses about factor effects. However, proofs are hard to find, and it is not always clear what a "correct" numerator SS would be.

This issue, whether and how to test for main effects in models that do not exclude interaction effects, continues to generate much discussion. See Searle (1994), Macnaughton (1998), Hector et al. (2010), Langsrud (2003), and Smith and Cribbie (2014). The books by Hocking (2013) and Khuri (2010) give detailed and comprehensive treatments of the topic. A cursory search of the internet will uncover dozens of discussions and lecture notes. Still, there is disagreement and some confusion on several points. I hope that the properties established in these chapters will resolve some and illuminate some. Some of the controversies are doctrinaire, though. It is unlikely that mathematical arguments will resolve those.

19.2 Constructing SSs as Variance Estimates

Yates's equation (A), (19.1) above, illustrates what was then a common approach to construct a numerator SS to test functions $G'\boldsymbol{\beta}$ in ANOVA settings. It was to identify independent linear statistics (like $u_1, \ldots, u_p$) all having the same mean if $G'\boldsymbol{\beta} = \mathbf{0}$, and then to define a weighted sum of squared deviations among them that estimates σ^2. That "[t]his estimate of the variance may be compared with the estimate of variance from the variation within subclasses by means of the z test" (Yates, 1934, p. 56) (the z-statistic was 1-to-1 with the logarithm of the F-statistic) required that the two sums of squares be independent, which followed if the u_is were linear functions of $\hat{\boldsymbol{y}}$.

That heuristic, and hence Yates's MWSM SS Q, is extended here to the general model $\mathrm{E}(\boldsymbol{Y}) = X\boldsymbol{\beta}$ to construct a numerator SS for a set $G'\boldsymbol{\beta}$ of estimable functions. It is shown that this SS is equivalent to the RMFM and GLH SSs. In the next section, it is shown how this plays out specifically in the two-factor ANOVA setting to coincide with Yates's MWSM SS.

That $G'\boldsymbol{\beta}$ is estimable in the model $X\boldsymbol{\beta}$ implies that there exists a matrix H with $\mathrm{sp}(H) \subset \mathrm{sp}(X)$ such that $X'H = G$.

Let A and C denote matrices such that A has linearly independent columns in $\mathrm{sp}(X)$ and $AC = H$. This guarantees that $D = A'A$ is positive-definite (pd) and hence has an inverse, and that $A'\boldsymbol{y} = A'\hat{\boldsymbol{y}}$ is a function of the estimated mean vector. Entries of $\boldsymbol{U} = A'\boldsymbol{Y}$ in this general setting correspond to Yates's marginal means of the subclass means in the two-factor setting he considered.

Matrices A and C satisfying these conditions exist in any case. Clearly, it is possible to choose A such that D is diagonal, or even the identity matrix. That may be a convenient choice in some situations, but in others it might require unnecessary computations. The only requirement here is that D be pd.

With $\mathrm{Var}(A'\boldsymbol{Y}) = D\sigma^2$, let $\boldsymbol{Z} = D^{-1/2}\boldsymbol{U} = D^{-1/2}A'\boldsymbol{Y}$, so that $\mathrm{E}(\boldsymbol{Z}) \in \mathrm{sp}(D^{-1/2}A'X)$ and $\mathrm{Var}(\boldsymbol{z}) = \sigma^2\mathrm{I}_c$. Let M be a matrix such that $\mathrm{sp}(M) = \mathrm{sp}(C)^{\perp}$. With the c columns of A linearly independent and in $\mathrm{sp}(X)$, it follows that $\mathrm{sp}(A'X) = \mathrm{sp}(A') = \Re^c$. [Clearly $\mathrm{sp}(A'X) \subset \mathrm{sp}(A')$;

and $\text{sp}(A') = \text{sp}(A'A) \subset \text{sp}(A'X)$; and $c = \text{rank}(\text{A}) = \text{rank}(\text{A}')$.] Then

$$\begin{aligned}\{A'X\boldsymbol{\beta} : \boldsymbol{\beta} \in \Re^k \text{ and } G'\boldsymbol{\beta} = \mathbf{0}\} &= \{A'X\boldsymbol{\beta} : \boldsymbol{\beta} \in \Re^k \text{ and } C'A'X\boldsymbol{\beta} = \mathbf{0}\} \\ &= \{\boldsymbol{\theta} \in \Re^c : C'\boldsymbol{\theta} = \mathbf{0}\}, \text{ where } \boldsymbol{\theta} = A'X\boldsymbol{\beta}, \\ &= \text{sp}(C)^\perp = \text{sp}(M).\end{aligned}$$

For $\boldsymbol{\beta} \in \text{sp}(G)^\perp$, the model for $\text{E}(\boldsymbol{Z})$ is $\text{sp}(D^{-1/2}M)$: this is the restricted model for $\text{E}(\boldsymbol{Z})$ under H_0. Thus MSE in this null model for $\boldsymbol{Z}$ is an unbiased estimator of σ^2 when $\text{H}_0 : G'\boldsymbol{\beta} = \mathbf{0}$ is true. SSE in this model is

$$\begin{aligned}SSE_{\boldsymbol{z}} &= \boldsymbol{z}'(\text{I} - \mathbf{P}_{D^{-1/2}M})\boldsymbol{z} \\ &= \boldsymbol{u}'(D^{-1} - D^{-1}M(M'D^{-1}M)^{-}M'D^{-1})\boldsymbol{u}. \qquad (19.3)\end{aligned}$$

This corresponds to (19.2) and is equivalent to Q in the setting that Yates (1934) considered, as shown in the next section.

Note further that

$$\begin{aligned}\boldsymbol{z}'(\text{I} - \mathbf{P}_{D^{-1/2}M})\boldsymbol{z} &= \boldsymbol{z}'\mathbf{P}_{D^{1/2}C}\boldsymbol{z}, \text{ by Exercise 1, p. 193,} \\ &= \boldsymbol{y}'AC(C'DC)^{-}C'A'\boldsymbol{y} \\ &= \boldsymbol{y}'\mathbf{P}_{AC}\boldsymbol{y} \\ &= \boldsymbol{y}'\mathbf{P}_{H}\boldsymbol{y} \text{ because } AC = H.\end{aligned}$$

The GLH form of the numerator SS to test $H'X\boldsymbol{\beta}$ is $\boldsymbol{y}'\mathbf{P}_{AC}\boldsymbol{y}$. Further, because $\text{sp}(H) \subset \text{sp}(X)$, it follows from Proposition 10.3 that $\mathbf{P}_H = \mathbf{P}_X - \mathbf{P}_{XN}$ (where $\text{sp}(N) = \text{sp}(G)^\perp$), and therefore that $SSE_{\boldsymbol{z}}$ is the RMFM SS for $G'\boldsymbol{\beta}$.

The development here shows that the vaguely-defined approach that Yates followed – find a quadratic form in $\hat{\boldsymbol{y}}$ that is an unbiased estimator of σ^2 if H_0 is true – can be extended to a general linear model and estimable functions $G'\boldsymbol{\beta}$. The SS that it produces is the RMFM SS, which is the unique SS $\boldsymbol{y}'P\boldsymbol{y}$ with $\text{sp}(P) \subset \text{sp}(X)$ that tests exclusively H_0. Further, although different choices of A are possible, all lead to the same SS, and so there is some room to choose an A that is convenient.

In Yates's formulation, this approach led to a closed-form algebraic expression for the SS, which made it practicable for computations at that time. That does not seem to be such an important consideration today. It is simpler and more direct to go straight to $\boldsymbol{y}'\mathbf{P}_H\boldsymbol{y}$.

19.3 SS for A Main Effects in the Two-Factor ANOVA Model

Consider the saturated two-factor model for the mean of the response under the ab FLCs of factors A and B. Notation is as described in Section 14.2. The model for the mean vector $\boldsymbol{\mu} = \mathrm{E}(\boldsymbol{Y})$ of the response is $\mathbb{K}\boldsymbol{\eta}$, corresponding to $X\boldsymbol{\beta}$ in the general formulation above. Yates (1934) required that all cells be filled, which renders the columns of $\mathbb{K}$ linearly independent, and so all linear functions of $\boldsymbol{\eta}$ are estimable. Note, however, that in the general development of the previous section, there is no such restriction on X, and so that construction could be extended to include empty cells.

Yates (1934) defined A main effects as differences among the A *population marginal means* $\bar{\eta}_{i\cdot} = (1/b)\sum_j \eta_{ij}$, $i = 1, \ldots, a$. The a-vector of A marginal means can be expressed as $\boldsymbol{\theta} = (1/b)(\mathrm{I}_a \otimes \mathbf{1}_b)'\boldsymbol{\eta}$. The hypothesis of equal A marginal means is $\mathrm{H}_0 : S_a\boldsymbol{\theta} = \mathbf{0}$, or, in terms of $\boldsymbol{\eta}$, $\mathrm{H}_0 : [(1/b)(\mathrm{I}_a \otimes \mathbf{1}_b)S_a]'\boldsymbol{\eta} = \mathbf{0}$. This takes the form $\mathrm{H}_0 : G'\boldsymbol{\beta} = \mathbf{0}$ with $\boldsymbol{\beta} = \boldsymbol{\eta}$ and $G = (1/b)(\mathrm{I}_a \otimes \mathbf{1}_b)S_a = (1/b)(S_a \otimes \mathbf{1}_b)$.

Let $D_{ab} = (\mathbb{K}'\mathbb{K})^{-1} = \mathrm{Diag}(1/n_{ij})$. To express the numerator SS in the form (19.3), $X = \mathbb{K}$ and $G = X'AC$ with

$$A = (1/b)\mathbb{K}D_{ab}(\mathrm{I}_a \otimes \mathbf{1}_b)$$

and $C = S_a$. Let $M = \mathbf{1}_a$ so that $\mathrm{sp}(M) = \mathrm{sp}(C)^{\perp}$. Then $\boldsymbol{\theta} = A'X\boldsymbol{\beta} = A'\mathbb{K}\boldsymbol{\eta} = (1/b)(\mathrm{I}_a \otimes \mathbf{1}_b)'\boldsymbol{\eta}$ is the a-vector of A population marginal means $\bar{\eta}_{i\cdot}$; and $\hat{\boldsymbol{\theta}} = A'\boldsymbol{y} = (\bar{\bar{y}}_{i\cdot} = (1/b)\sum_j \bar{y}_{ij})$ is the a-vector of averages, over levels of B, of the sample cell means $\bar{y}_{ij} = \sum_{\ell=1}^{n_{ij}} y_{ij\ell}/n_{ij}$ (which are the ab entries in $\hat{\boldsymbol{\eta}} = D_{ab}\mathbb{K}'\boldsymbol{y}$). Let

$$D_a = A'A = (1/b^2)(\mathrm{I}_a \otimes \mathbf{1}_b')D_{ab}(\mathrm{I}_a \otimes \mathbf{1}_b) = (1/b^2)\mathrm{Diag}\left(\sum_j (1/n_{ij})\right).$$

Diagonal entries of D_a are $1/w_i$ in (19.2). With these specifications, (19.3) is identical to (19.2), the MWSM numerator SS for A main effects. By the results in the last section, this is in turn equal to the RMFM SS for testing H_0.

With $H = AC = (1/b)\mathbb{K}D_{ab}(S_a \otimes \mathbf{1}_b)$, the ncp of Q is 0 iff $H'\mathbb{K}\boldsymbol{\eta} = \mathbf{0}$, and

$$\begin{aligned} H'\mathbb{K}\boldsymbol{\eta} &= \frac{1}{b}(S_a \otimes \mathbf{1}_b')D_{ab}\mathbb{K}'\mathbb{K}\boldsymbol{\eta} \\ &= S_a\left[\frac{1}{b}(\mathrm{I}_a \otimes \mathbf{1}_b')\boldsymbol{\eta}\right]. \end{aligned} \tag{19.4}$$

The quantity in brackets on the right is the a-vector of A marginal means, and the right-hand side is $\mathbf{0}$ is iff the a A marginal means $\bar{\eta}_{i\cdot}$ are all equal.

Defining matrices A and C in this way corresponds to the Yates (1934) formulation. This has the consequence that $M = \mathbf{1}_a$ is a column vector, which avoids matrix operations in (19.3). Another possible choice is $A = (1/b)\mathbb{K}D_{ab}$, so that $\boldsymbol{\theta} = \boldsymbol{\eta}$ and $C = S_a \otimes \mathbf{1}_b$. That would result in M having at least $ab - (a-1)$ columns. It is an alternative, and it would lead to the same SS, but evaluating (19.3) then would appear to be more burdensome than with A as defined above.

19.4 Summary

Although the MWSM dealt only with the numerator SS for testing main effects in two-factor models that permit interaction effects, it followed a rationale that was commonly used to justify SSs in a variety of settings: to find SSs in $\hat{\boldsymbol{y}}$ that estimated σ^2 if the hypothesis in question were true. That was the only property claimed by Yates. Documentation and proofs of the MWSM's properties as a test statistic only began to appear in the 1970s.

This chapter extends that heuristic to the general linear model and shows that the MWSM SS is implicit in that extension. This establishes at once that the MWSM is an RMFM SS and hence has all its properties.

19.5 MWSM Exercises

1. Let R be an $r \times c$ matrix, M a matrix such that $\mathrm{sp}(M) = \mathrm{sp}(R)^{\perp}$, D an $r \times r$ symmetric positive-definite (pd) matrix, $D^{1/2}$ a symmetric pd matrix such that $D^{1/2}D^{1/2} = D$, and $D^{-1/2} = (D^{1/2})^{-1}$.

   Prove that

   $$\mathbf{P}_{D^{1/2}R} = \mathrm{I} - \mathbf{P}_{D^{-1/2}M}.$$

2. Show that the estimable linear functions of the A marginal means are the linear functions of the A marginal means in rows that have no empty cells.

   It follows from this that the numerator SS for the estimable part of A main effects can be computed by Yates's MWSM Q on $\{y_{ijs} : i = 1, \ldots, a \text{ such that } n_{ij} > 0 \text{ for all } j = 1, \ldots b\}$.

20

Two-Factor Models with Proportional Subclass Numbers

Two-factor settings with proportional subclass numbers have long been cited as unbalanced designs where ANOVA SSs have the same properties as they do in balanced designs. This chapter presents results that show that this is not true, and instead that those SSs do not test ANOVA effects unless further conditions on the subclass numbers are satisfied, and that both A and B SSs test their eponymous effects only if the design is balanced.

20.1 Introduction

This chapter investigates properties of two-factor models with proportional subclass numbers (psn), as defined on p. 143. It is shown that in such settings, although the ANOVA SSs for A and B main effects are orthogonal, and hence that the AB interaction SS is a proper SS, nevertheless neither SS_A nor SS_B tests A or B main effects, respectively, unless further conditions hold, and that both test the named effects only if the model is balanced. Notation is as defined in Section 14.2.

Practically since the invention of ANOVA, it has been widely thought and taught that properties of two-factor models with psn were the same as if the model were balanced. Anderson and Bancroft (1952, p. 278) say that Snedecor (1946) "shows that if the numbers are proportional to the main effect totals, the usual analysis can be carried out." Snedecor and Cox (1935) note that, "The very excellence of the methods for analyzing tables with proportional subclass numbers has served to focus attention upon the disabilities of tables in which these numbers are disproportionate." Yates (1934) says that, if the subclass numbers are proportionate, "... the sum of squares appropriate to testing the A effects may be computed by the method ... for a single classification with unequal numbers in the various classes." Sokal and Rolf (1995, p. 358) give an example with proportional subclass numbers, noting that "[t]he computation of the sums of squares ... follows the familiar series of steps" for the balanced model.

Snedecor and Cochran (1980, p. 420) suggest that, if "the n_{ij} appear to be approximately proportional to their row and column totals" then "a good approximation to the least squares analysis is obtained by ... replacing the n_{ij} by proportional numbers $n'_{ij} = n_{i\cdot}n_{\cdot j}/n_{\cdot\cdot}$." Anderson and Bancroft (1952) say this method "is probably better than the method of unweighted means."

Longstanding convention and usage seem to indicate that the benefits of balanced models accrue as well to models with proportional subclass numbers. However, Snedecor and Cochran (1980, p. 416) distinguish three cases where "... the additive model can be fitted and the test for interactions can be made by simple methods," hinting that something might be different if the "n_{ij} are (i) equal, (ii) equal within any row or within any column, or (iii) proportional – that is, in the same proportion within any row." Snedecor and Cox (1935, p. 246) suggest that the MWSM "is especially appropriate if the postulated population has equal subclass numbers. ... [I]f the method is applied to a sample with equal subclass numbers it yields exactly the same results as the standard method for such numbers; but if it is applied to a sample with proportional (but not equal) subclass numbers the results do not coincide with those obtained from the standard method for proportional numbers." Kempthorne (1975, p. 479) observed that, with psn, the ANOVA SS does not test "the main effect of A with equal weights over the B levels." On a similarly discordant note, Hocking (2013, pp. 325–327) shows an example with proportional subclass numbers to illustrate that, although the sums of squares are orthogonal, they do not match those by "the marginal means method," and hence they do not test the ANOVA hypotheses defined in terms of the marginal means. Perhaps the "very excellence" of proportional subclass numbers is illusory after all.

Most of the results discussed here appear in LaMotte (2017, 2019). Propositions established there are given in Section 20.5. While their proofs are not entirely trivial, they are mostly technical and not very instructive, and so they are not included here.

20.2 "Excellence" of Proportional Subclass Numbers

If all the subclass numbers are equal, then the ANOVA SSs for A, B, and AB can be calculated as outlined on p. 175. This requires only SOSs. The symmetric matrices $Q_A = Q_{10}$, $Q_B = Q_{01}$, and $Q_{AB} = Q_{11}$ are idempotent and orthogonal. Consequently the SSs $SS_A = \boldsymbol{y}'Q_A\boldsymbol{y}$, SS_B, and SS_{AB} are distributed as proportional to independent chi-squared random variables. The ncp of SS_A is zero iff there are no A main effects, and the corresponding property holds for SS_B and SS_{AB}. ANOVA SSs are easy to compute, their null distributions are known, and they test meaningful and useful hypotheses. These properties are the foundation of the usefulness and importance of ANOVA in balanced models.

If the subclass numbers are proportional but not equal, the computations are only slightly more burdensome; they are practicable even for hand calculations. They follow exactly the definitions of subtotal SSs given on p. 175. It can be shown that the same orthogonality property holds for the three SSs, and hence that they are distributed as proportional to independent chi-squared random variables. It has long been tacitly assumed that they test the corresponding effects, too. Aside from the few doubts mentioned above, settings with psn have been regarded as equivalent to balanced settings in all respects.

20.3 Orthogonality of ANOVA SSs

Consider a model for two factors, A and B, with subclass numbers n_{ij}, $i = 1, \ldots, a$, $j = 1, \ldots, b$. For purposes of this discussion, the model for the mean vector $\boldsymbol{\mu}$ of the n-vector response $\boldsymbol{y}$ is $\text{sp}[\mathbb{K}(E_{00}, E_{10}, E_{01}, E_{11})] = \text{sp}(\mathbb{K})$, a dummy-variable model that includes A and B main effects and AB interaction effects.

ANOVA SSs are as defined on p. 175. Let

$$\begin{aligned} Q_A &= \mathbf{P}_{\mathbb{K}E_{10}} - \mathbf{P}_{1_n}, \\ Q_B &= \mathbf{P}_{\mathbb{K}E_{10}} - \mathbf{P}_{1_n}, \\ Q_{A,B} &= \mathbf{P}_{\mathbb{K}E_{11}} - \mathbf{P}_{1_n}, \text{ and} \\ Q_{AB} &= Q_{A,B} - Q_A - Q_B. \end{aligned} \tag{20.1}$$

Denote the resulting SSs as $SS_A = \boldsymbol{y}'Q_A\boldsymbol{y}$, $SS_B = \boldsymbol{y}'Q_B\boldsymbol{y}$, and $SS_{A,B} = \boldsymbol{y}'Q_{A,B}\boldsymbol{y}$.

Among what Snedecor and Cox (1935) call the "disabilities of tables in which [the subclass numbers n_{ij}] are disproportionate" is that "... the addition theorem for sums of squares does not apply." They illustrate with an example that Q_{AB} is not non-negative definite, which "... makes it impossible to compute directly the *interaction* sum of squares."

"The addition theorem" seems to be equivalent to "Q_{AB} is idempotent," so that $\boldsymbol{y}'Q_{AB}\boldsymbol{y}$ is a proper SS. With both $\text{sp}(Q_A)$ and $\text{sp}(Q_B)$ contained in $\text{sp}(Q_{A,B})$, it can be seen that Q_{AB} is idempotent iff $Q_AQ_B = 0$. Then "the addition theorem" is equivalent to orthogonality of the columns of Q_A to the columns of Q_B.

I have been unable to find a source where "orthogonality implies psn" is proved, while the implication "psn implies orthogonality" is fairly straightforward to show. Proposition 20.1 establishes that this orthogonality holds, and Q_{AB} is idempotent, if and only if the cell sample sizes n_{ij} are psn. In that case, Q_A, Q_B, and Q_{AB} computed in the sequence described on p. 175 yield proper SSs and $SS_{A,B} = SS_A + SS_B + SS_{AB}$.

It appears that this property, that Q_j in (18.2) is idempotent, holds as well for psn extended to f factors as if it were probabilistic independence, in the following sense. Let $p_{i\cdot} = n_{i\cdot}/n$, $p_{\cdot j} = n_{\cdot j}/n$, and $p_{ij} = n_{ij}/n$. Then the joint probability distribution p_{ij} is the product of its marginal distributions $p_{i\cdot}$ and $p_{\cdot j}$ iff $n_{ij} = n_{i\cdot}n_{\cdot j}/n$ for each i, j. Observing this, it appears that the psn property extends to f factors if the joint probability distribution $\{n_{\boldsymbol{\ell}}/n : \boldsymbol{\ell} \in \mathcal{L}\}$ is the product of its f marginal distributions. If all this is true, then psn for f factors implies that each $Q_{\boldsymbol{j}}$ defined in (18.2) is idempotent, and hence that $\{Q_{\boldsymbol{j}} : \boldsymbol{j} \in \mathcal{B}^f\}$ provides an orthogonal partition of $\mathbf{P}_{\mathbb{K}} - \mathbf{P}_{\mathbf{1}_n}$. However, as established here for two factors, more restrictive conditions than psn are required in order that these SSs test their corresponding ANOVA effects.

20.4 What Do They Test?

The set $(n_{ij}) = N_{21} = \begin{pmatrix} 6 & 4 & 4 \\ 3 & 2 & 2 \\ 3 & 2 & 2 \end{pmatrix}$ of subclass numbers in Table 18.1 is psn. The RMFM SS for Model 1 is SS_A, the ANOVA SS for A main effects ignoring B, also called Type I SS. The RMFM SS for Model 2 is Type II SS, for A main effects adjusted for B main effects. As a consequence of psn, these two SSs are the same, and they test exclusively A main effects both in model 1 and in model 2. In model 3, though, which includes interaction effects, SS_A tests the two contrasts

$$\begin{pmatrix} 6 & 4 & 4 \\ -3 & -2 & -2 \\ -3 & -2 & -2 \end{pmatrix} \text{ and } \begin{pmatrix} 0 & 0 & 0 \\ 3 & 2 & 2 \\ -3 & -2 & -2 \end{pmatrix} \tag{20.2}$$

on the cell means, neither of which is an A main effects contrast. It can be shown that the space they span intersects $\text{sp}(H_{10})$ only in $\mathbf{0}$. SS_A does not test any A main effects.

Confusing matters further, set N_{23} in Table 18.1 is not psn, and Type I and Type II SSs are different. Nevertheless, both test the same two of the three degrees of freedom for A main effects in the saturated model.

One of the criticisms of Type I and Type II SSs that prompted the creation of Type III SSs was that, in unbalanced models, the effects that they test depend on the distribution of the subclass numbers. That same deficit persists with proportional subclass numbers, as illustrated next. The two sets of subclass numbers

$$N_1 = \begin{pmatrix} 6 & 4 & 4 \\ 3 & 2 & 2 \\ 3 & 2 & 2 \end{pmatrix} \text{ and } N_2 = \begin{pmatrix} 1 & 7 & 2 \\ 2 & 14 & 4 \\ 3 & 21 & 6 \end{pmatrix} \tag{20.3}$$

are both psn. In both settings and under the saturated model, it can be seen that SS_A (which is both Type I and Type II) does not test any part of A main effects. Further, the effects tested by SS_A with N_1 and N_2, respectively, intersect only at $\mathbf{0}$. Not only does SS_A test different hypotheses under the two configurations, it does not test any A effects in either.

Proposition 20.2 establishes that Q_{AB} tests exclusively AB interaction effects iff the subclass numbers are psn. Proposition 20.3 establishes that Q_A tests exclusively A main effects iff the subclass numbers are *row balanced* such that $n_{ij} = n_{i\cdot}/b$, $i = 1, \ldots, a$. This is a more restrictive condition than psn. Proposition 20.4 establishes the corresponding condition for B main effects. Proposition 20.5 establishes that in the additive model, Q_A tests exclusively A main effects iff the subclass numbers are psn.

These facts are contrary to conventional wisdom about the "very excellence" of proportional subclass numbers. If we were restricted to doing computations by hand, then the simplification of computing SSs following (18.2) would be welcome. The problem is that then we lose control over the functions of the cell means that they test.

Today the best advice is "Don't do it by hand"; do all your statistical computations within a general-purpose statistical computing package. And teaching students that psn is the same as balanced is misleading, at best. There is no longer any need or benefit to teach or to know that some sums of squares can be computed as they are in balanced models if subclass numbers are psn.

20.5 Propositions

Following in this section are the propositions that are established in LaMotte (2018, 2019).

Proposition 20.1. $Q_A Q_B = 0$ *(equivalently, Q_{AB} is idempotent) iff* $n_{ij} = \frac{n_{i\cdot} n_{\cdot j}}{n_{\cdot\cdot}}$ *for all* $i = 1, \ldots, a$ *and* $j = 1, \ldots, b$.

Proposition 20.2. $Q_{AB} = \mathbf{P}_{\mathbb{K}} - \mathbf{P}_{\mathbb{K}(\mathrm{I}-H_{11})}$ *iff Q_{AB} is idempotent.*

Proposition 20.3. $Q_A = \mathbf{P}_{\mathbb{K}} - \mathbf{P}_{\mathbb{K}(\mathrm{I}-H_{10})}$ *iff* $n_{ij} = n_{i\cdot}/b$, $i = 1, \ldots, a$, $j = 1, \ldots, b$.

Proposition 20.4. $Q_B = \mathbf{P}_{\mathbb{K}} - \mathbf{P}_{\mathbb{K}(\mathrm{I}-H_{01})}$ *iff* $n_{ij} = n_{\cdot j}/a$, $i = 1, \ldots, a$, $j = 1, \ldots, b$.

Proposition 20.5. $Q_A = \mathbf{P}_{\mathbb{K}(\mathrm{I}-H_{11})} - \mathbf{P}_{\mathbb{K}(H_{00}+H_{01})}$ *iff* $n_{ij} = \frac{n_{i\cdot}n_{\cdot j}}{n_{\cdot\cdot}}$ *for all* $i = 1, \ldots, a$ *and* $j = 1, \ldots, b$.

20.6 PSN Exercises

At each FLC i, j of factors A and B at a and b levels, n_{ij} observations y_{ijs} are taken. The subclass numbers are psn.

Let

$$\begin{aligned} Q_A &= Q_{10} = ST_{10} = \mathbf{P}_{\mathbb{K}E_{10}} - \mathbf{P}_{\mathbf{1}_n}, \\ Q_B &= Q_{01} = ST_{01} = \mathbf{P}_{\mathbb{K}E_{01}} - \mathbf{P}_{\mathbf{1}_n}, \\ Q_{A,B} &= ST_{11} = \mathbf{P}_{\mathbb{K}} - \mathbf{P}_{\mathbf{1}_n}, \text{ and} \\ Q_{AB} &= Q_{11} = ST_{11} - ST_{10} - ST_{01} \\ &= Q_{A,B} - (Q_A + Q_B) \\ &= \mathbf{P}_{\mathbb{K}} - \mathbf{P}_{\mathbb{K}E_{10}} - \mathbf{P}_{\mathbb{K}E_{01}} + \mathbf{P}_{\mathbf{1}_n}. \end{aligned} \tag{20.4}$$

1. Prove: If L, M, and N are matrices with the same number of rows, and if $\text{sp}(M) \subset \text{sp}(L)$ and $\text{sp}(N) \subset \text{sp}(L)$, then $\mathbf{P}_L - (\mathbf{P}_M + \mathbf{P}_N)$ is idempotent iff $\mathbf{P}_M\mathbf{P}_N = 0$.

2. Prove: Q_{AB} is idempotent iff $Q_A Q_B = 0$.

3. Prove: $Q_A Q_B = 0$ iff $n_{ij} = n_{i\cdot}n_{\cdot j}/n$ for all $i = 1, \ldots, a$ and $j = 1, \ldots, b$.

4. The full model is $\text{sp}(\mathbb{K})$. Show that

$$\{\mathbb{K}\boldsymbol{\eta} : \boldsymbol{\eta} \in \Re^{ab} \text{ and } H_{10}\boldsymbol{\eta} = \mathbf{0}\} = \text{sp}[\mathbb{K}(\mathrm{I} - H_{10})].$$

   That is, the restricted model for $\mathrm{H}_0 : H_{10}\boldsymbol{\eta} = \mathbf{0}$, that there are no A main effects, is $\text{sp}[\mathbb{K}(\mathrm{I} - H_{10})]$.

5. Show that, if the model is balanced, so that $\mathbb{K} = \mathrm{I}_a \otimes \mathrm{I}_b \otimes \mathbf{1}_m$, then $\text{sp}[\mathbb{K}(\mathrm{I} - H_{10})] = \text{sp}(Q_A)$.

6. Suppose

$$N = (n_{ij}) = \begin{pmatrix} 6 & 4 & 4 \\ 3 & 2 & 2 \\ 3 & 2 & 2 \end{pmatrix}$$

   are the cell sample sizes for a two-factor setting with $a = b = 3$ levels of each factor. The following questions can be addressed numerically in terms of this specific setting.

(a) Describe $\mathbb{K}$, assuming that observations y_{ijs} are listed in lexicographic order on i, j, s.

(b) In the model $\text{sp}(\mathbb{K}) = \{\mathbb{K}\boldsymbol{\eta} : \boldsymbol{\eta} \in \Re^9\}$ for the mean vector $\boldsymbol{\mu}$ of $\boldsymbol{y}$, show that every linear function $\boldsymbol{g}'\boldsymbol{\eta}$ of the vector $\boldsymbol{\eta}$ of cell means is estimable.

(c) One representation of the additive-effects model is $\{\mathbb{K}\boldsymbol{\eta} : \boldsymbol{\eta} \in \Re^9 \text{ and } H_{11}\boldsymbol{\eta} = \mathbf{0}\}$. What linear functions of $\boldsymbol{\eta}$ are estimable in the additive-effects model?

(d) Find Q_{10}, Q_{01}, ST_{11}, and Q_{11} numerically.

(e) Identify vectors $\boldsymbol{r}$ and $\boldsymbol{c}$ such that $N = \boldsymbol{r}\boldsymbol{c}'$, and hence that this setting is psn.

(f) Verify that $Q_{10}Q_{01} = 0$ and that Q_{11} is idempotent.

(g) Show algebraically that

$$\begin{aligned} \boldsymbol{y}'Q_{10}\boldsymbol{y} &= \sum_{i=1}^{3}\sum_{j=1}^{3}\sum_{s=1}^{n_{ij}}(\bar{y}_{i\cdot} - \bar{y})^2 \text{ and} \\ \boldsymbol{y}'Q_{01}\boldsymbol{y} &= \sum_{i=1}^{3}\sum_{j=1}^{3}\sum_{s=1}^{n_{ij}}(\bar{y}_{\cdot j} - \bar{y})^2. \end{aligned}$$

That is, each is a sum of squared deviations from the average for levels of one factor, ignoring levels of the other.

(h) Find the dimension of $\text{sp}(Q_{10}) \cap \text{sp}(\mathbf{P}_{\mathbb{K}} - \mathbf{P}_{\mathbb{K}(\mathrm{I}-H_{10})})$. Is $Q_{10} = \mathbf{P}_{\mathbb{K}} - \mathbf{P}_{\mathbb{K}(\mathrm{I}-H_{10})}$?

(i) In terms of $\boldsymbol{\eta}$, what does $\boldsymbol{y}'Q_{10}\boldsymbol{y}$ test? With the full model being $\mathbb{K}\boldsymbol{\eta}$, $\boldsymbol{y}'Q_{10}\boldsymbol{y}$ tests exclusively $Q_{10}\mathbb{K}\boldsymbol{\eta} = \mathbf{0}$, which is equivalent to $G'\boldsymbol{\eta} = \mathbf{0}$ with $G = \mathbb{K}'Q_{10}$. Show that columns of G are contrasts, because $\mathbb{K}\mathbf{1}_{a_\times} = \mathbf{1}_n$. But are they contrasts on the A marginal means? Is $\text{sp}(G) \subset \text{sp}(H_{10})$? Find $\text{sp}(G) \cap \text{sp}(H_{10})$. What contrasts are in both $\text{sp}(G)$ and $\text{sp}(H_{10})$?

(j) Express G in terms of integer contrasts by rescaling. With more than one contrast, getting to recognizable contrasts entails finding linear transformations GT of the columns of G such that $\text{sp}(GT) = \text{sp}(G)$ and GT has nice structure. Another ploy is to make guesses, say, $C'\boldsymbol{\eta}$, and check whether some or all of $\text{sp}(C)$ is contained in $\text{sp}(G)$.

(k) Let $Q_{IIA} = \mathbf{P}_{\mathbb{K}(E_{10},E_{01})} - \mathbf{P}_{\mathbb{K}E_{01}}$. $SS_{IIA} = \boldsymbol{y}'Q_{IIA}\boldsymbol{y}$ is Type II SS for A main effects, in SAS's typology. As you can see, the full model includes (1), A, and B effects, and the restricted model excludes A effects. The full model for the cell means is $\text{sp}(H_{00} + H_{10} + H_{01})$, and the restricted model is $\text{sp}(H_{00} + H_{01})$, which is the correct restricted model for testing A main effects.

The additive model is $\text{sp}[\mathbb{K}(H_{00}+H_{10}+H_{01})] = \text{sp}[\mathbb{K}(E_{00}, E_{10}, E_{01})] = \text{sp}[\mathbb{K}(E_{10}, E_{01})]$.
Does SS_{IIA} test A main effects in the additive model? That is, is $Q_{IIA} = \mathbf{P}_{\mathbb{K}(H_{00}+H_{10}+H_{01})} - \mathbf{P}_{\mathbb{K}(H_{00}+H_{01})}$?
Does SS_{IIA} test A main effects in the full model sp($\mathbb{K}$)? That is, is $Q_{IIA} = \mathbf{P}_{\mathbb{K}} - \mathbf{P}_{\mathbb{K}(\mathrm{I}-H_{10})}$?

21

ANOVA Discussion and Comments

A numerator SS $\boldsymbol{y}'P\boldsymbol{y}$ for the conventional F-statistic, (10.1), tests exclusively that $PX\boldsymbol{\beta} = \mathbf{0}$. If our particular target of interest is $G'\boldsymbol{\beta}$, and if $G'\boldsymbol{\beta}$ is estimable, we have shown that then the RMFM, GLH, and Type III SSs are the same, and they test exclusively $\text{H}_0 : G'\boldsymbol{\beta} = \mathbf{0}$.

In practice, to rely on this fact requires verifying that $G'\boldsymbol{\beta}$ is estimable. If it is not, then the RMFM SS tests exclusively the estimable part, as established in Proposition 10.3. While that is reassuring, it does not identify specifically the estimable part. Interpreting results accurately then requires that one additional step.

When the target ANOVA effect $G'\boldsymbol{\beta}$ is not estimable, it was established here that both the GLH and Type III SSs test the estimable part, but they also test additional, unasked-for linear functions, up to ν_G degrees of freedom. Describing the estimable part and the lagniappe functions is required for meaningful interpretation.

If the specific target of inference is $G'\boldsymbol{\beta}$, then it seems that, whatever SS is chosen, estimability of $G'\boldsymbol{\beta}$ or its estimable part must be determined. Once that is done, then, testing it, all three SSs are the same, and whichever one is most convenient is just fine.

The extra linear functions tested by GLH and Type III SSs are useless if we're only interested in $G'\boldsymbol{\beta}$. If our interests are broader and less specific, though, as is often the case, those additional functions might highlight differences we hadn't considered. If that notion is appealing, and an overall objective is more exploratory than it is confirmatory, it might be useful to guide the exploration more. For example, in subclass numbers N_{13} in Table 18.1, Type III tests the one estimable A main effect contrast along with $2(\eta_{12} + \eta_{13}) - (\eta_{22} + \eta_{23}) - (\eta_{32} + \eta_{33})$. While that is not an A contrast in the 3×3 table, it is, in a 3×2 sub-table. It would be easy to think up and implement a systematic exploratory panel to look for differences like these.

I cannot think of a good reason to parameterize ANOVA effects in terms of dummy variables. While it has some natural appeal for additive effects, and does no harm there, its inter-dependencies in more complex models are unnecessary, and they have been responsible for much of the perception that you have to watch where you step when dealing with unbalanced ANOVA models. Those problems disappear when effects are formulated in terms of contrasts. Furthermore, with that formulation, any set of effects in the model can be tested with an RMFM test in which the restricted model is formed simply by deleting a set of columns from the X matrix.

Type III SSs and tests are a good substitute for dummy-variable formulations, in my opinion. They provide the same results for factor effects that are estimable, and they seem to provide results for reasonable sets of contrasts that partition potential effects better than the estimable ANOVA contrasts.

In a similar vein, I think formulas for balanced model sums of squares for two factors should be introduced to students only to illustrate Fisher's definitions of main effects and interaction effects. For that purpose, it is enough to show them as if $n_{ij} \equiv 1$.

The denouement of all this might be that, after a century of ANOVA, it is no longer useful to study or to teach about classical ANOVA at all, but to consider instead methods and constructions for modeling separate and joint effects of categorical factors within the broader context of a general linear model.

Appendices

A

Proofs and Solutions to Selected Exercises

This appendix provides proofs of most propositions. It provides also solutions to selected exercises.

A.2 Chapter 2 Solutions

1. $\hat{\sigma}^2 = 917.06456$, df$= 5$.

2. $\hat{\text{V}}\text{ar}(\hat{\beta}_2) = 2.99540^2$.

3. $c_2 = 2.99540^2/917.06456 = 0.00978$, and hence $\text{Var}(\hat{\beta}_2) = 0.00978\sigma^2$.

4. In Model 1, $\text{Var}(\hat{\beta}_2) = 5.96406^2/3916.26362\sigma^2 = 0.00908\sigma^2$. The variance of $\hat{\beta}_2$ is less in Model 1.

5. With Total SS $SST = 57102.50000$ and $SSE_2 = 4585.32278$ in Model 2, Regression SS is $SSR_2 = SST - SSE_2 = 57102.50000 - 4585.32278 = 52517.17722$, and $R^2 = SSR/SST = 0.91970$.

6. $|t| = |-14.95663/6.09869| = 2.45$. For df$= 5$, $p = \Pr(|T| \geqslant 2.45) = 0.058$.

7. For β_2 in Model 2, $t^2 = F = (19.49258/2.99540)^2 = 42.34762 = \frac{\Delta(SSE)}{\hat{\sigma}^2=917.06456}$, where $\Delta(SSE)$ is the increase in SSE due to deleting x_2 from Model 2. Then $\Delta(SSE) = 42.34762 \times 917.06456 = 38835.50082$. Then SSE after deleting x_2 from Model 2 would be $4585.32278 + 38835.50082 = 43420.82360$ (carrying all five decimal places in the calculations). Computed directly, 43420.89.

8. One way would be to compare the two least-squares estimates of β_2 in the two models: $\hat{\beta}_{2|2} - \hat{\beta}_{2|1} = 19.49258 - 15.29973 = 4.19285$. But then $\hat{\text{V}}\text{ar}(\hat{\beta}_{2|2} - \hat{\beta}_{2|1}) = c_* \times 917.06456$ is required for the denominator of the t-statistic. One way to determine c_* is by identifying vectors $\mathbf{c}_{2|2}$ and $\mathbf{c}_{2|1}$ such that $\hat{\beta}_{2|2} = \boldsymbol{c}'_{2|2}\boldsymbol{y}$ and $\hat{\beta}_{2|1} = c'_{2|1}\boldsymbol{y}$ and hence $\hat{\beta}_{2|2} - \hat{\beta}_{2|1} =$

$(\boldsymbol{c}_{2|2} - \boldsymbol{c}_{2|1})'\boldsymbol{y} = \boldsymbol{c}'\boldsymbol{y}$, and hence that $c_* = \boldsymbol{c}'\boldsymbol{c}$. These computations can be had readily using the Gram-Schmidt construction described in Chapter 6. This results in $\sqrt{\hat{\text{V}}\text{ar}(\hat{\beta}_{2|2} - \hat{\beta}_{2|1})} = 0.80189$, and then to $t = 5.23$ with 5 df and a p-value $p = 0.003$.

A.3 Chapter 3 Solutions

The following are R commands that can be used for Exercises 1–44, p. 28.

```
# Load packages matrix and psych
# Define these two functions:
fuzz<- function(x){x0<-x*(abs(x)>1.e-13); x0}
vec<-function(A) {vA<-matrix(A,ncol=1); vA}

A <- matrix(c(1,3,5,2,4,6),3,2)
a1<-matrix(c(1,3,5),3)
a2<-matrix(c(2,4,6),3)
AA<-cbind(a1,a2)
AA
a1
b1<-matrix(c(-3,1,1),3)
b2<-matrix(c(0,1,2),3)
b3<-matrix(c(3,2,-1),3)
B<- cbind(b1,b2,b3)
x<-matrix(c(3,7,11),3)
one3<-matrix(rep(1,3))
one3
one2<-matrix(rep(1,2))
y<-matrix(c(2,3))
U3<-(1/3)*one3%*%t(one3)
U3
I3 <- diag(3)
S3<- I3-U3
S3
ex_1 <- (1/2)*one2
ex_2 <- .5*y+.3*one2
ex_3 <- t(A)
ex_4 <- B%*%one3
ex_5 <- t(one3)%*%one3
ex_6 <- B%*%x
ex_7 <- t(A)%*%x
ex_8 <- B%*%a1
```

```
ex_9 <- one2%*%t(one2)
ex_10 <- I3%*%B
ex_11 <- t(A)%*%B
ex_12 <- A%*%t(A)
ex_13 <- t(A)%*%A
ex_14 <- B%*%B
ex_15 <- x*2
ex_16 <- 2*A
ex_17 <- B/3
ex_18 <- S3
ex_19 <- S3 %*% S3
ex_20 <- S3 %*% U3
ex_21 <- U3 %*% U3
ex_22 <- (1/3)*t(one3)%*%x
ex_23 <- U3 %*% x
ex_24 <- S3 %*% x
ex_25a <- B%*%a1
ex_25b <- B%*%a2
ex_25_c <- B %*% A
ex_26 <- x + 2*one3
ex_27 <- 2*A%*%t(A) + .5*B
ex_28 <- tr(B)
ex_29 <- tr(A%*%t(A))
ex_30 <- tr(t(A)%*%A)
ex_31 <- tr(S3)
ex_32 <- t(x)%*%S3%*%x
ex_33 <- vec(t(A))
ex_34 <- vec(B%*%A)
ex_35 <- A %x%one2
ex_36 <- A %x% x
ex_37 <- one3 %x% A
ex_38 <- Diagonal(x=x)
ex_39 <- bdiag(A, one2%*%t(one2),y)
C <- cbind(A, b1)
D <- rbind(t(A), t(one3))
ex_40 <- C %*% D
ex_41 <- max(abs(B - t(B)))
ex_42 <- max(abs(t(A)%*%A - t(t(A)%*%A)))
ex_43 <- max(abs(U3 - U3 %*% U3))
ex_44a <- max(abs(S3 - t(S3)))
ex_44b <- fuzz(max(abs(S3 %*% S3 - S3)))
S3pU3 <- S3 + U3
ex_44c <- max(abs(S3pU3%*%S3pU3 - S3pU3))
```

45. With $B = \begin{pmatrix} b_{11} & b_{12} \\ b_{21} & b_{22} \end{pmatrix}$, show that $AB = BA$ iff $b_{11} = b_{22}$ and $b_{12} = b_{21}$.

With $B = \begin{pmatrix} 1 & 0 \\ 1 & 2 \end{pmatrix}$, verify that $AB \neq BA$, and with $B = \begin{pmatrix} 1 & -1 \\ -1 & 1 \end{pmatrix}$, $AB = BA$.

46. Denote the c columns of A by $\boldsymbol{a}_1, \ldots, \boldsymbol{a}_c$ and the c columns of I_c by $\boldsymbol{e}_1, \ldots, \boldsymbol{e}_c$. Suppose $A\boldsymbol{y} = \mathbf{0}$ for every $\boldsymbol{y} \in \Re^c$. Then for $\boldsymbol{y} = \boldsymbol{e}_j$, $A\boldsymbol{y} = \boldsymbol{a}_j = \mathbf{0}$, and this holds for each $j = 1, \ldots, c$, hence all the columns of A are $\mathbf{0}$, and hence $A = 0_{r\times c}$.

47. $A = (\mathbf{1}_3, -\mathbf{1}_3)$ and $\boldsymbol{y} = \mathbf{1}_2$.

More generally, show that if $A = (\boldsymbol{a}_1, \boldsymbol{a}_2)$ and $\boldsymbol{a}_1 \neq \mathbf{0}$ or $\boldsymbol{a}_2 \neq \mathbf{0}$, then there exists a non-zero vector $\boldsymbol{y} = \binom{y_1}{y_2}$ such that $A\boldsymbol{y} = \mathbf{0}$ iff $\boldsymbol{a}_1$ is proportional to $\boldsymbol{a}_2$, that is, iff $\exists\ c \neq 0$ such that $\boldsymbol{a}_1 = c\boldsymbol{a}_2$.

48. That $\boldsymbol{e}_i' A \boldsymbol{e}_i = a_{ii} = 0$ for each i implies that $\boldsymbol{e}_i' A \boldsymbol{e}_j = a_{ii} + 2a_{ij} + a_{jj} = 2a_{ij} = 0$ for all $i \neq j$, hence that $A = (a_{ij}) = 0_{n\times n}$.

49. Let $\boldsymbol{z} = (z_1, \ldots, z_n)'$, so that $\boldsymbol{z}'\boldsymbol{z} = \sum_i z_i^2$. Clearly, if $\boldsymbol{z} = \mathbf{0}$, then $\boldsymbol{z}'\boldsymbol{z} = 0$. In any case, $z_i^2 \geqslant 0$ for each i. If $\boldsymbol{z}'\boldsymbol{z} = 0$, then $z_i^2 = -\sum_{j\neq i}^n z_j^2 \leqslant 0$, which together with $z_i^2 \geqslant 0$ implies that $z_i^2 = 0$, and hence $z_i = 0$. The same argument can be made for each i, and hence $\boldsymbol{z}'\boldsymbol{z} = 0$ implies that $\boldsymbol{z} = \mathbf{0}$.

50. Clearly $A = 0$ implies that $A'A = 0_{c\times c}$. In the other direction, if $A'A = 0$, then, for any c-vector $\boldsymbol{y}$, $\boldsymbol{y}'A'A\boldsymbol{y} = (A'\boldsymbol{y})'(A'\boldsymbol{y}) = 0$. That is, with $\boldsymbol{z} = A'\boldsymbol{y}$, $A'A = 0$ implies that $\boldsymbol{z}'\boldsymbol{z} = 0$, which, by 49., implies that $\boldsymbol{z} = A'\boldsymbol{y} = \mathbf{0}$ for each $\boldsymbol{y} \in \Re^c$, which, by 46., implies that $A' = 0$, hence that $A = (A')' = 0$.

51. The first assertion follows from (3.2) for the i,i-th entry of AB'. Then

$$\mathrm{tr}(AA') = \sum_{i=1}^r \sum_{j=1}^r (A_{ij})(A')_{ji} = \sum_{i=1}^r \sum_{j=1}^r a_{ij}a_{ij} = \sum_{i=1}^r \sum_{j=1}^r a_{ij}^2.$$

52. Follows from 51. and 49.

53. Solutions are left to the reader.

A.4 Chapter 4 Solutions

1. $\mathcal{A} + \mathcal{B} = \{\boldsymbol{a}_1 + \boldsymbol{b}_1, \boldsymbol{a}_1 + \boldsymbol{b}_2, \boldsymbol{a}_1 + \boldsymbol{b}_3, \boldsymbol{a}_2 + \boldsymbol{b}_1, \boldsymbol{a}_2 + \boldsymbol{b}_2, \boldsymbol{a}_2 + \boldsymbol{b}_3\}$.

2. Yes, $\boldsymbol{b}_1 = B\binom{1}{0}$.

3. Yes, $\boldsymbol{x} = A\binom{1}{1}$.

4. Yes, $\boldsymbol{a}_1 = A\boldsymbol{e}_1$, where $\boldsymbol{e}_1 = (1,0,0)'$ is the first column of the 3×3 identity matrix. With $\boldsymbol{z} = (-2,1,0)'$, $A'\boldsymbol{z} = \boldsymbol{e}_1$, and hence $AA'\boldsymbol{z} = A\boldsymbol{e}_1 = \boldsymbol{a}_1$.

5. $\boldsymbol{x} - \boldsymbol{a}_2 = (1,3,5)' = \boldsymbol{a}_1 \in \text{sp}(A)$.

6. $\boldsymbol{a}_1 + \mathbf{1}_3 = \boldsymbol{a}_2 \in \text{sp}(A)$. Note also that $\mathbf{1}_3 = \boldsymbol{a}_2 - \boldsymbol{a}_1 \in \text{sp}(A)$, and $\boldsymbol{a}_1 \in \text{sp}(A)$, and therefore $\boldsymbol{a}_1 + \mathbf{1}_3$ is in $\text{sp}(A)$ because $\text{sp}(A)$ is closed under linear combinations.

7. If $\boldsymbol{b}_1 \in \text{sp}(A)$, then there would exist a vector $\boldsymbol{z}$ such that $A\boldsymbol{z} = \boldsymbol{b}_1$, that is,

$$\begin{aligned} z_1 + 2z_2 &= -3, \\ 3z_1 + 4z_2 &= 1, \text{ and} \\ 5z_1 + 6z_2 &= 1. \end{aligned}$$

The last two say that $3(z_1 + z_2) + z_2 = 1$ and $5(z_1 + z_2) + z_2 = 1$, which imply that $z_1 + z_2 = 0$ and hence that $z_2 = 1$. But then, from the first equation, $z_1 + 2z_2 = (z_1 + z_2) + z_2 = -3$, hence $z_2 = -3$, contrary to $z_2 = 1$. Therefore, $\boldsymbol{b}_1$ is not in $\text{sp}(A)$.

We will show later that $\boldsymbol{v} \in \text{sp}(A)$ iff $\mathbf{P}_A\boldsymbol{v} = \boldsymbol{v}$, where $\mathbf{P}_A$ is the orthogonal projection matrix onto $\text{sp}(A)$. In this case, $\mathbf{P}_A\boldsymbol{b}_1 = (-2.\bar{3}, -0.\bar{3}, 1.\bar{6})' \neq \boldsymbol{b}_1$, and therefore $\boldsymbol{b}_1 \notin \text{sp}(A)$.

8. Yes, answered in 3.

9. No, answered in 7.

10. $\mathbf{c} = (1,-1,0)'$.

11. $\boldsymbol{c}'\boldsymbol{x} = 3c_1 + 7c_2 + 11c_3$. Then $\mathbf{c} = (7,-3,0)'$ satisfies $\boldsymbol{c}'\boldsymbol{x} = 0$.

12. Note that $A'\boldsymbol{z} = \binom{\boldsymbol{a}_1'\boldsymbol{z}}{\boldsymbol{a}_2'\boldsymbol{z}}$. By definition, $\boldsymbol{z} \in \text{sp}(A)^\perp$ iff $\boldsymbol{a}'\boldsymbol{z} = 0$ for every vector $\boldsymbol{a} \in \text{sp}(A)$, that is, $(A\boldsymbol{v})'\boldsymbol{z} = 0$ for every $\boldsymbol{v} \in \Re^2$.

Because $\boldsymbol{a}_1$ and $\boldsymbol{a}_2$ are both in $\text{sp}(A)$, then $\boldsymbol{z} \in \text{sp}(A)^\perp$ implies that $\boldsymbol{a}_1'\boldsymbol{z} = 0$ and $\boldsymbol{a}_2'\boldsymbol{z} = 0$.

In the other direction, suppose $A'\boldsymbol{z} = \mathbf{0}$. Let $\boldsymbol{w} = A\boldsymbol{v} \in \text{sp}(A)$. Then $\boldsymbol{w}'\boldsymbol{z} = (A\boldsymbol{v})'\boldsymbol{z} = \boldsymbol{v}'(A'\boldsymbol{z}) = \mathbf{0}$, and therefore $\boldsymbol{z} \in \text{sp}(A)^\perp$.

13. Left to the reader. Just follow the steps.

14. $(\boldsymbol{u} + \boldsymbol{v})'(\boldsymbol{u} + \boldsymbol{v}) = \boldsymbol{u}'\boldsymbol{u} + \boldsymbol{v}'\boldsymbol{v} + 2\boldsymbol{u}'\boldsymbol{v} = \boldsymbol{u}'\boldsymbol{u} + \boldsymbol{v}'\boldsymbol{v}$ because $\boldsymbol{u}'\boldsymbol{v} = 0$.

A.5 Chapter 5 Proofs

Proof of Propn. 5.1, p. 38. 5.1a: $\boldsymbol{z} \in \mathrm{sp}(Q) \Longrightarrow \exists\, \boldsymbol{w}$ such that $\boldsymbol{z} = Q\boldsymbol{w}$. Then $QQ'\boldsymbol{z} = QQ'(Q\boldsymbol{w}) = Q\boldsymbol{w} = \boldsymbol{z}$. In the other direction, $\boldsymbol{z} = QQ'\boldsymbol{z} = Q(Q'\boldsymbol{z}) \Longrightarrow \boldsymbol{z} \in \mathrm{sp}(Q)$.

5.1b: Clearly $\mathrm{sp}(QQ') \subset \mathrm{sp}(Q)$. And $Q = QQ'Q \Longrightarrow \mathrm{sp}(Q) \subset \mathrm{sp}(QQ')$. ■

Proof of Propn. 5.2, p. 38. If $\mathrm{sp}(Q) \subset \mathrm{sp}(R)$, then $RR'Q = Q$, by Prop. 5.1a, and so

$$\begin{aligned}(RR' - QQ')(RR' - QQ') &= RR' - QQ'RR' - RR'QQ' + QQ' \\ &= RR' - QQ'.\end{aligned}$$

If $\mathrm{sp}(Q) = \mathrm{sp}(R)$, then $QQ'R = R$ and $RR'Q = Q$, by Prop. 5.1a, and so

$$\begin{aligned}RR' - QQ' &= QQ'\,\overbrace{RR'Q}^{Q}\,Q' - QQ' \\ &= QQ' - QQ' = 0.\end{aligned}$$

In the other direction, suppose that

$$RR' - QQ' = (RR' - QQ')(RR' - QQ').$$

Noting that $Q'(RR' - QQ')Q = Q'RR'Q - \mathrm{I}$, we see also that

$$\begin{aligned}Q'(RR' - QQ')Q &= Q'(RR' - QQ')(RR' - QQ')Q \\ &= Q'RR'Q - Q'RR'QQ'Q - Q'QQ'RR'Q + \mathrm{I} \\ &= \mathrm{I} - Q'RR'Q = -Q'(RR' - QQ')Q,\end{aligned}$$

which implies that $Q'(RR' - QQ')Q = 0$. This, together with $(RR' - QQ')(RR' - QQ') = (RR' - QQ')$, implies that

$$[(RR' - QQ')Q]'[(RR' - QQ')Q] = 0,$$

which implies that $(RR' - QQ')Q = 0$, and therefore $RR'Q = Q$, and therefore $\mathrm{sp}(Q) \subset \mathrm{sp}(R)$, by Prop. 5.1a.

If $RR' = QQ'$, then $\mathrm{sp}(R) = \mathrm{sp}(RR') = \mathrm{sp}(QQ') = \mathrm{sp}(Q)$, by Prop. 5.1b. ■

Proof of Propn. 5.3, p. 38. Suppose $\mathrm{sp}(Q) \subset \mathrm{sp}(R)$, and let $Z = RR' - QQ'$. By Prop. 5.2, $Z = ZZ = Z'Z$, and hence $\mathrm{tr}(Z) = \eta - \nu = \sum_{i,j} z_{ij}^2 \geqslant 0$. If $\eta = \nu$, then $\sum_{i,j} z_{ij}^2 = 0$ implies that all $z_{ij} = 0$, hence that $RR' = QQ'$, which implies that $\mathrm{sp}(R) = \mathrm{sp}(Q)$, by Prop. 5.2.

If $\mathrm{sp}(Q) = \mathrm{sp}(R)$, then $RR' = QQ'$, by the last part of Proposition 5.2, and therefore $\mathrm{tr}(RR') = \eta = \mathrm{tr}(QQ') = \nu$. ■

Proof of Propn. 5.4, p. 38. Let $\boldsymbol{v}_1$ be a non-zero vector in $\mathcal{S}$, $\boldsymbol{q}_1 = (1/||\boldsymbol{v}_1||)\boldsymbol{v}_1$, and Q_1 be the matrix with $\boldsymbol{q}_1$ as its one column. For $i = 1, 2, \ldots$, at the i-th step Q_i is a matrix such that $Q_i'Q_i = \mathrm{I}_i$ and $\mathrm{sp}(Q_i) \subset \mathcal{S}$. There are then three possibilities: (1) $\mathrm{sp}(Q_i) = \mathcal{S}$; (2) $i = n$, and therefore $\mathrm{sp}(Q_i) = \mathcal{S} = \Re^n$, by Prop. 5.3; or (3) there exists a vector in $\mathcal{S}$ that is not in $\mathrm{sp}(Q_i)$. If either (1) or (2) is true, then existence is proved with $Q = Q_i$ and $\nu = i$. If (3) is true, then take another step: let $\boldsymbol{v}_{i+1}$ be a vector in $\mathcal{S}$ that is not in $\mathrm{sp}(Q_i)$, and let $\boldsymbol{q}_{i+1} = c(\boldsymbol{v}_{i+1} - Q_iQ_i'\boldsymbol{v}_{i+1})$, with c such that $\boldsymbol{q}_{i+1}'\boldsymbol{q}_{i+1} = 1$. Since $\mathrm{sp}(Q_i) \subset \mathcal{S}$ and $\boldsymbol{v}_{i+1} \in \mathcal{S}$, $\boldsymbol{q}_{i+1} \in \mathcal{S}$ because $\mathcal{S}$ is a linear subspace. Append $\boldsymbol{q}_{i+1}$ to the right end of Q_i to form $Q_{i+1} = (Q_i, \boldsymbol{q}_{i+1})$. By this construction, $\mathrm{sp}(Q_{i+1}) \subset \mathcal{S}$ and $Q_{i+1}'Q_{i+1} = \mathrm{I}_{i+1}$. By Prop. 5.3, at most n steps can be taken, and so the steps must conclude with (1) or (2). ■

Proof of Propn. 5.5, p. 39. Uniqueness follows directly from the definition of OP in a linear subspace: see Exercise 3a, p. 47. Existence is established as follows. By Propn. 5.4, let Q be a matrix such that $Q'Q = \mathrm{I}$ and $\mathrm{sp}(Q) = \mathcal{S}$. Let $\boldsymbol{y}$ be an n-vector. Then $QQ'\boldsymbol{y} \in \mathcal{S}$; and $Q'(\boldsymbol{y} - QQ'\boldsymbol{y}) = \boldsymbol{0}$, which implies that $QQ'\boldsymbol{y}$ is the OP of $\boldsymbol{y}$ in $\mathcal{S}$. ■

Proof of Propn. 5.6, p. 39. 1. That $\mathcal{S} \subset (\mathcal{S}^\perp)^\perp$ was established in Exercise 25c, p. 35. To show that $(\mathcal{S}^\perp)^\perp \subset \mathcal{S}$, let $\boldsymbol{z} \in (\mathcal{S}^\perp)^\perp$. As noted just above, there exist $\boldsymbol{z}_1 \in \mathcal{S}$ and $\boldsymbol{z}_2 \in \mathcal{S}^\perp$ such that $\boldsymbol{z} = \boldsymbol{z}_1 + \boldsymbol{z}_2$. Then $\boldsymbol{z}_2'\boldsymbol{z}_1 = 0$, and $\boldsymbol{z}_2'\boldsymbol{z} = 0$, because $\boldsymbol{z} \in (\mathcal{S}^\perp)^\perp$ and $\boldsymbol{z}_2 \in \mathcal{S}^\perp$, and therefore $\boldsymbol{z}_2'\boldsymbol{z}_2 = 0$, which implies that $\boldsymbol{z}_2 = \boldsymbol{0}$, and therefore $\boldsymbol{z} = \boldsymbol{z}_1 \in \mathcal{S}$. Therefore $(\mathcal{S}^\perp)^\perp \subset \mathcal{S}$, and hence $(\mathcal{S}^\perp)^\perp = \mathcal{S}$.

2. We have already shown (Exercise 25c, p. 35) that $\mathcal{S} \subset \mathcal{T} \Longrightarrow \mathcal{T}^\perp \subset \mathcal{S}^\perp$. Substitute $\mathcal{T}^\perp$ for $\mathcal{S}$ and $\mathcal{S}^\perp$ for $\mathcal{T}$ so that, by 5.6a, $\mathcal{T}^\perp \subset \mathcal{S}^\perp \Longrightarrow \mathcal{S} = (\mathcal{S}^\perp)^\perp \subset (\mathcal{T}^\perp)^\perp = \mathcal{T}$. Note that 5.6.2 implies that $\mathcal{S} = \mathcal{T}$ iff $\mathcal{S}^\perp = \mathcal{T}^\perp$.

3. We have already shown in Exercise 16c, p. 34, that $(\mathcal{S}+\mathcal{T})^\perp = \mathcal{S}^\perp \cap \mathcal{T}^\perp$. Replace both linear subspaces by their orthogonal complements, which implies that $(\mathcal{S}^\perp + \mathcal{T}^\perp)^\perp = (\mathcal{S}^\perp)^\perp \cap (\mathcal{T}^\perp)^\perp = \mathcal{S} \cap \mathcal{T}$, by 5.6.1. Then by 5.6.1, the result follows. ■

A.7 Chapter 7 Solutions

7.1.

(a) Suppose $\mathcal{S} = \mathcal{T}$ and $\boldsymbol{s}_0 - \boldsymbol{t}_0 \in \mathcal{S}$. Then $\{\boldsymbol{s}_0\} + \mathcal{S} = \{\boldsymbol{t}_0 + (\boldsymbol{s}_0 - \boldsymbol{t}_0)\} + \mathcal{S} = \{\boldsymbol{t}_0\} + \mathcal{T}$ because $\boldsymbol{s}_0 - \boldsymbol{t}_0 \in \mathcal{S}$ and $\mathcal{S} = \mathcal{T}$.

In the other direction, suppose $\{\boldsymbol{s}_0\} + \mathcal{S} = \{\boldsymbol{t}_0\} + \mathcal{T}$. Then $\exists\ \boldsymbol{t}_*$ such that $\boldsymbol{s}_0 + \boldsymbol{0} = \boldsymbol{t}_0 + \boldsymbol{t}_* \Longrightarrow \boldsymbol{s}_0 - \boldsymbol{t}_0 \in \mathcal{T}$. And $\exists\ \boldsymbol{s}_*$ such that $\boldsymbol{t}_0 = \boldsymbol{s}_0 + \boldsymbol{s}_* \Longrightarrow \boldsymbol{t}_0 - \boldsymbol{s}_0 = \boldsymbol{s}_* \in \mathcal{S} \Longrightarrow \boldsymbol{s}_0 - \boldsymbol{t}_0 \in \mathcal{S}$ because $\mathcal{S}$ is a linear subspace. Therefore, $\boldsymbol{s}_0 - \boldsymbol{t}_0 \in \mathcal{S}$ and $\boldsymbol{s}_0 - \boldsymbol{t}_0 \in \mathcal{T}$. Let $\boldsymbol{s} \in \mathcal{S}$.

Then $\exists\, \boldsymbol{t} \in \mathcal{T}$ such that $\boldsymbol{s}_0+\boldsymbol{s} = \boldsymbol{t}_0+\boldsymbol{t} \Longrightarrow \boldsymbol{s} = \boldsymbol{t}_0-\boldsymbol{s}_0+\boldsymbol{t} \in \mathcal{T}$ because $\boldsymbol{t}_0 - \boldsymbol{s}_0 \in \mathcal{T}$. Similarly, $\boldsymbol{t} \in \mathcal{T} \Longrightarrow \exists\, \boldsymbol{s} \in \mathcal{S}$ such that $\boldsymbol{t}_0 + \boldsymbol{t} = \boldsymbol{s}_0 + \boldsymbol{s}$ $\Longrightarrow \boldsymbol{t} = \boldsymbol{s}_0 - \boldsymbol{t}_0 + \boldsymbol{s} \in \mathcal{S}$. Therefore, $\mathcal{S} = \mathcal{T}$.

(b) $\boxed{\mathcal{S} = \mathcal{S}_0 + \mathcal{S}_0^\perp \cap \mathcal{S} \Longrightarrow \mathcal{S}_0 \subset \mathcal{S}:}$ $\boldsymbol{s}_0 \in \mathcal{S}_0$ and $\boldsymbol{0} \in \mathcal{S}_0^\perp \cap \mathcal{S} \Longrightarrow \boldsymbol{s}_0 = \boldsymbol{s}_0 + \boldsymbol{0} \in \mathcal{S} \Longrightarrow \mathcal{S}_0 \subset \mathcal{S}$.

$\boxed{\mathcal{S}_0 \subset \mathcal{S} \Longrightarrow \mathcal{S} = \mathcal{S}_0 + \mathcal{S}_0^\perp \cap \mathcal{S}:}$ Clearly $\mathcal{S}_0 + \mathcal{S}_0^\perp \cap \mathcal{S} \subset \mathcal{S}$. To show that $\mathcal{S} \subset \mathcal{S}_0 + \mathcal{S}_0^\perp \cap \mathcal{S}$, suppose $\boldsymbol{s} \in \mathcal{S}$ and let $\hat{\boldsymbol{s}}_0 = \mathbf{P}_{\mathcal{S}_0}\boldsymbol{s}$, so that $\boldsymbol{s} = \hat{\boldsymbol{s}}_0 + (\boldsymbol{s} - \hat{\boldsymbol{s}}_0)$. Both $\boldsymbol{s}$ and $\hat{\boldsymbol{s}}_0$ are in $\mathcal{S}$, and so $(\boldsymbol{s} - \hat{\boldsymbol{s}}_0)$ is in $\mathcal{S}$. And, by definition of orthogonal projection, $(\boldsymbol{s} - \hat{\boldsymbol{s}}_0)$ is in $\mathcal{S}_0^\perp$. Therefore, $(\boldsymbol{s} - \hat{\boldsymbol{s}}_0) \in \mathcal{S}_0^\perp \cap \mathcal{S}$, and therefore $\boldsymbol{s} \in \mathcal{S}_0 + \mathcal{S}_0^\perp \cap \mathcal{S}$. It follows that $\mathcal{S} \subset \mathcal{S}_0 + \mathcal{S}_0^\perp \cap \mathcal{S}$ and hence that $\mathcal{S} = \mathcal{S}_0 + \mathcal{S}_0^\perp \cap \mathcal{S}$.

7.2.

$\mathrm{sp}(G) = \mathrm{sp}(G_*)$ iff $\mathrm{sp}(G)^\perp = \mathrm{sp}(G_*)^\perp$. Then $\boldsymbol{b} - \boldsymbol{b}_0 \in \mathrm{sp}(G)^\perp$ iff $\boldsymbol{b} - \boldsymbol{b}_0 \in \mathrm{sp}(G_*)^\perp$.

7.3.

(a) That $\hat{\boldsymbol{y}}$ and $\tilde{\boldsymbol{y}}$ are both OPs of $\boldsymbol{y}$ in $\mathcal{S}$ means that both are in $\mathcal{S}$ and both $\boldsymbol{y} - \hat{\boldsymbol{y}}$ and $\boldsymbol{y} - \tilde{\boldsymbol{y}}$ are in $\mathcal{S}^\perp$. Then $(\boldsymbol{y} - \hat{\boldsymbol{y}}) - (\boldsymbol{y} - \tilde{\boldsymbol{y}}) = \tilde{\boldsymbol{y}} - \hat{\boldsymbol{y}}$ is both in $\mathcal{S}$ and in $\mathcal{S}^\perp$, which implies that $\tilde{\boldsymbol{y}} - \hat{\boldsymbol{y}} = \boldsymbol{0}$.

(b) If $\boldsymbol{y}$ is the OP of $\boldsymbol{y}$ in $\mathcal{S}$, then $\boldsymbol{y} \in \mathcal{S}$, from the definition of OP.
If $\boldsymbol{y} \in \mathcal{S}$, then $\boldsymbol{y} - \boldsymbol{y} = \boldsymbol{0} \in \mathcal{S}^\perp \Longrightarrow \boldsymbol{y}$ is the OP of $\boldsymbol{y}$ in $\mathcal{S}$.

(c) If $\boldsymbol{y} \in \mathcal{S}^\perp$, then $\boldsymbol{0} \in \mathcal{S}$ and $\boldsymbol{y} - \boldsymbol{0} = \boldsymbol{y} \in \mathcal{S}^\perp$, which implies that $\boldsymbol{0}$ is the OP of $\boldsymbol{y}$ in $\mathcal{S}$.
If $\boldsymbol{0}$ is the OP of $\boldsymbol{y}$ in $\mathcal{S}$, then $\boldsymbol{y} - \boldsymbol{0} = \boldsymbol{y} \in \mathcal{S}^\perp$,.

(d) If $\hat{\boldsymbol{y}}$ is the OP of $\boldsymbol{y}$ in $\mathcal{S}$, then $\boldsymbol{y} - \hat{\boldsymbol{y}} \in \mathcal{S}^\perp$ and $\boldsymbol{y} - (\boldsymbol{y} - \hat{\boldsymbol{y}}) = \hat{\boldsymbol{y}} \in \mathcal{S} = (\mathcal{S}^\perp)^\perp \Longrightarrow \boldsymbol{y} - \hat{\boldsymbol{y}}$ is the OP of $\boldsymbol{y}$ in $\mathcal{S}^\perp$.
If $\boldsymbol{y} - \hat{\boldsymbol{y}}$ is the OP of $\boldsymbol{y}$ in $\mathcal{S}^\perp$, then $\boldsymbol{y} - \hat{\boldsymbol{y}} \in \mathcal{S}^\perp$ and $\boldsymbol{y} - (\boldsymbol{y} - \hat{\boldsymbol{y}}) = \hat{\boldsymbol{y}} \in (\mathcal{S}^\perp)^\perp = \mathcal{S} \Longrightarrow \hat{\boldsymbol{y}}$ is the OP of $\boldsymbol{y}$ in $\mathcal{S}$.

(e) Because $\mathcal{S}$ is a linear subspace and $\hat{\boldsymbol{y}}_1$ and $\hat{\boldsymbol{y}}_2$ are in $\mathcal{S}$, $c_1\hat{\boldsymbol{y}}_1 + c_2\hat{\boldsymbol{y}}_2$ is in $\mathcal{S}$. And

$$\begin{aligned}(c_1\boldsymbol{y}_1 + c_2\boldsymbol{y}_2) &- (c_1\hat{\boldsymbol{y}}_1 + c_2\hat{\boldsymbol{y}}_2)\\ &= c_1(\boldsymbol{y}_1 - \hat{\boldsymbol{y}}_1) + c_2(\boldsymbol{y}_2 - \hat{\boldsymbol{y}}_2)\end{aligned}$$

is in $\mathcal{S}^\perp$ because both $c_1(\boldsymbol{y}_1 - \hat{\boldsymbol{y}}_1)$ and $c_2(\boldsymbol{y}_2 - \hat{\boldsymbol{y}}_2)$ are in $\mathcal{S}^\perp$, and $\mathcal{S}^\perp$ is a linear subspace.

(f) Clearly $\hat{\boldsymbol{y}}_s + \hat{\boldsymbol{y}}_t$ is in $\mathcal{S} + \mathcal{T}$. So we must prove that (I) $\boldsymbol{y} - \hat{\boldsymbol{y}}_s - \hat{\boldsymbol{y}}_t$ is in $(\mathcal{S} + \mathcal{T})^\perp = \mathcal{S}^\perp \cap \mathcal{T}^\perp$ iff (II) $\hat{\boldsymbol{y}}_{st} = \hat{\boldsymbol{y}}_{ts} = \boldsymbol{0}$. $\boxed{\text{I} \Longrightarrow \text{II.}}$ Note that

$$\begin{aligned}\hat{\boldsymbol{y}}_s &= \overbrace{\hat{\boldsymbol{y}}_{ts}}^{\in\mathcal{T}} + \overbrace{(\hat{\boldsymbol{y}}_s - \hat{\boldsymbol{y}}_{ts})}^{\in\mathcal{T}^\perp} \text{ and}\\ \hat{\boldsymbol{y}}_t &= \underbrace{\hat{\boldsymbol{y}}_{st}}_{\in\mathcal{S}} + \underbrace{(\hat{\boldsymbol{y}}_t - \hat{\boldsymbol{y}}_{st})}_{\in\mathcal{S}^\perp}.\end{aligned}$$

If $\hat{\boldsymbol{y}}_{s+t} = \hat{\boldsymbol{y}}_s + \hat{\boldsymbol{y}}_t$, then

$$\begin{aligned}\boldsymbol{y} - \hat{\boldsymbol{y}}_s - \hat{\boldsymbol{y}}_t &= \boldsymbol{y} - \hat{\boldsymbol{y}}_{ts} - \hat{\boldsymbol{y}}_{st} - (\hat{\boldsymbol{y}}_s - \hat{\boldsymbol{y}}_{ts}) - (\hat{\boldsymbol{y}}_t - \hat{\boldsymbol{y}}_{st}) \\ &= \underbrace{\hat{\boldsymbol{y}} - \hat{\boldsymbol{y}}_s - (\hat{\boldsymbol{y}}_t - \hat{\boldsymbol{y}}_{st})}_{\in \mathcal{S}^\perp} - \underbrace{\hat{\boldsymbol{y}}_{st}}_{\in \mathcal{S}}\end{aligned}$$

is in $\mathcal{S}^\perp$, which implies that $\hat{\boldsymbol{y}}_{st} = \mathbf{0}$.
Similarly,

$$\begin{aligned}\boldsymbol{y} - \hat{\boldsymbol{y}}_s - \hat{\boldsymbol{y}}_t &= \boldsymbol{y} - \hat{\boldsymbol{y}}_{ts} - \hat{\boldsymbol{y}}_{st} - (\hat{\boldsymbol{y}}_s - \hat{\boldsymbol{y}}_{ts}) - (\hat{\boldsymbol{y}}_t - \hat{\boldsymbol{y}}_{st}) \\ &= \underbrace{\hat{\boldsymbol{y}} - \hat{\boldsymbol{y}}_t - (\hat{\boldsymbol{y}}_s - \hat{\boldsymbol{y}}_{ts})}_{\in \mathcal{T}^\perp} - \underbrace{\hat{\boldsymbol{y}}_{ts}}_{\in \mathcal{T}}\end{aligned}$$

is in $\mathcal{T}^\perp$, which implies that $\hat{\boldsymbol{y}}_{ts} = \mathbf{0}$. ■

$\boxed{\text{II} \Longrightarrow \text{I.}}$ If $\hat{\boldsymbol{y}}_{st} = \hat{\boldsymbol{y}}_{ts} = \mathbf{0}$, then $\hat{\boldsymbol{y}}_t - \hat{\boldsymbol{y}}_{st} = \hat{\boldsymbol{y}}_t$ is in $\mathcal{S}^\perp$ and $\hat{\boldsymbol{y}}_s - \hat{\boldsymbol{y}}_{ts} = \hat{\boldsymbol{y}}_s$ is in $\mathcal{T}^\perp$. Then $\boldsymbol{y} - \hat{\boldsymbol{y}}_s - \hat{\boldsymbol{y}}_t$ is in $\mathcal{S}^\perp$ because both $\boldsymbol{y} - \hat{\boldsymbol{y}}_s$ and $\hat{\boldsymbol{y}}_t$ are in $\mathcal{S}^\perp$, and it is in $\mathcal{T}^\perp$ because both $\boldsymbol{y} - \hat{\boldsymbol{y}}_t$ and $\hat{\boldsymbol{y}}_s$ are in $\mathcal{T}^\perp$. ■

7.4.

(a) Clearly, $C\boldsymbol{y} = y_1\hat{\boldsymbol{e}}_1 + \cdots + y_n\hat{\boldsymbol{e}}_n$ is in $\mathcal{S}$. And

$$C'(\mathrm{I} - C) = [C'(\boldsymbol{e}_1 - \hat{\boldsymbol{e}}_1), \ldots, C'(\boldsymbol{e}_n - \hat{\boldsymbol{e}}_n)] = 0$$

because each $\boldsymbol{e}_i - \hat{\boldsymbol{e}}_i$ is in $\mathcal{S}^\perp$. Therefore, $\boldsymbol{y} - C\boldsymbol{y}$ is in $\mathcal{S}^\perp$, and hence $C\boldsymbol{y}$ is the orthogonal projection of $\boldsymbol{y}$ on $\mathcal{S}$.

(b) Clearly $\mathrm{sp}(C) \subset \mathcal{S}$.
Let $\boldsymbol{s} \in \mathcal{S}$. Then, by 3b, $\boldsymbol{s}$ is the orthogonal projection of $\boldsymbol{s}$ in $\mathcal{S}$; by 4a, $C\boldsymbol{s}$ is the orthogonal projection of $\boldsymbol{s}$ in $\mathcal{S}$; and, by 3a, $C\boldsymbol{s} = \boldsymbol{s}$. Therefore $\mathcal{S} \subset \mathrm{sp}(C)$, and hence $\mathrm{sp}(C) = \mathcal{S}$.

(c) Because each $\hat{\boldsymbol{e}}_i$ is the orthogonal projection of $\boldsymbol{e}_i$ in $\mathcal{S}$, each column $\boldsymbol{e}_i - \hat{\boldsymbol{e}}_i$ is in $\mathcal{S}^\perp = \mathrm{sp}(C)^\perp$, and therefore $C'(\mathrm{I} - C), = 0_{n\times n}$. Therefore, $C' = C'C$, which is symmetric, and hence C is symmetric, and therefore $C'C = CC$, that is, C is symmetric and idempotent.

(d) By 3a, $M\boldsymbol{y} = C\boldsymbol{y}\ \forall\ \boldsymbol{y} \in \Re^n$, which implies that $C = M$.

(e) Sufficiency is established in 4a. Necessity is clear: if every vector in $\Re^n$ has an OP in $\mathcal{S}$, then each $\boldsymbol{e}_i$, being a vector in $\Re^n$, has an OP in $\mathcal{S}$.

7.5.

Clearly $\mathrm{sp}(\hat{\boldsymbol{w}}_1, \ldots, \hat{\boldsymbol{w}}_m) \subset \mathcal{S}$ because $\hat{\boldsymbol{w}}_1, \ldots, \hat{\boldsymbol{w}}_m$ are in $\mathcal{S}$. Let $\boldsymbol{s} \in \mathcal{S}$. Then there exist scalars $a_1, \ldots, a_m$ such that $\boldsymbol{s} = a_1\boldsymbol{w}_1 + \cdots + a_m\boldsymbol{w}_m$. Further, because $\boldsymbol{s} \in \mathcal{S}$, it is equal to its OP in $\mathcal{S}$. Exercise 3a showed that orthogonal projection is linear, and therefore $\boldsymbol{s} = a_1\hat{\boldsymbol{w}}_1 + \cdots + a_m\hat{\boldsymbol{w}}_m$, and therefore $\mathcal{S} \subset \mathrm{sp}(\hat{\boldsymbol{w}}_1, \ldots, \hat{\boldsymbol{w}}_m)$.

7.6.

(a) $Q'Q = (\boldsymbol{q}_i'\boldsymbol{q}_j) = (\delta_{ij}) = \mathrm{I}$ because $\boldsymbol{q}_1, \ldots, \boldsymbol{q}_k$ are norm-1 and pairwise orthogonal.

(b) Because it is a linear combination of $\boldsymbol{q}_1, \ldots, \boldsymbol{q}_k$, $\hat{\boldsymbol{y}}$ is in $\mathcal{S}$. And

$$Q'(\boldsymbol{y} - \hat{\boldsymbol{y}}) = (\boldsymbol{q}_i'(\boldsymbol{y} - \boldsymbol{q}_i\boldsymbol{q}_i'\boldsymbol{y})) = \mathbf{0},$$

hence $\boldsymbol{y} - \hat{\boldsymbol{y}} \in \mathcal{S}^{\perp}$. Therefore, $\hat{\boldsymbol{y}}$ is the orthogonal projection of $\boldsymbol{y}$ in $\mathcal{S}$.

7.7.

(a) $(A'A)G'(A'A) = [(A'A)G(A'A)]' = A'A$.

(b) $A'(AGA'A - A) = 0 \Longrightarrow \mathrm{sp}(AGA'A - A) \subset \mathrm{sp}(A)^{\perp} \cap \mathrm{sp}(A) \Longrightarrow AGA'A = A$.

(c) Follows from 7b.

(d) Follows from 7b.

(e) Clearly $\mathrm{sp}(AGA') \subset \mathrm{sp}(A)$. Suppose $\boldsymbol{z} \in \mathrm{sp}(A)$, so that $\boldsymbol{z} = A\boldsymbol{x}$ for some r-vector $\boldsymbol{x}$. Then $AGA'\boldsymbol{z} = AGA'A\boldsymbol{x} = A\boldsymbol{x} = \boldsymbol{z} \Longrightarrow \mathrm{sp}(A) \subset \mathrm{sp}(AGA')$.

(f) Follows from 7b and $A'(AG_1A' - AG_2A') = 0$.

(g) $AGA'\boldsymbol{y} \in \mathrm{sp}(A)$ and $A'(\boldsymbol{y} - AGA'\boldsymbol{y}) = \mathbf{0}$.

7.8.

(a) $\forall\ \boldsymbol{x} \in \Re^n\ \mathbf{P}_{\mathcal{S}}\boldsymbol{x} \in \mathcal{S} \Longrightarrow \mathrm{sp}(\mathbf{P}_{\mathcal{S}}) \subset \mathcal{S}$. And $\forall\ \boldsymbol{s} \in \mathcal{S},\ \boldsymbol{s} = \mathbf{P}_{\mathcal{S}}\boldsymbol{s} \Longrightarrow \mathcal{S} \subset \mathrm{sp}(\mathbf{P}_{\mathcal{S}})$.

(b) $\mathcal{S} = \mathcal{T} \Longrightarrow \mathbf{P}_{\mathcal{S}} = \mathbf{P}_{\mathcal{T}}$. $\mathbf{P}_{\mathcal{S}} = \mathbf{P}_{\mathcal{T}} \Longrightarrow \mathcal{S} = \mathrm{sp}(\mathbf{P}_{\mathcal{S}}) = \mathrm{sp}(\mathbf{P}_{\mathcal{T}}) = \mathcal{T}$.

7.9.

(a) For any $\boldsymbol{y} \in \Re^n$, $QQ'\boldsymbol{y} \in \mathrm{sp}(Q) = \mathrm{sp}(A)$ and $Q'(\boldsymbol{y} - QQ'\boldsymbol{y}) = \mathbf{0} \Longrightarrow QQ'\boldsymbol{y}$ is the OP of $\boldsymbol{y}$ in $\mathrm{sp}(A)$, that is, $QQ'\boldsymbol{y} = \mathbf{P}_A\boldsymbol{y}\ \forall\ \boldsymbol{y} \in \Re^n$, hence $QQ' = \mathbf{P}_A$.

(b) $Q\boldsymbol{x} = \mathbf{0} \Longrightarrow Q'Q\boldsymbol{x} = \mathbf{0} = \boldsymbol{x}$.

(c) $T\boldsymbol{x} = \mathbf{0} \Longrightarrow AT\boldsymbol{x} = Q\boldsymbol{x} = \mathbf{0} \Longrightarrow \boldsymbol{x} = \mathbf{0}$.

(d) The GS construction is such that no $\boldsymbol{b}_{i+1}$ is a linear combination of its predecessors. Then $\sum_1^{\nu} x_i\boldsymbol{b}_i = \mathbf{0} \Longrightarrow x_{\nu} = 0$, for otherwise $\boldsymbol{b}_{\nu}$ would be a linear combination of its predecessors; that in turn implies that $x_i = 0$, $i = (\nu - 1) : 2$. Finally, $\boldsymbol{q}_{j_1}$ is proportional to $\boldsymbol{b}_1 = \boldsymbol{a}_{j_1}$, hence $\boldsymbol{a}_{j_1} \neq \mathbf{0}$, and therefore $x_1\boldsymbol{b}_1 = \mathbf{0} \Longrightarrow x_1 = 0$. Thus $\sum_1^{\nu} x_i\boldsymbol{b}_i = \mathbf{0} \Longrightarrow x_i = 0$, $i = 1 : \nu$, and therefore $\boldsymbol{b}_i = \boldsymbol{a}_{j_i}$, $i = 1 : \nu$, are linearly independent.

(e) For any $\boldsymbol{y} \in \Re^n$, $A\boldsymbol{y} \in \text{sp}(A)$ and $\boldsymbol{y} - A\boldsymbol{y}$ is in $\text{sp}(A)^\perp$, because $A'(\boldsymbol{y} - A\boldsymbol{y}) = A'(\text{I} - A)\boldsymbol{y} = (A - AA)\boldsymbol{y} = \boldsymbol{0}$, because $A = A' = AA$, hence $\boldsymbol{y} - A\boldsymbol{y} \in \text{sp}(A)^\perp$. That is, $A\boldsymbol{y}$ is the OP of $\boldsymbol{y}$ in $\text{sp}(A)$. The OP matrix is unique, so $A = \mathbf{P}_A = QQ'$.

(f) The columns contributed to Q are the same for both GS on A and GS on B.

(g) Follows from GS on B.

(h) At most m columns of A can contribute columns to Q.

(i) Follows from Proposition 5.3 because $\text{sp}(Q) \subset \text{sp}(\boldsymbol{e}_1, \ldots, \boldsymbol{e}_n)$.

(j) Let $V = (\boldsymbol{v}_1, \ldots, \boldsymbol{v}_\nu)$. Let R be an orthonormal basis from GS on V, so that $\text{sp}(V) = \text{sp}(R)$. Because the columns of V are linearly independent, each contributes a column to R, and $\text{sp}(R) \subset \text{sp}(V)$, which implies that $\text{sp}(V) = \text{sp}(A)$, by Proposition 5.3, p. 38.

7.10.

(a) Let $\{\boldsymbol{u}_1, \ldots, \boldsymbol{u}_\nu\}$ and $\{\boldsymbol{v}_1, \ldots, \boldsymbol{v}_\eta\}$ be bases of $\mathcal{S}$. Then GS on $\{\boldsymbol{u}_1, \ldots, \boldsymbol{u}_\nu\}$ yields Q such that $\text{sp}(Q) = \mathcal{S}$ and $Q'Q = \text{I}_\nu$; and Q has ν columns because $\{\boldsymbol{u}_1, \ldots, \boldsymbol{u}_\nu\}$ are linearly independent. And GS on $\{\boldsymbol{v}_1, \ldots, \boldsymbol{v}_\eta\}$ yields R such that $R'R = \text{I}_\eta$, R has η columns, and $\text{sp}(R) = \mathcal{S}$. By Proposition 5.3, $\eta = \nu$.

(b) Let R be a matrix such that $\text{sp}(R) = \mathcal{S}$ and $R'R = \text{I}_\nu$. Let $W = (\boldsymbol{w}_1, \ldots, \boldsymbol{w}_\nu)$ be a matrix with linearly independent columns such that $\text{sp}(W) \subset \mathcal{S}$. Let $Q = (\boldsymbol{q}_1, \ldots, \boldsymbol{q}_\nu)$ be the result of GS on W (ν because the columns of W are linearly independent) such that $\text{sp}(Q) = \text{sp}(W)$ and $Q'Q = \text{I}_\nu$. By Proposition 5.3, $\text{sp}(Q) = \text{sp}(R)$, and therefore $\text{sp}(W) = \mathcal{S}$.

7.11. Let $A = (\boldsymbol{a}_1, \ldots, \boldsymbol{a}_n)$ be a square matrix.

$\boxed{\text{Columns of } A \text{ are LI} \Longrightarrow A \text{ has an inverse:}}$ That the n columns of the $n \times n$ matrix A are linearly independent $\Longrightarrow \text{sp}(A) = \Re^n = \text{sp}(\boldsymbol{e}_1, \ldots, \boldsymbol{e}_n)$, by Proposition 5.3. Then, for each $i = 1 : n$, there exists a vector $\boldsymbol{c}_i$ such that $A\boldsymbol{c}_i = \boldsymbol{e}_i$. That is, $AC = \text{I}_n$ with $C = (\boldsymbol{c}_1, \ldots, \boldsymbol{c}_n)$. And $A(CA) = A$ implies that $C\boldsymbol{a}_i = \boldsymbol{e}_i$ because the columns of A are linearly independent and therefore $\boldsymbol{e}_i$ is the unique vector such that $A\boldsymbol{e}_i = \boldsymbol{a}_i$. Thus $AC = CA = \text{I}$, that is, $C = A^{-1}$.

$\boxed{A \text{ has an inverse} \Longrightarrow \text{the columns of } A \text{ are LI:}}$ Suppose that C is a matrix such that $CA = AC = \text{I}$. Then $A\boldsymbol{x} = \boldsymbol{0} \Longrightarrow CA\boldsymbol{x} = \boldsymbol{x} = \boldsymbol{0}$.

7.12. These require only straightforward algebraic reexpression. For example,

$$\begin{aligned}(\text{I} + AB)[\text{I} - A(\text{I} + BA)^{-1}B] &= \text{I} - A(\text{I} + BA)^{-1}B - ABA(\text{I} - BA)^{-1}B \\ &= \text{I} - A[\text{I} - (\text{I} + BA) - BA](\text{I} + BA)^{-1}B \\ &= \text{I}.\end{aligned}$$

7.13.

(a) Columns of A are LI $\Longrightarrow$ columns of $A'A$ are LI: By Exercise 23b, p. 34, $A'A\boldsymbol{x} = \mathbf{0} \iff A\boldsymbol{x} = \mathbf{0}$. If the columns of A are LI, then $A'A\boldsymbol{x} = \mathbf{0} \Longrightarrow A\boldsymbol{x} = \mathbf{0} \Longrightarrow \boldsymbol{x} = \mathbf{0}$.

Columns of $A'A$ are LI $\Longrightarrow$ columns of A are LI: $A\boldsymbol{x} = \mathbf{0} \Longrightarrow A'A\boldsymbol{x} = \mathbf{0} \Longrightarrow \boldsymbol{x} = \mathbf{0}$.

(b) $\forall\ \boldsymbol{y} \in \Re^n$, $\boldsymbol{z} = A(A'A)^{-1}A'\boldsymbol{y}$ is in $\text{sp}(A)$ and $A'(\boldsymbol{y} - \boldsymbol{z}) = \mathbf{0} \Longrightarrow \boldsymbol{y} - \boldsymbol{z}$ is in $\text{sp}(A)^{\perp}$, and hence $A(A'A)^{-1}A' = \mathbf{P}_A$.

7.14.

(a) For any $\boldsymbol{y} \in \Re^n$, let $\boldsymbol{z} = (\mathbf{P}_A - \mathbf{P}_{AB})\boldsymbol{y}$. Clearly $\boldsymbol{z} \in \text{sp}(A)$; and, for any $AB\boldsymbol{x} \in \text{sp}(AB)$, $(AB\boldsymbol{x})'\boldsymbol{z} = \boldsymbol{x}'B'A'(\mathbf{P}_A - \mathbf{P}_{AB})\boldsymbol{y} = \mathbf{0} \Longrightarrow \boldsymbol{z} \in \text{sp}(A) \cap \text{sp}(AB)^{\perp}$.

(b) Let Q be a matrix such that $Q'Q = \text{I}_\nu$ and $\text{sp}(Q) = \text{sp}(A)$. Then $\mathbf{P}_A = QQ'$ and $\nu_A = \nu$.

$\text{sp}(A') = \text{sp}(A'Q)$: Clearly, $\text{sp}(A'Q) \subset \text{sp}(A')$. And, noting that $QQ'A = A$, it follows that $\text{sp}(A') \subset \text{sp}(A'Q)$, and therefore $\text{sp}(A') = \text{sp}(A'Q)$.

The columns of $A'Q$ are linearly independent: If $A'Q\boldsymbol{x} = \mathbf{0}$, then $Q\boldsymbol{x} = \mathbf{0}$ because $Q\boldsymbol{x} \in \text{sp}(A) \cap \text{sp}(A)^{\perp} = \{\mathbf{0}\}$. And $Q\boldsymbol{x} = \mathbf{0} \Longrightarrow \boldsymbol{x} = Q'(Q\boldsymbol{x}) = \mathbf{0}$. Therefore, the columns of $A'Q$ are linearly independent. That is, the ν_A columns of $A'Q$ are linearly independent and they span $\text{sp}(A')$. Therefore, the column rank of A' is ν_A, the same as the column rank of A.

(c) $\text{sp}(AB) \subset \text{sp}(A) \Longrightarrow \mathbf{P}_A - \mathbf{P}_{AB}$ is the orthogonal projection matrix onto $\text{sp}(A) \cap \text{sp}(AB)^{\perp}$, and hence is symmetric and idempotent. Therefore, $\text{tr}(\mathbf{P}_A - \mathbf{P}_{AB}) = \text{tr}[(\mathbf{P}_A - \mathbf{P}_{AB})^2] \geqslant 0 \Longrightarrow \nu_A \geqslant \nu_{AB}$. Correspondingly, $\nu_{B'A'} \leqslant \nu_{B'}$, and then, by 14b, $\nu_{AB} \leqslant \nu_B$.

7.15.

(a) $\text{sp}(A) = \text{sp}(Q) \Longrightarrow \exists\ U$ such that $A = QU \Longrightarrow QQ'A = QQ'QU = QU = A$.

(b) $\boldsymbol{y} \in \Re^n$ and $\text{sp}(Q) = \text{sp}(A) \Longrightarrow QQ'\boldsymbol{y} \in \text{sp}(A)$; and $Q'(\boldsymbol{y} - QQ'\boldsymbol{y}) = \mathbf{0} \Longrightarrow \boldsymbol{y} - QQ'\boldsymbol{y} \in \text{sp}(Q)^{\perp} = \text{sp}(A)^{\perp}$. Therefore, $QQ' = \mathbf{P}_A$.

(c) Follows from $QQ'A = A$ and $Q'Q = \text{I}$.

(d) $(A'A)(TT')(A'A) = A'QQ'A = A'A$ and $(TT')(A'A)(TT') = TQ'QT' = TT'$.

(e) $ATQ' = QQ' = \mathbf{P}_A = A(A'A)^{-1}A'$, by Exercise 13, $\Longrightarrow A[TQ' - (A'A)^{-1}A'] = 0 \Longrightarrow TQ' = (A'A)^{-1}A'$ because the columns of A are LI.

(f) If A is square and non-singular, then it has an inverse and its columns are LI. By Exercise 15e, $TQ' = (A'A)^{-1}A' = A^{-1}(A')^{-1}A' = A^{-1}$.

(g) $AN = A - ATQ'A = A - QQ'A = A - A = 0 \Longrightarrow \text{sp}(N) \subset \text{sp}(A')^{\perp}$. $A\boldsymbol{x} = \mathbf{0} \Longrightarrow \boldsymbol{x} = \boldsymbol{x} - TQ'A\boldsymbol{x} \in \text{sp}(N) \Longrightarrow \text{sp}(A')^{\perp} \subset \text{sp}(N)$.

(h) With $\boldsymbol{b} = Q\boldsymbol{z} \in \text{sp}(A)$, $A\boldsymbol{x}_* = ATQ'\boldsymbol{b} = QQ'Q\boldsymbol{z} = Q\boldsymbol{z} = \boldsymbol{b}$.

(i) If $\boldsymbol{x}$ is such that $A\boldsymbol{x} = \boldsymbol{b}$, then $\boldsymbol{x} = \boldsymbol{x}_* + (\boldsymbol{x} - \boldsymbol{x}_*)$, and $A(\boldsymbol{x} - \boldsymbol{x}_*) = \mathbf{0}$ $\Longrightarrow \boldsymbol{x} - \boldsymbol{x}_* \in \text{sp}(A')^{\perp} = \text{sp}(N)$. Conversely, if $\boldsymbol{x} = \boldsymbol{x}_* + \boldsymbol{z}$ and $A\boldsymbol{z} = \mathbf{0}$, then $A\boldsymbol{x} = A\boldsymbol{x}_* = \boldsymbol{b}$, and hence $\boldsymbol{x} \in \{\boldsymbol{u} \in \Re^m : A\boldsymbol{u} = \boldsymbol{b}\}$.

7.16. $\forall\ \boldsymbol{y} \in \Re^n$, $P\boldsymbol{y} \in \text{sp}(P)$ and $P(\boldsymbol{y} - P\boldsymbol{y}) = \mathbf{0}$ and so $\boldsymbol{y} - P\boldsymbol{y} \in \text{sp}(P)^{\perp}$. That is, $P\boldsymbol{y}$ is the orthogonal projection of $\boldsymbol{y}$ in sp(P). Let Q be such that sp$(Q) = $ sp(P) and $Q'Q = \text{I}_{\nu}$. Then $QQ' = P$ because the OP matrix is unique; and Q is an orthonormal basis for sp(P), hence dim[sp(P)] $= \nu =$ tr(P).

7.17. Note first that, because of the left-right progression of GS, sp$(Q_1) =$ sp(A_1).

| $\text{sp}(Q_2) \subset \text{sp}(A) \cap \text{sp}(A_1)^{\perp}$: | $\text{sp}(Q_2) \subset \text{sp}(Q_1, Q_2) = \text{sp}(A)$.

If $Q_2\boldsymbol{z}_2 \in \text{sp}(Q_2)$, then $Q_1'(Q_2\boldsymbol{z}_2) = \mathbf{0} \Longrightarrow Q_2\boldsymbol{z}_2 \in \text{sp}(Q_1)^{\perp} = \text{sp}(A_1)^{\perp}$. Together these imply that $\text{sp}(Q_2) \subset \text{sp}(A) \cap \text{sp}(A_1)^{\perp}$.

| $\text{sp}(A) \cap \text{sp}(A_1)^{\perp} \subset \text{sp}(Q_2)$: | If $\boldsymbol{y} \in \text{sp}(A) \cap \text{sp}(A_1)^{\perp} = \text{sp}(Q) \cap \text{sp}(Q_1)^{\perp}$, then $\boldsymbol{y} = Q_1\boldsymbol{z}_1 + Q_2\boldsymbol{z}_2$ and $Q_1'\boldsymbol{y} = \boldsymbol{z}_1 = \mathbf{0} \Longrightarrow \boldsymbol{y} = Q_2\boldsymbol{z}_2 \in \text{sp}(Q_2)$.

7.18. | (1)$\Longrightarrow$(2): | $\text{sp}(M) \subset \text{sp}(A) \Longrightarrow \exists\ B$ such that $M = AB$. Then $A'(\text{I} - M) = 0 \Longrightarrow A' = A'M = A'AB \Longrightarrow M' = B'A' = B'A'AB$, which is symmetric, and therefore $M = M'$. Then $A'(\text{I}-M) = 0 \Longrightarrow (\text{I}-M)A = 0 \Longrightarrow (\text{I} - M)AB = 0 \Longrightarrow M = MM$. That $MA = A \Longrightarrow \text{sp}(A) \subset \text{sp}(M)$; coupled with $\text{sp}(M) \subset \text{sp}(A)$, this implies that $\text{sp}(A) = \text{sp}(M)$. $\boldsymbol{z} \in \text{sp}(A)^{\perp} = \text{sp}(M)^{\perp} \Longrightarrow M'\boldsymbol{z} = M\boldsymbol{z} = \mathbf{0}$ and hence $\boldsymbol{z} = \boldsymbol{z} - M\boldsymbol{z} \in \text{sp}(\text{I}-M)$; therefore, $\text{sp}(A)^{\perp} \subset \text{sp}(\text{I} - M)$, and therefore $\text{sp}(\text{I} - M) = \text{sp}(A)^{\perp}$.

| (2)$\Longrightarrow$ (1): | Clear.

7.19. | (1) $\Longrightarrow$ (2): | Define columns of M by $\boldsymbol{m}_i = \hat{\boldsymbol{e}}_i$, $i = 1:n$.

| (2) $\Longrightarrow$ (3): | Note that $\text{sp}(M) \subset \mathcal{S}$ and $\text{sp}(\text{I}-M) \subset \mathcal{S}^{\perp} \Longrightarrow M'(\text{I}-M) = 0 = M' - M'M \Longrightarrow M' = M'M$ is symmetric and hence $M = MM$. $\boldsymbol{s} \in \mathcal{S}$ $\Longrightarrow (\text{I} - M)\boldsymbol{s} = \mathbf{0} \Longrightarrow \boldsymbol{s} \in \text{sp}(M)$, hence $\text{sp}(M) = \mathcal{S}$. And $\boldsymbol{z} \in \mathcal{S}^{\perp} \Longrightarrow$ $M\boldsymbol{z} = \mathbf{0} \Longrightarrow \boldsymbol{z} = \boldsymbol{z} + M\boldsymbol{z} = (\text{I} - M)\boldsymbol{z} \in \text{sp}(\text{I} - M) \Longrightarrow \text{sp}(\text{I} - M) = \mathcal{S}^{\perp}$. GS on M yields Q such that $\text{sp}(Q) = \text{sp}(M) = \mathcal{S}$ and $Q'Q = \text{I}$.

| (3) $\Longrightarrow$ (1): | $\forall\ \boldsymbol{y} \in \Re^n$, $\hat{\boldsymbol{y}} = QQ'\boldsymbol{y} \in \text{sp}(Q) = \mathcal{S}$ and $\boldsymbol{y} - \hat{\boldsymbol{y}} \in \text{sp}(Q)^{\perp} = \mathcal{S}^{\perp}$.

7.20.

(a) Suppose $\mathbf{P}_{\mathcal{A}+\mathcal{B}} = \mathbf{P}_{\mathcal{A}}+\mathbf{P}_{\mathcal{B}}$. Then $\mathbf{P}_{\mathcal{A}}+\mathbf{P}_{\mathcal{B}}$ is idempotent, which implies that

$$\mathbf{P}_{\mathcal{A}}\mathbf{P}_{\mathcal{B}} + \mathbf{P}_{\mathcal{B}}\mathbf{P}_{\mathcal{A}} = 0,$$

hence, multiplying by $\mathbf{P}_{\mathcal{A}}$ on the left,

$$\mathbf{P}_{\mathcal{A}}\mathbf{P}_{\mathcal{B}} + \mathbf{P}_{\mathcal{A}}\mathbf{P}_{\mathcal{B}}\mathbf{P}_{\mathcal{A}} = 0,$$

implying that $\mathbf{P}_{\mathcal{A}}\mathbf{P}_{\mathcal{B}}$ is symmetric, and therefore $\mathbf{P}_{\mathcal{A}}\mathbf{P}_{\mathcal{B}} = 0$.
In the other direction, for any n-vector $\boldsymbol{z}$, note first that $(\mathbf{P}_{\mathcal{A}}+\mathbf{P}_{\mathcal{B}})\boldsymbol{z} \in \mathcal{A}+\mathcal{B}$. And, if $\mathbf{P}_{\mathcal{A}}\mathbf{P}_{\mathcal{B}} = 0$, then $\mathbf{P}_{\mathcal{B}}\mathbf{P}_{\mathcal{A}} = 0$ because $0_{n\times n}$ is symmetric. Then

$$\begin{aligned} \mathbf{P}_{\mathcal{A}}[\boldsymbol{z} - (\mathbf{P}_{\mathcal{A}} + \mathbf{P}_{\mathcal{B}})\boldsymbol{z}] &= 0 \text{ and} \\ \mathbf{P}_{\mathcal{B}}[\boldsymbol{z} - (\mathbf{P}_{\mathcal{A}} + \mathbf{P}_{\mathcal{B}})\boldsymbol{z}] &= 0. \end{aligned}$$

Thus $\boldsymbol{z}-(\mathbf{P}_{\mathcal{A}}+\mathbf{P}_{\mathcal{B}})\boldsymbol{z} \in \mathcal{A}^{\perp} \cap \mathcal{B}^{\perp} = (\mathcal{A}+\mathcal{B})^{\perp}$. Together, these establish that $\mathbf{P}_{\mathcal{A}} + \mathbf{P}_{\mathcal{B}}$ is the orthogonal projection matrix onto $\mathcal{A} + \mathcal{B}$ iff $\mathbf{P}_{\mathcal{A}}\mathbf{P}_{\mathcal{B}} = 0$.

(b) Follows immediately from (a).

7.21. Note that $\mathbf{P}_{\mathcal{S}} - \mathbf{P}_{\mathcal{S}_0}$ is symmetric and idempotent, and so it is the OP matrix onto its span. It remains to show that $\text{sp}(\mathbf{P}_{\mathcal{S}} - \mathbf{P}_{\mathcal{S}_0}) = \mathcal{S} \cap \mathcal{S}_0^{\perp}$.
Let $\boldsymbol{z} = (\mathbf{P}_{\mathcal{S}} - \mathbf{P}_{\mathcal{S}_0})\boldsymbol{w} \in \text{sp}(\mathbf{P}_{\mathcal{S}} - \mathbf{P}_{\mathcal{S}_0})$. It is clear that, because $\mathcal{S}_0 \subset \mathcal{S}$, $\boldsymbol{z} \in \mathcal{S}$. And

$$\mathbf{P}_{\mathcal{S}_0}\boldsymbol{z} = \mathbf{P}_{\mathcal{S}_0}(\mathbf{P}_{\mathcal{S}} - \mathbf{P}_{\mathcal{S}_0})\boldsymbol{w} = (\mathbf{P}_{\mathcal{S}_0} - \mathbf{P}_{\mathcal{S}_0})\boldsymbol{w} = \mathbf{0}$$

$\Longrightarrow \boldsymbol{z} \in \mathcal{S}_0^{\perp} \Longrightarrow \text{sp}(\mathbf{P}_{\mathcal{S}} - \mathbf{P}_{\mathcal{S}_0}) \subset \mathcal{S} \cap \mathcal{S}_0^{\perp}$.
Let $\boldsymbol{z} \in \mathcal{S} \cap \mathcal{S}_0^{\perp}$. Then $\mathbf{P}_{\mathcal{S}}\boldsymbol{z} = \boldsymbol{z}$, $\mathbf{P}_{\mathcal{S}_0}\boldsymbol{z} = \mathbf{0}$, and so

$$\boldsymbol{z} = \mathbf{P}_{\mathcal{S}}\boldsymbol{z} - \mathbf{P}_{\mathcal{S}_0}\boldsymbol{z} = (\mathbf{P}_{\mathcal{S}} - \mathbf{P}_{\mathcal{S}_0})\boldsymbol{z} \in \text{sp}(\mathbf{P}_{\mathcal{S}} - \mathbf{P}_{\mathcal{S}_0})$$

$\Longrightarrow \mathcal{S} \cap \mathcal{S}_0^{\perp} \subset \text{sp}(\mathbf{P}_{\mathcal{S}} - \mathbf{P}_{\mathcal{S}_0})$.

7.22. Let $\boldsymbol{x}_* \in \mathcal{S}$. Then $A\boldsymbol{x}_* \in A\mathcal{S}$ is in

$$\begin{aligned} \{A\boldsymbol{x} : B\boldsymbol{x} = \mathbf{0}\} &= A[\text{sp}(A') \cap \text{sp}(B')]^{\perp} \\ &= A[\text{sp}(A')^{\perp} + \text{sp}(B')^{\perp}] \end{aligned}$$

$\Longrightarrow \exists\ \boldsymbol{x}_1 \in \text{sp}(A')^{\perp}$ and $\boldsymbol{x}_2 \in \text{sp}(B')^{\perp}$ such that

$$A\boldsymbol{x}_* = A(\boldsymbol{x}_1 + \boldsymbol{x}_2) = A\boldsymbol{x}_2,$$

which implies that $\boldsymbol{x}_* - \boldsymbol{x}_2 \in \text{sp}(A')^{\perp}$, which implies that

$$\boldsymbol{x}_* = (\boldsymbol{x}_* - \boldsymbol{x}_2) + \boldsymbol{x}_2 \in \text{sp}(A')^{\perp} + \text{sp}(B')^{\perp}.$$

7.23. Note that

$$\{A\boldsymbol{x} : B\boldsymbol{x} = \mathbf{0}\} = A\{\text{sp}(A')^{\perp} + \text{sp}(B')^{\perp}\},$$

and the same holds substituting C for B.

$\boxed{\text{I.} \Longrightarrow \text{II.}}$ If I. is true, then

$$\text{sp}(A')^{\perp} + \text{sp}(B')^{\perp} = \text{sp}(A')^{\perp} + \text{sp}(C')^{\perp}$$

$\Longrightarrow$

$$\begin{aligned} \{A\boldsymbol{x} : B\boldsymbol{x} = \mathbf{0}\} &= A\{\text{sp}(A')^{\perp} + \text{sp}(B')^{\perp}\} \\ &= A\{\text{sp}(A')^{\perp} + \text{sp}(C')^{\perp}\} \\ &= \{A\boldsymbol{x} : C\boldsymbol{x} = \mathbf{0}\}. \end{aligned}$$

$\boxed{\text{II.} \Longrightarrow \text{I.}}$ If $\boldsymbol{z}_1 \in \text{sp}(A')^{\perp} + \text{sp}(B')^{\perp}$, then $\exists\ \boldsymbol{z}_{1A} \in \text{sp}(A')^{\perp}$ and $\boldsymbol{z}_{1B} \in \text{sp}(B')^{\perp}$ such that $\boldsymbol{z}_1 = \boldsymbol{z}_{1A} + \boldsymbol{z}_{1B}$. Then, because

$$A\{\text{sp}(A')^{\perp} + \text{sp}(B')^{\perp}\} = A\{\text{sp}(A')^{\perp} + \text{sp}(C')^{\perp}\},$$

there exists $\boldsymbol{z}_2 = \boldsymbol{z}_{2A} + \boldsymbol{z}_{2C}$ with $\boldsymbol{z}_{2A} \in \text{sp}(A')^{\perp}$ and $\boldsymbol{z}_{2C} \in \text{sp}(C')^{\perp}$ such that $A\boldsymbol{z}_1 = A\boldsymbol{z}_2$. Then $A\boldsymbol{z}_1 = A\boldsymbol{z}_{1B} = A\boldsymbol{z}_2 = A\boldsymbol{z}_{2C}$, which implies that $\boldsymbol{z}_{1B} - \boldsymbol{z}_{2C} \in \text{sp}(A')^{\perp}$. Then

$$\begin{aligned} \boldsymbol{z}_1 &= \boldsymbol{z}_{1A} + \boldsymbol{z}_{1B} \\ &= (\boldsymbol{z}_{1A} + \boldsymbol{z}_{1B} - \boldsymbol{z}_{2C}) + \boldsymbol{z}_{2C} \\ &\in \text{sp}(A')^{\perp} + \text{sp}(C')^{\perp}. \end{aligned}$$

That is, $\text{sp}(A')^{\perp} + \text{sp}(B')^{\perp} \subset \text{sp}(A')^{\perp} + \text{sp}(C')^{\perp}$. The same argument works switching B and C, reversing the inclusion. Thus $\text{sp}(A')^{\perp} + \text{sp}(B')^{\perp} = \text{sp}(A')^{\perp} + \text{sp}(C')^{\perp}$.

7.24. $\boxed{\Longrightarrow :}$ This part is due to Denise Danos, née Moore. Suppose $\boldsymbol{z} \in \text{sp}(A') \cap \text{sp}(B')$. Then $\exists\ \boldsymbol{\ell}_1$ and $\boldsymbol{\ell}_2$ such that $\boldsymbol{z} = A'\boldsymbol{\ell}_1 = B'\boldsymbol{\ell}_2$. Then

$$\begin{pmatrix} A \\ B \end{pmatrix}' \begin{pmatrix} \boldsymbol{\ell}_1 \\ -\boldsymbol{\ell}_2 \end{pmatrix} = A'\boldsymbol{\ell}_1 - B'\boldsymbol{\ell}_2 = \mathbf{0},$$

which implies that

$$\mathbf{P}_{\binom{A}{B}} \begin{pmatrix} \boldsymbol{\ell}_1 \\ -\boldsymbol{\ell}_2 \end{pmatrix} = \mathbf{0},$$

and hence, by the supposition,

$$\mathbf{P}_{\binom{A}{B}} \begin{pmatrix} \boldsymbol{\ell}_1 \\ -\boldsymbol{\ell}_2 \end{pmatrix} = \begin{pmatrix} \mathbf{P}_A\boldsymbol{\ell}_1 \\ -\mathbf{P}_B\boldsymbol{\ell}_2 \end{pmatrix} = \begin{pmatrix} \mathbf{0} \\ \mathbf{0} \end{pmatrix},$$

so $\mathbf{P}_A\boldsymbol{\ell}_1 = \mathbf{0} \Longrightarrow \boldsymbol{z} = A'\boldsymbol{\ell}_1 = \mathbf{0}$. Therefore, $\mathrm{sp}(A') \cap \mathrm{sp}(B') = \{\mathbf{0}\}$.

$\boxed{\Longleftarrow :}$ Suppose $\mathrm{sp}(A') \cap \mathrm{sp}(B') = \{\mathbf{0}\}$. Let $\boldsymbol{y} = \binom{\boldsymbol{y}_1}{\boldsymbol{y}_2} \in \Re^{a+b}$. Note that this implies that $\mathrm{sp}(A') + \mathrm{sp}(B')$ is a direct sum (see 29, p. 35). Then

$$\boldsymbol{y} = \begin{pmatrix} \boldsymbol{y}_1 \\ \boldsymbol{y}_2 \end{pmatrix} = \mathbf{P}_{\binom{A}{B}} \begin{pmatrix} \boldsymbol{y}_1 \\ \boldsymbol{y}_2 \end{pmatrix} + \begin{pmatrix} \boldsymbol{y}_{21} \\ \boldsymbol{y}_{22} \end{pmatrix} = \begin{pmatrix} A \\ B \end{pmatrix} \boldsymbol{b} + \begin{pmatrix} \boldsymbol{y}_{21} \\ \boldsymbol{y}_{22} \end{pmatrix}$$

for some $\boldsymbol{b}$ and $\binom{\boldsymbol{y}_{21}}{\boldsymbol{y}_{22}} \in \mathrm{sp}\binom{A}{B}^{\perp}$. Then $\mathbf{0} = \binom{A}{B}'\binom{\boldsymbol{y}_{21}}{\boldsymbol{y}_{22}} = A'\boldsymbol{y}_{21} + B'\boldsymbol{y}_{22}$, hence $A'\boldsymbol{y}_{21} = -B'\boldsymbol{y}_{22}$, which implies that both $A'\boldsymbol{y}_{21}$ and $B'\boldsymbol{y}_{22}$ are in $\mathrm{sp}(A') \cap \mathrm{sp}(B')$, and hence both are $\mathbf{0}$. [Alternatively, they are both $\mathbf{0}$ because of the direct sum.] This implies that $\boldsymbol{y}_{21} \in \mathrm{sp}(A)^{\perp}$ and $\boldsymbol{y}_{22} \in \mathrm{sp}(B)^{\perp}$. Therefore, $\boldsymbol{y}_1 = A\boldsymbol{b} + \boldsymbol{y}_{21}$, with $A\boldsymbol{b} \in \mathrm{sp}(A)$ and $\boldsymbol{y}_1 - A\boldsymbol{b} = \boldsymbol{y}_{21} \in \mathrm{sp}(A)^{\perp}$, and hence $A\boldsymbol{b}$ is the orthogonal projection of $\boldsymbol{y}_1$ in $\mathrm{sp}(A)$; and similarly $B\boldsymbol{b} = \mathbf{P}_B\boldsymbol{y}_2$.

7.25. Switch A and B for proofs of the second part of each assertion.

(a) $\boxed{\subset:}$ $\boldsymbol{z} \in \mathrm{sp}(A') \cap \mathrm{sp}(B') \Longrightarrow \exists\, \boldsymbol{a}, \boldsymbol{b}$ such that $\boldsymbol{z} = A'\boldsymbol{a} = B'\boldsymbol{b} \Longrightarrow$

$$A(\mathrm{I} - \mathbf{P}_{B'})\boldsymbol{z} = A(\mathrm{I} - \mathbf{P}_{B'})A'\boldsymbol{a} = A(\mathrm{I} - \mathbf{P}_{B'})B'\boldsymbol{b} = \mathbf{0}$$

$\Longrightarrow \boldsymbol{z} \in \mathrm{sp}[(\mathrm{I} - \mathbf{P}_{B'})A']^{\perp} \Longrightarrow \mathrm{sp}(A') \cap \mathrm{sp}(B') \subset \mathrm{sp}(A') \cap [\mathrm{sp}(\mathrm{I} - \mathbf{P}_{B'})A']^{\perp}$.

$\boxed{\supset:}$ $\boldsymbol{z} \in \mathrm{sp}(A') \cap \mathrm{sp}[(\mathrm{I} - \mathbf{P}_{B'})A']^{\perp} \Longrightarrow \exists\, \boldsymbol{a}$ such that $\boldsymbol{z} = A'\boldsymbol{a}$ and $A(\mathrm{I} - \mathbf{P}_{B'})\boldsymbol{z} = A(\mathrm{I} - \mathbf{P}_{B'})A'\boldsymbol{a} = A(\mathrm{I} - \mathbf{P}_{B'})(\mathrm{I} - \mathbf{P}_{B'})A'\boldsymbol{a} = \mathbf{0} \Longrightarrow (\mathrm{I} - \mathbf{P}_{B'})A'\boldsymbol{a} = \mathbf{0} \Longrightarrow A'\boldsymbol{a} = \boldsymbol{z} \in \mathrm{sp}(B') \Longrightarrow \boldsymbol{z} \in \mathrm{sp}(A') \cap \mathrm{sp}(B')$.

(b) $\boxed{\mathrm{sp}(A')^{\perp} \cap \mathrm{sp}[(\mathrm{I} - \mathbf{P}_{B'})A'] = \mathbf{0}:}$ $\boldsymbol{z} = (\mathrm{I} - \mathbf{P}_{B'})A'\boldsymbol{a}$ and $A\boldsymbol{z} = \mathbf{0} \Longrightarrow \mathbf{0} = A(\mathrm{I}-\mathbf{P}_{B'})A'\boldsymbol{a} = A(\mathrm{I}-\mathbf{P}_{B'})(\mathrm{I}-\mathbf{P}_{B'})A'\boldsymbol{a} \Longrightarrow (\mathrm{I}-\mathbf{P}_{B'})A'\boldsymbol{a} = \boldsymbol{z} = \mathbf{0}$.

(c) By Exercise 22,

$$\begin{aligned} \{A\boldsymbol{x} : B\boldsymbol{x} = \mathbf{0}\} &= A\{\mathrm{sp}(A')^{\perp} + \mathrm{sp}(B')^{\perp}\} \\ &= A\{\mathrm{sp}(B')^{\perp}\} \\ &= A\{\mathrm{sp}(A')^{\perp} + \mathrm{sp}[(\mathrm{I} - \mathbf{P}_{B'})A']\}, \text{ by 7.2,} \\ &= A\{\mathrm{sp}[(\mathrm{I} - \mathbf{P}_{B'})A']\}. \end{aligned}$$

7.26.

(a) $\boxed{\Longrightarrow:}$ $\mathrm{sp}(B') \subset \mathrm{sp}(A') \Longrightarrow (\mathrm{I} - \mathbf{P}_{A'})B' = 0$.
$\boxed{\Longleftarrow:}$ $(\mathrm{I} - \mathbf{P}_{A'})B' = 0 \Longrightarrow \mathrm{sp}(B') \subset \mathrm{sp}(A')$.

(b) $\boxed{\Longrightarrow:}$ If $\boldsymbol{z}_*$ is such that $A\boldsymbol{z}_* = \mathbf{0}$ and $B\boldsymbol{z}_* \neq \mathbf{0}$, then $\boldsymbol{z}_* \in \mathrm{sp}(A')^{\perp}$ and $\boldsymbol{z}_* \notin \mathrm{sp}(B')^{\perp}$, hence $\mathrm{sp}(A')^{\perp} \not\subset \mathrm{sp}(B')^{\perp}$.

$\boxed{\Longleftarrow:}$ If $\mathrm{sp}(A')^{\perp} \not\subset \mathrm{sp}(B')^{\perp}$, then there exists $\boldsymbol{z}_*$ such that $\boldsymbol{z}_* \in \mathrm{sp}(A')^{\perp}$ and $\boldsymbol{z}_* \notin \mathrm{sp}(B')^{\perp}$. That is, $A\boldsymbol{z}_* = \mathbf{0}$ and $B\boldsymbol{z}_* \neq \mathbf{0}$.

(c) Let $\boldsymbol{w}$ be such that $\boldsymbol{z} = (\mathrm{I}-\mathbf{P}_{A'})B'\boldsymbol{w}$. Then $B\boldsymbol{z} = B(\mathrm{I}-\mathbf{P}_{A'})B'\boldsymbol{w} = \mathbf{0}$ $\Longrightarrow \boldsymbol{z} = (\mathrm{I} - \mathbf{P}_{A'})B'\boldsymbol{w} = \mathbf{0}$.

(d) $\boxed{\Longrightarrow:}$ If $\text{sp}(A') \subset \text{sp}(B')$, then $\exists\ C$ such that $A' = B'C$. Then

$$AB^-B = C'BB^-B = C'B = A.$$

$\boxed{\Longleftarrow:}$ $AB^-B = A \Longrightarrow A' = B'(B^-)'A' \Longrightarrow \text{sp}(A') \subset \text{sp}(B')$.

7.27.

(a) $\boxed{\text{sp}(H) \subset \text{sp}(A)}$

$\text{sp}(H) = \text{sp}(\mathbf{P}_H) = \text{sp}(\mathbf{P}_A - \mathbf{P}_{AN}) = \text{sp}(A) \cap \text{sp}(AN)^\perp \subset \text{sp}(A)$.

$\boxed{\text{sp}(H) \subset \text{sp}(A) \Longrightarrow \text{sp}(H, A|H) = \text{sp}(A)}$

Clearly,$\text{sp}(H, A|H) \subset \text{sp}(A)$. $A\boldsymbol{x} \in \text{sp}(A) \Longrightarrow \mathbf{P}_H(A\boldsymbol{x}) + (\text{I} - \mathbf{P}_H)A\boldsymbol{x} \in \text{sp}(H, A|H)$. Therefore, $\text{sp}(A) \subset \text{sp}(H, A|H)$, and hence $\text{sp}(A) = \text{sp}(H, A|H)$.

(b) $\boxed{\text{sp}(AN) = \{A\boldsymbol{x} : B\boldsymbol{x} = \mathbf{0}\}}$

$\boxed{\subset:}$ $\boldsymbol{z} \in \text{sp}(AN) \Longrightarrow \exists\ \boldsymbol{w}$ such that $\boldsymbol{z} = A(N\boldsymbol{w}) \in \{A\boldsymbol{x} : B\boldsymbol{x} = \mathbf{0}\}$.

$\boxed{\supset:}$ $\boldsymbol{z} \in \{A\boldsymbol{x} : B\boldsymbol{x} = \mathbf{0}\} \Longrightarrow \exists\ \boldsymbol{x}_0$ such that $B\boldsymbol{x}_0 = \mathbf{0}$ and $\boldsymbol{z} = A\boldsymbol{x}_0$. Then $\boldsymbol{x}_0 \in \text{sp}(B')^\perp = \text{sp}(N) \Longrightarrow \exists \boldsymbol{w}$ such that $\boldsymbol{x}_0 = N\boldsymbol{w} \Longrightarrow A\boldsymbol{x}_0 = AN\boldsymbol{w} \in \text{sp}(AN)$.

$\boxed{\text{sp}(AN) = \text{sp}(A) \cap \text{sp}(H)^\perp}$

$\boxed{\subset:}$ $\boldsymbol{z} = AN\boldsymbol{x} \subset \text{sp}(A) \Longrightarrow \mathbf{P}_H\boldsymbol{z} = (\mathbf{P}_A - \mathbf{P}_{AN})\boldsymbol{z} = \boldsymbol{z} - \boldsymbol{z} = \mathbf{0} \Longrightarrow \boldsymbol{z} \in \text{sp}(H)^\perp \Longrightarrow \boldsymbol{z} \in \text{sp}(A) \cap \text{sp}(H)^\perp$.

$\boxed{\supset:}$ $\boldsymbol{z} \in \text{sp}(A) \cap \text{sp}(H)^\perp \Longrightarrow \mathbf{P}_H\boldsymbol{z} = \mathbf{0} \Longrightarrow \mathbf{P}_A\boldsymbol{z} = \boldsymbol{z} = \mathbf{P}_{AN}\boldsymbol{z} \Longrightarrow \boldsymbol{z} \in \text{sp}(AN)$.

$\boxed{\text{sp}(A) \cap \text{sp}(H)^\perp = \{A\boldsymbol{x} : H'A\boldsymbol{x} = \mathbf{0}\}}$ Clear.

$\boxed{\{A\boldsymbol{x} : H'A\boldsymbol{x} = \mathbf{0}\} = \text{sp}(A|H)}$

$\boxed{\subset:}$ $\boldsymbol{z} \in \{A\boldsymbol{x} : H'A\boldsymbol{x} = \mathbf{0}\} \Longrightarrow \boldsymbol{z} = A\boldsymbol{x}$ and $\mathbf{P}_H\boldsymbol{z} = \mathbf{0} = (\mathbf{P}_A - \mathbf{P}_{AN})A\boldsymbol{x} = \boldsymbol{z} - \mathbf{P}_{AN}\boldsymbol{z} \Longrightarrow \boldsymbol{z} = \mathbf{P}_{AN}\boldsymbol{z} = (\mathbf{P}_A - \mathbf{P}_H)\boldsymbol{z} = A\boldsymbol{x} - \mathbf{P}_H A\boldsymbol{x} \Longrightarrow \boldsymbol{z} = (\text{I} - \mathbf{P}_H)A\boldsymbol{x} \in \text{sp}(A|H)$.

$\boxed{\supset:}$ Note that $\text{sp}(A|H) \subset \text{sp}(A)$ because $\text{sp}(H) \subset \text{sp}(A)$. Then $\boldsymbol{z} \in \text{sp}(A|H) \Longrightarrow \boldsymbol{z} \in \text{sp}(A)$ and $\boldsymbol{z} = (\text{I} - \mathbf{P}_H)A\boldsymbol{x} \Longrightarrow H'\boldsymbol{z} = H'(\text{I} - \mathbf{P}_H)A\boldsymbol{x} = \mathbf{0} \Longrightarrow \boldsymbol{z} \in \{A\boldsymbol{x} : H'A\boldsymbol{x} = \mathbf{0}\}$.

(c) $\boxed{\text{sp}(A'H)^\perp = \text{sp}(A')^\perp + \text{sp}(B')^\perp}$

$\boxed{\subset:}$ $\boldsymbol{z} \in \text{sp}(A'H)^\perp \Longrightarrow \mathbf{0} = \mathbf{P}_H A\boldsymbol{z} = (\mathbf{P}_A - \mathbf{P}_{AN})A\boldsymbol{z} = A\boldsymbol{z} - A\boldsymbol{x} = A(\boldsymbol{z} - \boldsymbol{x})$, with $B\boldsymbol{x} = \mathbf{0}$. Thus $\boldsymbol{z} = (\boldsymbol{z} - \boldsymbol{x}) + \boldsymbol{x}$, with $\boldsymbol{z} - \boldsymbol{x} \in \text{sp}(A')^\perp$ and $\boldsymbol{x} \in \text{sp}(B')^\perp$.

$\boxed{\supset:}$ $\boldsymbol{z} = \boldsymbol{z}_A + \boldsymbol{z}_B$ with $\boldsymbol{z}_A \in \text{sp}(A')^\perp$ and $\boldsymbol{z}_B \in \text{sp}(B')^\perp \Longrightarrow A\boldsymbol{z} = A\boldsymbol{z}_B$. Then $\mathbf{P}_H A\boldsymbol{z} = (\mathbf{P}_A - \mathbf{P}_{AN})A\boldsymbol{z}_B = \mathbf{0}$ because $A\boldsymbol{z}_B \in \text{sp}(AN)$. Therefore, $H'A\boldsymbol{z} = \mathbf{0}$ and hence $\boldsymbol{z} \in \text{sp}(A'H)^\perp$.

(d) $\boxed{\{A\boldsymbol{x} : B\boldsymbol{x} = \mathbf{0}\} = \text{sp}(A) \iff \text{sp}(A') \cap \text{sp}(B') = \{\mathbf{0}\}}$

$\{A\boldsymbol{x} : B\boldsymbol{x} = \mathbf{0}\} = A[\text{sp}(A') \cap \text{sp}(B')]^{\perp}$; if $\text{sp}(A') \cap \text{sp}(B') = \{\mathbf{0}\}$, then $[\text{sp}(A') \cap \text{sp}(B')]^{\perp} = \{\mathbf{0}\}^{\perp} = \Re^c$, and $A\Re^c = \text{sp}(A)$.

$\{A\boldsymbol{x} : B\boldsymbol{x} = \mathbf{0}\} = \text{sp}(AN)$; if $\{A\boldsymbol{x} : B\boldsymbol{x} = \mathbf{0}\} = \text{sp}(A)$, then $\text{sp}(AN) = \text{sp}(A)$; and then $\mathbf{P}_{AN} = \mathbf{P}_A - \mathbf{P}_H$ implies that $\mathbf{P}_H = 0$, which implies that $H = 0$, and hence that $\text{sp}(A') \cap \text{sp}(B') = \text{sp}(A'H) = \{\mathbf{0}\}$.

(e) $\boxed{\{A\boldsymbol{x} : B\boldsymbol{x} = \mathbf{0}\} = \{\mathbf{0}\} \iff \text{sp}(A') \subset \text{sp}(B')}$

If $\text{sp}(A') \subset \text{sp}(B')$, then $\{A\boldsymbol{x} : B\boldsymbol{x} = \mathbf{0}\} = A\text{sp}(B')^{\perp} = \{\mathbf{0}\}$ because $\text{sp}(B')^{\perp} \subset \text{sp}(A')^{\perp}$.

If $\{A\boldsymbol{x} : B\boldsymbol{x} = \mathbf{0}\} = \{\mathbf{0}\}$, then $\text{sp}(AN) = \{\mathbf{0}\}$, and, because $\mathbf{P}_H = \mathbf{P}_A - \mathbf{P}_{AN}$, that implies that $\mathbf{P}_H = \mathbf{P}_A$ and hence that $\text{sp}(H) = \text{sp}(A)$ and hence that $\text{sp}(A') \cap \text{sp}(B') = \text{sp}(A'H) = \text{sp}(A'A) = \text{sp}(A')$. That in turn implies that $\text{sp}(A') \subset \text{sp}(B')$.

(f) $\boxed{\mathbf{P}_{A|H} = \mathbf{P}_A - \mathbf{P}_H = \mathbf{P}_{AN}}$ follows directly from (b), that $\text{sp}(AN) = \text{sp}(A|H)$.

(g) $\boxed{\text{sp}[(A|H)'] \cap \text{sp}(A'H) = \{\mathbf{0}\}}$

$\boldsymbol{z} \in \text{sp}[(A|H)'] \cap \text{sp}(A'H) \Longrightarrow \exists\ \boldsymbol{\ell}, \boldsymbol{c}$ such that $\boldsymbol{z} = A'(\text{I} - \mathbf{P}_H)\boldsymbol{\ell} = A'H\boldsymbol{c}$. This implies that $(\text{I} - \mathbf{P}_H)\boldsymbol{\ell} - H\boldsymbol{c} \in \text{sp}(A)^{\perp} \subset \text{sp}(H)^{\perp}$, which implies that $\mathbf{P}_H H\boldsymbol{c} = H\boldsymbol{c} = \mathbf{0}$, which implies that $\boldsymbol{z} = \mathbf{0}$.

(h) $\boxed{\text{sp}[(A|H)'] \cap \text{sp}(B') = \{\mathbf{0}\}}$

It suffices to show that $\text{sp}[(A|H)'] \cap \text{sp}(B') = \text{sp}[(A|H)'] \cap \text{sp}(A'H)$.

$$
\begin{aligned}
\{\text{sp}[(A|H)'] \cap \text{sp}(A'H)\}^{\perp} &= \text{sp}[(A|H)']^{\perp} + \text{sp}(A'H)^{\perp} \\
&= \text{sp}[(A|H)']^{\perp} + [\text{sp}(A') \cap \text{sp}(B')]^{\perp} \\
&= \text{sp}[(A|H)']^{\perp} + \text{sp}(A')^{\perp} + \text{sp}(B')^{\perp} \\
&= \text{sp}[(A|H)']^{\perp} + \text{sp}(B')^{\perp} \\
&\quad \text{because } \text{sp}(A')^{\perp} \subset \text{sp}[(A|H)']^{\perp} \\
&= \{\text{sp}[(A|H)'] \cap \text{sp}(B')\}^{\perp}.
\end{aligned}
$$

(i) $\boxed{\text{sp}[(A|H)'] \cap \text{sp}(A')^{\perp} = (\mathbf{0})}$

$\boldsymbol{z} \in \text{sp}[(A|H)'] \Longrightarrow \exists\ \boldsymbol{x}$ such that $\boldsymbol{z} = A'(\text{I} - \mathbf{P}_H)\boldsymbol{x}$; and $\boldsymbol{z} \in \text{sp}(A')^{\perp} \Longrightarrow A\boldsymbol{z} = \mathbf{0}$. Then $\boldsymbol{z}'\boldsymbol{z} = \boldsymbol{x}'(\text{I} - \mathbf{P}_H)A\boldsymbol{z} = \mathbf{0} \Longrightarrow \boldsymbol{z} = \mathbf{0}$.

(j) From 24 and that $\text{sp}[(A|H)'] \cap \text{sp}(B') = \mathbf{0}$, it follows that

$$
\mathbf{P}_{\binom{A|H}{B}} = \begin{pmatrix} \mathbf{P}_{A|H} & 0 \\ 0 & \mathbf{P}_B \end{pmatrix},
$$

and hence that

$$\mathbf{P}_{\binom{A|H}{B}}\begin{pmatrix}\boldsymbol{z}\\ \mathbf{0}\end{pmatrix} = \begin{pmatrix}\mathbf{P}_{A|H}\boldsymbol{z}\\ \mathbf{P}_B\mathbf{0}\end{pmatrix}.$$

To show that $(A|H)\boldsymbol{x} = A\boldsymbol{x}$: $B\boldsymbol{x} = \mathbf{0} \Longrightarrow \boldsymbol{x} \in \text{sp}(B')^{\perp} \subset \text{sp}(B')^{\perp} + \text{sp}(A')^{\perp} = [\text{sp}(A'H)]^{\perp} \Longrightarrow \mathbf{P}_H A\boldsymbol{x} = \mathbf{0}$. Therefore, $\mathbf{P}_{A|H}\boldsymbol{z} = A\boldsymbol{x}$ with $B\boldsymbol{x} = \mathbf{0}$.

7.28. $A'H = A'LB \Longrightarrow \text{sp}(H - LB) \subset \text{sp}(A)^{\perp} \cap \text{sp}(A) = \{\mathbf{0}\} \Longrightarrow \text{sp}(H) \subset \text{sp}(L)$. And $A'L = A'HC \Longrightarrow \text{sp}(L - HC) \subset \text{sp}(A)^{\perp} \cap \text{sp}(A) = \{\mathbf{0}\} \Longrightarrow \text{sp}(L) \subset \text{sp}(H)$.

7.29.

Let N be a matrix such that $\text{sp}(N) = \text{sp}(G)^{\perp} = \text{sp}(G_*)^{\perp}$. Let $L = \{\boldsymbol{b} \in \Re^r : G'\boldsymbol{b} = G'\boldsymbol{b}_0\}$ and $R = \{\boldsymbol{b} \in \Re^r : G'_*\boldsymbol{b} = G'_*\boldsymbol{b}_0\}$. If $\boldsymbol{b}_L \in L$, then $G'\boldsymbol{b}_L = G'\boldsymbol{b}_0 \Longrightarrow \exists\ \boldsymbol{z}_L$ such that $\boldsymbol{b}_L = \boldsymbol{b}_0 + N\boldsymbol{z}_L$. Then $G'_*\boldsymbol{b}_L = G'_*\boldsymbol{b}_0$, which implies that $\boldsymbol{b}_L \in R$.

If $\boldsymbol{b}_R \in R$, then $G'_*\boldsymbol{b}_R = G'_*\boldsymbol{b}_0 \Longrightarrow \exists\ \boldsymbol{z}_R$ such that $\boldsymbol{b}_R = \boldsymbol{b}_0 + N\boldsymbol{z}_R$. Then $G'\boldsymbol{b}_R = G'\boldsymbol{b}_0 = \boldsymbol{c}_0 \Longrightarrow \boldsymbol{b}_R \in L$.

7.30. Patience and precision are sufficient to establish all parts. For (e), establish that $\mathbf{P}_A \otimes \mathbf{P}_B$ is symmetric and idempotent and that $\text{sp}(A \otimes B) = \text{sp}(\mathbf{P}_A \otimes \mathbf{P}_B)$.

7.31.

(a)

$$\begin{aligned}(\boldsymbol{y} - \boldsymbol{s})'(\boldsymbol{y} - \boldsymbol{s}) &= (\boldsymbol{y} - \hat{\boldsymbol{y}} + \hat{\boldsymbol{y}} - \boldsymbol{s})'(\boldsymbol{y} - \hat{\boldsymbol{y}} + \hat{\boldsymbol{y}} - \boldsymbol{s})\\ &= (\boldsymbol{y} - \hat{\boldsymbol{y}})'(\boldsymbol{y} - \hat{\boldsymbol{y}}) + (\hat{\boldsymbol{y}} - \boldsymbol{s})'(\hat{\boldsymbol{y}} - \boldsymbol{s})\\ &\geqslant (\boldsymbol{y} - \hat{\boldsymbol{y}})'(\boldsymbol{y} - \hat{\boldsymbol{y}})\end{aligned}$$

because $(\boldsymbol{y} - \hat{\boldsymbol{y}}) \in \mathcal{S}^{\perp}$ and $(\hat{\boldsymbol{y}} - \boldsymbol{s}) \in \mathcal{S} \Longrightarrow (\boldsymbol{y} - \hat{\boldsymbol{y}})'(\hat{\boldsymbol{y}} - \boldsymbol{s}) = 0$. Equality holds iff $(\hat{\boldsymbol{y}} - \boldsymbol{s})'(\hat{\boldsymbol{y}} - \boldsymbol{s}) = 0$ iff $\hat{\boldsymbol{y}} = \boldsymbol{s}$.

(b) Clear.

(c)
i. As shown previously, the SS is minimized by $\check{\boldsymbol{b}}$ iff $X\check{\boldsymbol{b}} = \mathbf{P}_X\boldsymbol{y}$, that is, iff $\check{\boldsymbol{b}}$ is a LS solution.
ii. This equation is equivalent to "$X\check{\boldsymbol{b}} = \mathbf{P}_X\boldsymbol{y}$," because $X\check{\boldsymbol{b}} \in \text{sp}(X)$ and $\boldsymbol{y} - X\check{\boldsymbol{b}} \in \text{sp}(X)^{\perp}$. That is, it is equivalent to "$\check{\boldsymbol{b}}$ is a LS solution."
iii. $XTQ'\boldsymbol{y} = QQ'\boldsymbol{y} = \mathbf{P}_X\boldsymbol{y}$.
iv. $X(X'X)^{-}X' = \mathbf{P}_X$.
v. Immediately apparent.

vi. If $\boldsymbol{g} \in \mathrm{sp}(X')$, then $\exists\ \boldsymbol{\ell}$ such that $\boldsymbol{g} = X'\boldsymbol{\ell} = X'\boldsymbol{\ell}_*$, where $\boldsymbol{\ell}_* = \mathbf{P}_X\boldsymbol{\ell}$ is the unique vector in $\mathrm{sp}(X)$ such that $X'\boldsymbol{\ell}_* = \boldsymbol{g}$. Then, for any $\check{\boldsymbol{b}}$ such that $X\check{\boldsymbol{b}}(\boldsymbol{y}) = \mathbf{P}_X\boldsymbol{y}$ for all $\boldsymbol{y} \in \Re^n$, $\boldsymbol{g}'\check{\boldsymbol{b}} = \boldsymbol{\ell}_*'\boldsymbol{y}$.
If $\check{\boldsymbol{b}}(\boldsymbol{y}) = A'\boldsymbol{y}$ is a linear LS solution, then $XA' = \mathbf{P}_X$. That $XA' = \mathbf{P}_X$ is equivalent to $A' = A_0' + (\mathrm{I} - \mathbf{P}_{X'})Z$ for some Z of appropriate dimensions. Then "$\boldsymbol{g}'A'$ is the same for all solutions A' to $XA' = \mathbf{P}_X$" is true iff $(\mathrm{I}-\mathbf{P}_{X'})\boldsymbol{g} = \mathbf{0}$, that is, iff $\boldsymbol{g} \in \mathrm{sp}(X')$.

7.32.

Let $\boldsymbol{z} = \boldsymbol{y} - \boldsymbol{m}_0$ and $X\boldsymbol{d} = \boldsymbol{m} - \boldsymbol{m}_0$ (because $\boldsymbol{m} - \boldsymbol{m}_0 \in \mathrm{sp}(X)$). Then, as a result of Exercise 31,

$$\begin{aligned}(\boldsymbol{y} - \boldsymbol{m})'(\boldsymbol{y} - \boldsymbol{m}) &= (\boldsymbol{z} - X\boldsymbol{d})'(\boldsymbol{z} - X\boldsymbol{d}) \\ &\geqslant \boldsymbol{z}'(\mathrm{I} - \mathbf{P}_X)\boldsymbol{z} \\ &= (\boldsymbol{y} - \boldsymbol{m}_0)'(\mathrm{I} - \mathbf{P}_X)(\boldsymbol{y} - \boldsymbol{m}_0),\end{aligned}$$

and equality holds iff $X\boldsymbol{d} = \mathbf{P}_X\boldsymbol{z}$, which is equivalent to $\boldsymbol{m} = \boldsymbol{m}_0 + \mathbf{P}_X(\boldsymbol{y} - \boldsymbol{m}_0)$.

7.33.

(a) $\boxed{\mathcal{M} \subset \{X\boldsymbol{b}_0\} + \mathrm{sp}(XN):}$ If $\boldsymbol{m} \in \mathcal{M}$, then there exists $\boldsymbol{b}$ such that $\boldsymbol{m} = X\boldsymbol{b}$ and $R'\boldsymbol{b} = \boldsymbol{r}_0$. This implies that there exists $\boldsymbol{z}$ such that $\boldsymbol{b} = \boldsymbol{b}_0 + N\boldsymbol{z}$, and hence that $\boldsymbol{m} = X\boldsymbol{b}_0 + XN\boldsymbol{z}$, which is a member of $\{X\boldsymbol{b}_0\} + \mathrm{sp}(XN)$.

$\boxed{\{X\boldsymbol{b}_0\} + \mathrm{sp}(XN) \subset \mathcal{M}:}$ If $\boldsymbol{m} \in \{X\boldsymbol{b}_0\} + \mathrm{sp}(XN)$, then $\exists\ \boldsymbol{z}$ such that $\boldsymbol{m} = X\boldsymbol{b}_0 + XN\boldsymbol{z} = X(\boldsymbol{b}_0 + N\boldsymbol{z})$. Because $\mathrm{sp}(N) \subset \mathrm{sp}(R)^\perp + \mathrm{sp}(X')^\perp$, there exist vectors $\boldsymbol{u} \in \mathrm{sp}(R)^\perp$ and $\boldsymbol{v} \in \mathrm{sp}(X')^\perp$ such that $N\boldsymbol{z} = \boldsymbol{u} + \boldsymbol{v}$. Then $XN\boldsymbol{z} = X\boldsymbol{u}$, and hence $\boldsymbol{m} = X(\boldsymbol{b}_0 + \boldsymbol{u})$, and $R'(\boldsymbol{b}_0 + \boldsymbol{u}) = R'\boldsymbol{b}_0 = \boldsymbol{r}_0$. Therefore, $\boldsymbol{m} \in \mathcal{M}$.

(b) If $\boldsymbol{m} \in \mathcal{M}$, then $\exists\ \boldsymbol{c}$ such that $\boldsymbol{m} = X\boldsymbol{b}_0 + XN\boldsymbol{c}$. Let $\boldsymbol{z} = \boldsymbol{y} - X\boldsymbol{b}_0$. Then, for any $\boldsymbol{y}$ and any $X\boldsymbol{b} \in \mathcal{M}$,

$$\begin{aligned}(\boldsymbol{y} - X\boldsymbol{b})'(\boldsymbol{y} - X\boldsymbol{b}) &= (\boldsymbol{z} - XN\boldsymbol{c})'(\boldsymbol{z} - XN\boldsymbol{c}) \\ &\geqslant \boldsymbol{z}'(\mathrm{I} - \mathbf{P}_{XN})\boldsymbol{z}\end{aligned}$$

with equality iff $XN\boldsymbol{c} = \mathbf{P}_{XN}\boldsymbol{z}$. Thus $\hat{\boldsymbol{c}}$ is an RLS solution in the model $XN\boldsymbol{c}$ for $\boldsymbol{z}$ iff $XN\hat{\boldsymbol{c}} = \mathbf{P}_{XN}\boldsymbol{z}$. In that case, $\tilde{\boldsymbol{b}} = \boldsymbol{b}_0 + N\hat{\boldsymbol{c}}$ is such that $X\tilde{\boldsymbol{b}}$ is in $\mathcal{M}$, and among members of $\mathcal{M}$ it minimizes $(\boldsymbol{y} - X\boldsymbol{b})'(\boldsymbol{y} - X\boldsymbol{b})$. We may say that $\tilde{\boldsymbol{b}}$ is an RLS solution in the model $\mathcal{M}$ for $\boldsymbol{y}$ iff $X\tilde{\boldsymbol{b}} = X\boldsymbol{b}_0 + \mathbf{P}_{XN}(\boldsymbol{y} - X\boldsymbol{b}_0)$.

(c) One set of computational steps:

i. Find $\boldsymbol{b}_0$: $\mathrm{GS}(R') \to Q, T$; $\boldsymbol{b}_0 = TQ'\boldsymbol{r}_0$.

ii. Find N: $\mathrm{GS}(R, \mathrm{I}) \to (Q_1, Q_2), T$: $N = Q_2$, then $\mathrm{sp}(N) = \mathrm{sp}(R)^\perp$.

iii. Find $\mathbf{P}_{XN}$: $\text{GS}(XN) \to Q, T$; $\mathbf{P}_{XN} = QQ'$.
iv. Find $\hat{\boldsymbol{z}} = \mathbf{P}_{XN}(\boldsymbol{y} - X\boldsymbol{b}_0)$, then $\hat{\boldsymbol{c}} = TQ'\hat{\boldsymbol{z}}$.
v. $\tilde{\boldsymbol{b}} = \boldsymbol{b}_0 + N\hat{\boldsymbol{c}}$.

(d)

$$\begin{aligned}(\boldsymbol{y} - X\tilde{\boldsymbol{b}})'(\boldsymbol{y} - X\tilde{\boldsymbol{b}}) &= [\boldsymbol{y} - X\boldsymbol{b}_0 - \mathbf{P}_{XN}(\boldsymbol{y} - X\boldsymbol{b}_0)]' \\ &\quad \times [\boldsymbol{y} - X\boldsymbol{b}_0 - \mathbf{P}_{XN}(\boldsymbol{y} - X\boldsymbol{b}_0)] \\ &= (\boldsymbol{y} - X\boldsymbol{b}_0)'(\mathrm{I} - \mathbf{P}_{XN})(\boldsymbol{y} - X\boldsymbol{b}_0).\end{aligned}$$

A.8 Chapter 8 Proofs and Solutions

Proof of Propn. 8.1, p. 63.

1. Because $\boldsymbol{g} \in \text{sp}(X')$, $\exists\ \boldsymbol{\ell}_0$ such that $X'\boldsymbol{\ell}_0 = \boldsymbol{g}$. Thus $\mathcal{L}$ is not empty. For $\boldsymbol{\ell} \in \mathcal{L}$, let $\boldsymbol{\ell}_* = \mathbf{P}_X\boldsymbol{\ell}$. Then $\boldsymbol{\ell}_* \in \text{sp}(X)$ and $X'\boldsymbol{\ell}_* = X'\mathbf{P}_X\boldsymbol{\ell} = X'\boldsymbol{\ell} = \boldsymbol{g}$. If $\boldsymbol{\ell}_{**} \in \text{sp}(X)$ and $X'(\boldsymbol{\ell}_* - \boldsymbol{\ell}_{**}) = \mathbf{0}$, then $\boldsymbol{\ell}_* - \boldsymbol{\ell}_{**} \in \text{sp}(X) \cap \text{sp}(X)^\perp$ and therefore $\boldsymbol{\ell}_* = \boldsymbol{\ell}_{**}$. That is, $\boldsymbol{\ell}_* \in \text{sp}(X)$ such that $X'\boldsymbol{\ell}_* = \boldsymbol{g}$ is unique.

2. If $\boldsymbol{\ell} \in \mathcal{L}$, then $X'\boldsymbol{\ell} = \boldsymbol{g}$ and $X'(\mathbf{P}_X\boldsymbol{\ell}) = \boldsymbol{g}$, which implies by (1) that $\mathbf{P}_X\boldsymbol{\ell} = \boldsymbol{\ell}_*$. If $\mathbf{P}_X\boldsymbol{\ell} = \boldsymbol{\ell}_*$, then $X'\boldsymbol{\ell} = X'\mathbf{P}_X\boldsymbol{\ell} = X'\boldsymbol{\ell}_* = \boldsymbol{g}$, and therefore $\boldsymbol{\ell} \in \mathcal{L}$.

3. If $\boldsymbol{\ell} \in \mathcal{L}$, then

$$\boldsymbol{\ell}'\boldsymbol{\ell} - \boldsymbol{\ell}_*'\boldsymbol{\ell}_* = [(\mathrm{I} - \mathbf{P}_X)\boldsymbol{\ell}]'[(\mathrm{I} - \mathbf{P}_X)\boldsymbol{\ell}] \geqslant 0,$$

and equality holds iff $(\mathrm{I} - \mathbf{P}_X)\boldsymbol{\ell} = \mathbf{0}$, that is, iff $\boldsymbol{\ell} = \boldsymbol{\ell}_*$.

Proof of Propn. 8.2, p. 63. Let $\boldsymbol{\ell}_*$ be the vector in $\text{sp}(X)$ such that $\boldsymbol{g} = X'\boldsymbol{\ell}_*$. Then

$$\begin{aligned}\boldsymbol{g}'\check{\boldsymbol{b}}(\boldsymbol{Y}) &= \boldsymbol{\ell}_*'X\check{\boldsymbol{b}}(\boldsymbol{Y}) \\ &= \boldsymbol{\ell}_*'\mathbf{P}_X\boldsymbol{Y} \\ &= \boldsymbol{\ell}_*'\boldsymbol{Y} \text{ because } \boldsymbol{\ell}_* \in \text{sp}(X).\end{aligned}$$

Therefore, for any LS solution $\check{\boldsymbol{b}}(\boldsymbol{Y})$ (some of which may be non-linear), $\boldsymbol{g}'\check{\boldsymbol{b}}(\boldsymbol{Y})$ is a linear estimator. And, for any $\boldsymbol{\beta} \in \Re^{k+1}$ and $\mathrm{E}(\boldsymbol{Y}) = X\boldsymbol{\beta}$,

$$\begin{aligned}\mathrm{E}(\boldsymbol{g}'\check{\boldsymbol{b}}(\boldsymbol{Y})) &= \boldsymbol{\ell}_*'\mathrm{E}(\boldsymbol{Y}) \\ &= \boldsymbol{\ell}_*'X\boldsymbol{\beta} \\ &= \boldsymbol{g}'\boldsymbol{\beta}.\end{aligned}$$

That is, $\boldsymbol{g}'\check{\boldsymbol{b}}(\boldsymbol{Y})$ is an unbiased linear estimator of $\boldsymbol{g}'\boldsymbol{\beta}$.

Furthermore, because $\boldsymbol{g}'\check{\boldsymbol{b}}(\boldsymbol{Y}) \equiv \boldsymbol{\ell}_*'\boldsymbol{Y}$, by Propn. 8.1, its variance $\sigma^2\boldsymbol{\ell}_*'\boldsymbol{\ell}_*$ is least among variances $\sigma^2\boldsymbol{\ell}'\boldsymbol{\ell}$ of linear unbiased estimators $\boldsymbol{\ell}'\boldsymbol{Y}$. ■

Proof of Propn. 8.3, p. 64. 1. and 2. follow directly from Propn. 8.1 applied to each column of L and L_*. 3. follows from 1. and 2., because then

$$\begin{aligned} L'L - L_*'L_* &= L'(\mathrm{I} - \mathbf{P}_X)L \\ &= [(\mathrm{I} - \mathbf{P}_X)L]'[(\mathrm{I} - \mathbf{P}_X)L], \end{aligned}$$

which is nnd; and it is 0 iff $(\mathrm{I} - \mathbf{P}_X)L = L - \mathbf{P}_X L = L - L_* = 0$, that is, iff $L = L_*$. ■

A.9 Chapter 9 Proofs and Solutions

Proof of Propn. 9.1, p. 70.

$\star \iff 1$: Straightforward.

$1 \iff 2$: From 27e, p. 53,

$$\{G'\boldsymbol{b} : \boldsymbol{b} \in \Re^{k+1} \text{ and } X\boldsymbol{b} = X\boldsymbol{b}_*\} = \{G'\boldsymbol{b}_*\}$$

iff $\mathrm{sp}(G) \subset \mathrm{sp}(X')$.

$1 \Longrightarrow 3$: If $X\boldsymbol{b}_0 \in \mathcal{M}_0$, then there exists a vector $\boldsymbol{b}_*$ such that $G'\boldsymbol{b}_* = \mathbf{0}$ and $X\boldsymbol{b}_0 = X\boldsymbol{b}_*$. Then $G'\boldsymbol{b}_0 = L'X\boldsymbol{b}_0 = L'X\boldsymbol{b}_* = G'\boldsymbol{b}_* = \mathbf{0}$.

$3 \Longrightarrow 1$: $\boldsymbol{z} \in \mathrm{sp}(X')^\perp \Longrightarrow X\boldsymbol{z} = \mathbf{0} \in \mathcal{M}_0 \Longrightarrow G'\boldsymbol{z} = \mathbf{0} \Longrightarrow \mathrm{sp}(X')^\perp \subset \mathrm{sp}(G)^\perp \Longrightarrow \mathrm{sp}(G) \subset \mathrm{sp}(X')$.

$1 \Longrightarrow 4$: $\mathrm{sp}(G) \subset \mathrm{sp}(X') \Longrightarrow \exists\, L$ such that $G = X'L \Longrightarrow X'AG = X'AX'L = X'\mathbf{P}_X L = X'L = G$.

$4 \Longrightarrow 1$: $G = X'(AG) \Longrightarrow \mathrm{sp}(G) \subset \mathrm{sp}(X')$.

$1 \iff 5$: Let $\boldsymbol{g} \in \mathrm{sp}(G)$ and $\boldsymbol{g} \neq \mathbf{0}$. By 27d, p. 53,

$$\{X\boldsymbol{b} : \boldsymbol{b} \in \Re^{k+1} \text{ and } \boldsymbol{g}'\boldsymbol{b} = 0\} = \mathcal{M}$$

if and only if $\mathrm{sp}(\boldsymbol{g}) \cap \mathrm{sp}(X') = \{\mathbf{0}\}$. Therefore, 5 holds iff $\mathrm{sp}(\boldsymbol{g}) \cap \mathrm{sp}(X') \neq \{\mathbf{0}\}$, which is true iff $\boldsymbol{g} \in \mathrm{sp}(X')$ for each $\boldsymbol{g} \in \mathrm{sp}(G)$, and hence iff $\mathrm{sp}(G) \subset \mathrm{sp}(X')$.

$1 \iff 6$: 6 is equivalent to $\mathrm{sp}(G) = \mathrm{sp}(X'\mathbf{P}_H)$. If 6 is true, then $\mathrm{sp}(G) \subset \mathrm{sp}(X')$, that is, $6 \Longrightarrow 1$. If 1 is true, then $\mathrm{sp}(G) = \mathrm{sp}(G) \cap \mathrm{sp}(X') = \mathrm{sp}(X'H) \Longrightarrow 6$. ■

$7 \iff 1$: Let A_0' be a matrix such that $XA_0' = \mathbf{P}_X$. The set of all matrices A' such that $XA' = \mathbf{P}_X$ is

$$\mathcal{S} = \{A' = A_0' + (\mathrm{I} - \mathbf{P}_{X'})Z : Z \text{ is a } (k+1) \times n \text{ matrix}\}.$$

(7.) is equivalent to $A\boldsymbol{g} = A_0\boldsymbol{g}$ for each $\boldsymbol{g} \in \mathrm{sp}(G)$ and every $(k+1) \times n$ matrix Z. That is, (7.) is equivalent to

$$Z'(\mathrm{I} - \mathbf{P}_{X'})\boldsymbol{g} = \mathbf{0}$$

for each $\boldsymbol{g} \in \mathrm{sp}(G)$ and every $(k+1)\times n$ matrix Z. That in turn is equivalent to $\mathrm{sp}(G) \subset \mathrm{sp}(X')$. ∎

Chapter 9 Solutions

8. See Exercise 22, p. 52.

9. See Exercise 25, p. 52.

10. See Exercise 22, p. 52.

11. (a) See Exercise 27, p. 53.

13. Note that $\Delta_0 = \boldsymbol{b}_0 + \mathrm{sp}(G)^\perp$ and $\Delta_{0*} = \boldsymbol{b}_0 + \mathrm{sp}(X'H)^\perp$.

    (a) Because $\mathrm{sp}(X'H) = \mathrm{sp}(X') \cap \mathrm{sp}(G)$ (Exercise 27, p. 53), it follows that $\mathrm{sp}(X'H) = \mathrm{sp}(G) \cap \mathrm{sp}[(\mathrm{I} - \mathbf{P}_{X'})G]^\perp$, by Exercise 25, p. 52. Then

    $$\begin{aligned}\Delta_{0*} &= \boldsymbol{b}_0 + \mathrm{sp}(X'H)^\perp \\ &= \boldsymbol{b}_0 + \mathrm{sp}(G)^\perp + \mathrm{sp}[(\mathrm{I} - \mathbf{P}_{X'})G] \\ &= \Delta_0 + \mathrm{sp}[(\mathrm{I} - \mathbf{P}_{X'})G].\end{aligned}$$

    And $\mathrm{sp}(G)^\perp \cap \mathrm{sp}[(\mathrm{I} - \mathbf{P}_{X'})G] = \{\mathbf{0}\}$, by Exercise 25, p. 52.

    (b) If $\mathrm{sp}[(\mathrm{I} - \mathbf{P}_{X'})G] = \{\mathbf{0}\}$, then $\Delta_0 = \Delta_{0*}$.
    If $\Delta_0 = \Delta_{0*}$, let $\boldsymbol{z} \in \mathrm{sp}[(\mathrm{I} - \mathbf{P}_{X'})G]$. Let $\boldsymbol{c}_1 \in \mathrm{sp}(G)^\perp$. Because $\Delta_0 = \Delta_{0*}$, $\boldsymbol{b}_0 + \boldsymbol{c}_1 + \boldsymbol{z} \in \Delta_0$, and so there exists $\boldsymbol{c}_2 \in \mathrm{sp}(G)^\perp$ such that $\boldsymbol{b}_0 + \boldsymbol{c}_2 = \boldsymbol{b}_0 + \boldsymbol{c}_1 + \boldsymbol{z}$. Then $\boldsymbol{c}_1 - \boldsymbol{c}_2 = -\boldsymbol{z}$, and hence both are in $\mathrm{sp}[(\mathrm{I} - \mathbf{P}_{X'})G] \cap \mathrm{sp}(G)^\perp = \{\mathbf{0}\}$.

    (c) $(\mathrm{I} - \mathbf{P}_{X'})G = 0 \iff G = \mathbf{P}_{X'}G \iff \mathrm{sp}(G) \subset \mathrm{sp}(X')$.

14. (b) Proof: If $X_1\boldsymbol{\beta}_1$ is estimable in X, then there exists a matrix L such that $X'L = \binom{X_1'}{X_2'}L = \binom{X_1'}{0}$. Then $\mathrm{sp}(L) \subset \mathrm{sp}(N_2)$, because $X_2'L = 0$, and hence $\exists\ M$ such that $L = N_2M$. Then $X_1' = X_1'N_2M$, which implies that $\mathrm{sp}(X_1') \subset \mathrm{sp}(X_1'N_2)$. In the other direction, clearly $\mathrm{sp}(X_1'N_2) \subset \mathrm{sp}(X_1')$.
    If $\mathrm{sp}(X_1') = \mathrm{sp}(X_1'N_2)$, then $\exists\ Z$ such that $X_1' = X_1'N_2Z$. With $L = N_2Z$, then, $\binom{X_1'}{X_2'}L = \binom{X_1'}{0}$, and so $\mathrm{sp}\binom{X_1'}{0} \subset \mathrm{sp}(X')$, and hence $X_1\boldsymbol{\beta}_1$ is estimable in X.

(d) Proof: If $X_1\boldsymbol{\beta}_1$ is estimable in X, then there exists a matrix L such that $X'L = \binom{X_1'L}{X_2'L} = \binom{X_1'}{0}$. Suppose $\boldsymbol{z} \in \text{sp}(X_1) \cap \text{sp}(X_2)$, so that $\boldsymbol{z} = X_1\boldsymbol{b}_1 = X_2\boldsymbol{b}_2$. Then $X_1\boldsymbol{b}_1 = L'X_1\boldsymbol{b}_1 = L'X_2\boldsymbol{b}_2 = \mathbf{0}$ because $L'X_2 = 0$. Therefore, $\text{sp}(X_1) \cap \text{sp}(X_2) = \mathbf{0}$.
If $\text{sp}(X_1) \cap \text{sp}(X_2) = \mathbf{0}$, then

$$\mathbf{P}_{\binom{X_1'}{X_2'}} = \begin{pmatrix} \mathbf{P}_{X_1'} & 0 \\ 0 & \mathbf{P}_{X_2'} \end{pmatrix},$$

and hence

$$\begin{aligned} \mathbf{P}_{\binom{X_1'}{X_2'}} \begin{pmatrix} X_1' \\ 0 \end{pmatrix} &= \begin{pmatrix} X_1'L \\ X_2'L \end{pmatrix} \\ &= \begin{pmatrix} \mathbf{P}_{X_1'}X_1' \\ \mathbf{P}_{X_2'}0 \end{pmatrix} \\ &= \begin{pmatrix} X_1' \\ 0 \end{pmatrix}, \end{aligned}$$

and therefore $X_1\boldsymbol{\beta}_1$ is estimable in (X_1, X_2).

A.10 Chapter 10 Proofs and Solutions

Proof of Proposition 10.1, p. 83.

If $\text{sp}(G) \subset \text{sp}(X')$, then there exists a matrix B such that $G = X'B$. Let $P = \mathbf{P}_B$, which is symmetric and idempotent. Then $\text{sp}(B) = \text{sp}(P)$ and $\text{sp}(X'B) = \text{sp}(X'P)$.

If there exists a symmetric idempotent matrix P such that $\text{sp}(G) \subset \text{sp}(X'P)$, then for any $\boldsymbol{x}$ there exists a $\boldsymbol{z}$ such that $G\boldsymbol{x} = (X'P)\boldsymbol{z} = X'(P\boldsymbol{z}) \in \text{sp}(X')$. ■

Proof of Proposition 10.2, p. 83.

1. $\boxed{\text{sp}(\mathbf{P}_X L) \subset \text{sp}(H)\text{:}}$ $\boldsymbol{z} = \mathbf{P}_X L\boldsymbol{a} \in \text{sp}(\mathbf{P}_X L) \Longrightarrow \exists\, \boldsymbol{b}$ such that $X'\boldsymbol{z} = X'H\boldsymbol{b}$ $\Longrightarrow \boldsymbol{z} - H\boldsymbol{b} \in \text{sp}(X) \cap \text{sp}(X)^\perp \Longrightarrow \boldsymbol{z} = H\boldsymbol{b} \in \text{sp}(H)$.

   $\boxed{\text{sp}(H) \subset \text{sp}(\mathbf{P}_X L)\text{:}}$ $\boldsymbol{z} = H\boldsymbol{b} \in \text{sp}(H)$ (hence $\boldsymbol{z} \in \text{sp}(X)$) $\Longrightarrow \exists\, \boldsymbol{a}$ such that $X'\boldsymbol{z} = X'H\boldsymbol{b} = X'L\boldsymbol{a} = X'\mathbf{P}_X L\boldsymbol{a} \Longrightarrow \boldsymbol{z} - \mathbf{P}_X L\boldsymbol{a} \in \text{sp}(X) \cap \text{sp}(X)^\perp \Longrightarrow$ $\boldsymbol{z} \in \text{sp}(\mathbf{P}_X L)$. Therefore, $\text{sp}(\mathbf{P}_X L) = \text{sp}(H)$.

2. If $\text{sp}(X'L) = \text{sp}(X'H)$, then $L\boldsymbol{a} \in \text{sp}(L) \Longrightarrow \exists\, \boldsymbol{b}$ such that $X'L\boldsymbol{a} = X'H\boldsymbol{b}$ $\Longrightarrow L\boldsymbol{a} - H\boldsymbol{b} \in \text{sp}(X)^\perp \Longrightarrow L\boldsymbol{a} \in \text{sp}(H) + \text{sp}(X)^\perp$.

3. Because $\text{sp}(H)$ and $\text{sp}(X)^{\perp}$ are orthogonal, the orthogonal projection matrix onto $\text{sp}(H) + \text{sp}(X)^{\perp}$ is $\mathbf{P}_H + (\mathrm{I} - \mathbf{P}_X)$. And, because $\text{sp}(L) \subset \text{sp}(H) + \text{sp}(X)^{\perp}$, $Q = \mathbf{P}_H + (\mathrm{I} - \mathbf{P}_X) - \mathbf{P}_L$ is symmetric and idempotent. It follows then that, for any vector $\boldsymbol{z}$,

$$\begin{aligned} \boldsymbol{z}'(X'\mathbf{P}_H X - X'\mathbf{P}_L X)\boldsymbol{z} &= \boldsymbol{z}'X'QX\boldsymbol{z} \\ &= (Q'X\boldsymbol{z})'(Q'X\boldsymbol{z}) \geqslant 0, \end{aligned}$$

that is, that $X'\mathbf{P}_H X - X'\mathbf{P}_L X$ is nnd.

4. Because $Q = QQ'$, $\text{tr}(Q) = \text{tr}(QQ') \geqslant 0$, and hence $\nu_L = \text{tr}(\mathbf{P}_L) \leqslant \text{tr}(\mathbf{P}_H) + \text{tr}(\mathrm{I} - \mathbf{P}_X) = \nu_H + n - \nu_X$. Because $\text{sp}(H) = \text{sp}(\mathbf{P}_X L)$, $\nu_H = \nu_{\mathbf{P}_X L} \leqslant \nu_L$. ■

Proof of Proposition 10.2(5), p. 83. This proof is longer and more complicated than the proofs of the other parts of this proposition, and so it is set apart. As you can see, this proof establishes that $\text{sp}(L) \supset \text{sp}(H)$ and $\text{sp}(L)$ includes vectors that are not in $\text{sp}(H)$. Is it possible to show this without all the fancy construction with non-singular T that diagonalizes both $A'H'HA$ and $B'(\mathrm{I} - \mathbf{P}_X)B$?

By Prop. 10.2(2), there exist matrices A and B such that $L = HA + (\mathrm{I} - \mathbf{P}_X)B$. It follows that $\text{sp}(X'HA) = \text{sp}(X'H)$, which implies that $\text{sp}(HA) = \text{sp}(H)$ and hence that $\mathbf{P}_{HA} = \mathbf{P}_H$.

Following LaMotte et al. (2020), let T be a non-singular matrix such that $T'(A'H'HA)T = \Delta_1 = \text{Diag}(\delta_{1i})$ and $T'[B'(\mathrm{I} - \mathbf{P}_X)B]T = \Delta_2 = \text{Diag}(\delta_{2i})$. These imply that δ_{1i} and δ_{2i} are $\geqslant 0$ for all i. Re-order columns of T as $T = (T_{++}, T_{+0}, T_{0+}T_{00})$ corresponding to: $\delta_{1i} > 0$ and $\delta_{2i} > 0$ (++); $\delta_{1i} > 0$ and $\delta_{2i} = 0$ (+0); $\delta_{1i} = 0$ and $\delta_{2i} > 0$ (0+); and $\delta_{1i} = 0$ and $\delta_{2i} = 0$ (00). Omit categories that are void.

Then

$$\begin{aligned} \Delta_1 &= \text{Diag}(\Delta_{1++}, \Delta_{1+0}, 0, 0) \text{ and} \\ \Delta_2 &= \text{Diag}(\Delta_{2++}, 0, \Delta_{20+}, 0). \end{aligned}$$

Recall that $\mathbf{P}_L = L(L'L)^{-}L'$ is the same for any generalized inverse of $L'L$. Note that

$$\begin{aligned} X'\mathbf{P}_L X &= X'HA[A'H'HA + B'(\mathrm{I} - \mathbf{P}_X)B]^{+}A'H'X \\ &= X'HAT(\Delta_1 + \Delta_2)^{+}T'A'H'X, \end{aligned}$$

and

$$X'\mathbf{P}_H X = X'\mathbf{P}_{HA}X = X'HAT\Delta_1^{+}T'A'H'X.$$

It follows that

$$\begin{aligned} X'\mathbf{P}_H X - X'\mathbf{P}_L X &= X'HAT[\Delta_1^{+} - (\Delta_1 + \Delta_2)^{+}]T'A'H'X \\ &= X'HAT[\Delta_2\Delta_1^{+}(\Delta_1 + \Delta_2)^{+}]T'A'H'X. \end{aligned}$$

Note that

$$\Delta_2\Delta_1^+(\Delta_1+\Delta_2)^+ = \text{Diag}\{\Delta_{2++}\Delta_{1++}^{-1}(\Delta_{1++}+\Delta_{2++})^{-1},0,0,0\}.$$

Then

$$X'\mathbf{P}_HX - X'\mathbf{P}_LX = X'HAT_{++}\Delta_{2++}\Delta_{1++}^{-1}(\Delta_{1++}+\Delta_{2++})^{-1}T'_{++}A'H'X,$$

and $X'\mathbf{P}_HX - X'\mathbf{P}_LX = 0$ implies that $X'HAT_{++} = 0$, which implies that $HAT_{++} = 0$, because $\text{sp}(HAT_{++}) \subset \text{sp}(H) \subset \text{sp}(X)$.

Then if $X'\mathbf{P}_HX - X'\mathbf{P}_LX = 0$, it follows that there is no ++ category, that is, there is no i such that both δ_{1i} and δ_{2i} are > 0. Then $T = (T_{+0}, T_{0+}, T_{00})$, and

$$\begin{aligned} T'A'H'HAT &= \text{Diag}(\Delta_{1+0},0,0) \text{ and} \\ T'B'(\mathrm{I}-\mathbf{P}_X)BT &= \text{Diag}(0,\Delta_{20+},0), \end{aligned}$$

from which $HAT_{0+} = 0$, $HAT_{00} = 0$, $(\mathrm{I}-\mathbf{P}_X)BT_{+0} = 0$, and $(\mathrm{I}-\mathbf{P}_X)BT_{00} = 0$. Then

$$\begin{aligned} LT &= [HA + (\mathrm{I}-\mathbf{P}_X)B][T_{+0}, T_{0+}, T_{00}] \\ &= (HAT_{+0},0,0) + (0,(\mathrm{I}-\mathbf{P}_X)BT_{0+},0) \\ &= [HAT_{+0},(\mathrm{I}-\mathbf{P}_X)BT_{0+},0]. \end{aligned}$$

Because T is non-singular, $\text{sp}(HAT) = \text{sp}(HAT_{+0}) = \text{sp}(HA) = \text{sp}(H)$; and $\text{sp}[(\mathrm{I}-\mathbf{P}_X)BT] = \text{sp}[(\mathrm{I}-\mathbf{P}_X)BT_{0+}]$.

Given that $X'\mathbf{P}_HX - X'\mathbf{P}_LX = 0$, the ++ category is void. Then $\mathbf{P}_L \neq \mathbf{P}_H$ implies that the 0+ category is non-void, which implies that the columns of $(\mathrm{I}-\mathbf{P}_X)BT_{0+}$ are non-zero and orthogonal.

With T non-singular, $\text{sp}(LT) = \text{sp}(L)$; and thus

$$\begin{aligned} \mathbf{P}_L &= \mathbf{P}_{LT} \\ &= \mathbf{P}_{HAT_{+0}} + \mathbf{P}_{[(\mathrm{I}-\mathbf{P}_X)BT_{0+}]} \\ &= \mathbf{P}_H + \mathbf{P}_{[(\mathrm{I}-\mathbf{P}_X)BT_{0+}]}. \end{aligned}$$

It follows that $\nu_L > \nu_H$. ■

Proof of Proposition 10.3, p. 85.

$\text{sp}(H) \subset \text{sp}(X)$ and $\text{sp}(X'H) = \text{sp}(G) \cap \text{sp}(X') \Longrightarrow \mathbf{P}_H = \mathbf{P}_X - \mathbf{P}_{XN}$:

Show that $\text{sp}(H) \subset \text{sp}(\mathbf{P}_X - \mathbf{P}_{XN})$:

$$\begin{aligned} \text{sp}(X'H) &= \text{sp}(X') \cap \text{sp}(G) \subset \text{sp}(N)^\perp \\ \Longrightarrow N'X'H &= (XN)'H = \mathbf{0} \Longrightarrow P_{XN}H = 0 \\ \Longrightarrow (\mathbf{P}_X - \mathbf{P}_{XN})H &= \mathbf{P}_XH = H, \text{ because } \text{sp}(H) \subset \text{sp}(X), \\ \Longrightarrow \text{sp}(H) &\subset \text{sp}(\mathbf{P}_X - \mathbf{P}_{XN}). \end{aligned}$$

Show that $\text{sp}(\mathbf{P}_X - \mathbf{P}_{XN}) \subset \text{sp}(H)$: $\boldsymbol{z} \in \text{sp}(\mathbf{P}_X - \mathbf{P}_{XN}) \Longrightarrow (XN)'\boldsymbol{z} = \mathbf{0} \Longrightarrow X'\boldsymbol{z} \in \text{sp}(X') \cap \text{sp}(N)^{\perp} \subset \text{sp}(X') \cap \text{sp}(G) = \text{sp}(X'H) \Longrightarrow \exists\, \boldsymbol{x}$ such that $X'\boldsymbol{z} = X'H\boldsymbol{x}$; since both $\boldsymbol{z}$ and $H\boldsymbol{x}$ are in $\text{sp}(X)$, it follows that $\boldsymbol{z} = H\boldsymbol{x} \in \text{sp}(H)$. Therefore, $\text{sp}(\mathbf{P}_X - \mathbf{P}_{XN}) \subset \text{sp}(H)$, and therefore $\text{sp}(H) = \text{sp}(\mathbf{P}_X - \mathbf{P}_{XN})$. Both $\mathbf{P}_H$ and $\mathbf{P}_X - \mathbf{P}_{XN}$ are orthogonal projection matrices onto the same linear subspace, and therefore $\mathbf{P}_H = \mathbf{P}_X - \mathbf{P}_{XN}$.

$\mathbf{P}_H = \mathbf{P}_X - \mathbf{P}_{XN} \Longrightarrow \text{sp}(X'H) = \text{sp}(G) \cap \text{sp}(X')$:

If $\boldsymbol{z} \in \text{sp}(X'H)$, then $\exists\, \boldsymbol{x}$ such that $\boldsymbol{z} = X'H\boldsymbol{x}$; then $N'\boldsymbol{z} = (XN)'H\boldsymbol{x} = (XN)'\mathbf{P}_H H\boldsymbol{x} = \mathbf{0} \Longrightarrow \boldsymbol{z} \in \text{sp}(N)^{\perp} \subset \text{sp}(G)$, and hence $\boldsymbol{z} \in \text{sp}(X') \cap \text{sp}(G)$. Therefore $\text{sp}(X'H) \subset \text{sp}(X') \cap \text{sp}(G)$.

If $\boldsymbol{z} \in \text{sp}(G) \cap \text{sp}(X')$, which is contained in $\text{sp}(N)^{\perp}$, then $N'\boldsymbol{z} = \mathbf{0}$; and $\exists\, \boldsymbol{h} \in \text{sp}(X)$ such that $\boldsymbol{z} = X'\boldsymbol{h}$, hence $N'X'\boldsymbol{h} = \mathbf{0} \Longrightarrow \boldsymbol{h} \in \text{sp}(X) \cap \text{sp}(XN)^{\perp} = \text{sp}(\mathbf{P}_H) = \text{sp}(H)$, and therefore $\boldsymbol{z} = X'\boldsymbol{h}$ is in $\text{sp}(X'H)$. Therefore, $\text{sp}(G) \cap \text{sp}(X') \subset \text{sp}(X'H)$, and therefore $\text{sp}(X'H) = \text{sp}(G) \cap \text{sp}(X')$. ■

Proof of Propn. 10.4, p. 86 (Corollary to Proposition 10.2).

1. $\delta_L^2 = 0$ iff $\mathbf{P}_L \mathbb{K} M_2 \boldsymbol{\beta}_2 = 0$, equivalent to $L'\mathbb{K}M_2\boldsymbol{\beta}_2 = L'\mathbb{K}\mathbf{P}_{M_2}(M_2\boldsymbol{\beta}_2) = G'(M_2\boldsymbol{\beta}_2) = \mathbf{0}$.

2. Follows from (1), replacing M_2 by M_1 and noting that $\mathbf{P}_{M_1}\mathbf{P}_{M_2} = \mathbf{P}_{M_1}$ because $\text{sp}(M_1) \subset \text{sp}(M_2)$.

3. This follows from Proposition 10.2, with $H = H_2$ and $X = X_2 = \mathbb{K}M_2$ upon noting that $X_2'H_2 = X_2'\mathbf{P}_{X_2}L = X_2'L$.

4. This follows from Proposition 10.2, with $H = H_1$ and $X = X_1$ upon noting that $X_1'H_1 = X_1'\mathbf{P}_{X_1}L = X_1'L$.

5. This follows from Proposition 10.2, with $L = H_2$, $X = X_1$, and $H = H_1$. ■

Proof of Propn. 10.5, p. 87.

1. By (11, p. 74), $\text{sp}(X'H) = \text{sp}(X') \cap \text{sp}(G) \subset \text{sp}(G) \Longrightarrow \text{sp}(H) = \text{sp}(\mathbf{P}_X H) = \text{sp}(AX'H) \subset \text{sp}(AG)$, which in turn implies that $\text{sp}(X'H) \subset \text{sp}(X'AG)$.

2. If $\text{sp}(G) \subset \text{sp}(X')$, then there exists a matrix L such that $G = X'L$, and hence $X'AG = X'AX'L = X'\mathbf{P}_X L = X'L = G$. In the other direction, clearly $\text{sp}(X'AG) = \text{sp}(G)$ implies that $\text{sp}(G) \subset \text{sp}(X')$. ■

Chapter 10 Solutions

19. (a) $\text{sp}(M) = \text{sp}(\mathbf{1}_a, \text{I}_a) = \text{sp}(\mathbf{1}_a) + \text{sp}(\text{I}_a) = \text{sp}(\text{I}_a) = \Re^a$, so $\text{sp}(X) = \mathbb{K}\{\text{sp}(M)\} = \mathbb{K}\{\Re^a\} = \text{sp}(\mathbb{K})$.

(b) $\mathbb{K}'\mathbb{K} = \text{Diag}(\mathbf{1}'_{n_i})\text{Diag}(\mathbf{1}_{n_i}) = \text{diag}(n_i) = N$. Then $\mathbf{P}_{\mathbb{K}} = \mathbb{K}N^{-1}\mathbb{K}'$.

(c) Reexpression.

(d) $(\text{I} - \mathbf{P}_X)\boldsymbol{y} = \boldsymbol{y} - \mathbf{P}_{\mathbb{K}}\boldsymbol{y} = \boldsymbol{y} - \mathbb{K}\bar{\boldsymbol{y}} = (y_{ij} - \bar{y}_{i\cdot})$. Then SSE has $\text{tr}(\text{I} - \mathbf{P}_X) = n - a$ df; and $(\text{I} - \mathbf{P}_X)X = 0$, so its ncp is 0.

(e) Let $A = (\mathbf{0}_n, \mathbb{K}N^{-1})$. Then

$$XA' = \mathbb{K}(\mathbf{1}_a, \text{I}_a)\begin{pmatrix} \mathbf{0}'_n \\ N^{-1}\mathbb{K}' \end{pmatrix} = \mathbb{K}N^{-1}\mathbb{K}' = \mathbf{P}_X.$$

(f) Note that $\text{sp}(\mathbb{K}') = \Re^a$. If $\boldsymbol{g} \in \text{sp}(X')$, then $\boldsymbol{g} \in \text{sp}(M')$, and $\exists$ $\boldsymbol{m}$ such that $\boldsymbol{g} = (\mathbf{1}'_a\boldsymbol{m}, \boldsymbol{m}')'$, and hence $g_0 = \mathbf{1}'_a\boldsymbol{g}_1$. In the other direction, if $\boldsymbol{g} = (g_0, \boldsymbol{g}'_1)'$ and $g_0 = \mathbf{1}'_a\boldsymbol{g}_1$, let $\boldsymbol{\ell} = \mathbb{K}N^{-1}\boldsymbol{g}_1$. Then $X'\boldsymbol{\ell} = (\mathbf{1}'_a\boldsymbol{g}_1, \boldsymbol{g}'_1)' = (g_0, \boldsymbol{g}'_1)' = \boldsymbol{g}$, and therefore $\boldsymbol{g} \in \text{sp}(X')$.
With $g_0 = 1$ and $\boldsymbol{g}_1 = \mathbf{0}_a$, $\boldsymbol{g}'\boldsymbol{\beta} = \beta_0$ is not estimable, because $\mathbf{1}'_a\boldsymbol{g}_1 \neq 1$. And $\boldsymbol{\beta}_1 = G'\boldsymbol{\beta}$ with $G = \begin{pmatrix} \mathbf{0}'_a \\ \text{I}_a \end{pmatrix}$ is not estimable: in fact, no column of G is estimable.

(g) Show that $\text{sp}(C_1) = \text{sp}(S_a)$: the $a - 1$ columns of C_1 are linearly independent members of $\text{sp}(\mathbf{1}_a)^\perp = \text{sp}(S_a)$, and therefore $\text{sp}(C_1) = \text{sp}(S_a)$, by Exercise 9j, p. 49. Then $\text{sp}(C_1)^\perp = \text{sp}(S_a)^\perp = \text{sp}(\mathbf{1}_a)$. That is, $C'\boldsymbol{\beta} = C'_1\boldsymbol{\beta}_1 = \mathbf{0}$ iff $\boldsymbol{\beta}_1 \in \text{sp}(\mathbf{1}_a)$, that is, $\boldsymbol{\beta} = b\mathbf{1}_a$ for some b.

(h) Each column of C has the form $\boldsymbol{c} = (0, \boldsymbol{c}'_1)'$, with $\mathbf{1}'\boldsymbol{c}_1 = 0$, and therefore, by (19f) above, $\boldsymbol{c}'\boldsymbol{\beta}$ is estimable. It follows that $C'\boldsymbol{\beta}$ is estimable.

(i) Because $\text{sp}(C) \subset \text{sp}(X')$, $\exists$ L such that $C = X'L$. Then, for any A such that $XA' = \mathbf{P}_X$, $AC = AX'L = \mathbf{P}_XL$. That is, $A_1C = \mathbf{P}_XL$ and $A_2C = \mathbf{P}_XL$, and hence $A_1C = A_2C$.

(j) Note that $\mathbb{K}\mathbf{1}_a = \mathbf{1}_n$. With $A = (\mathbf{0}, \mathbb{K}N^{-1})$, $AC = \mathbb{K}N^{-1}C_1$. Then $\text{sp}(AC) \subset \text{sp}(\mathbb{K})$. And $\mathbf{1}'_n\mathbb{K}N^{-1}C_1 = \mathbf{1}'_aC_1 = \mathbf{0}$ implies that $\text{sp}(AC) \subset \text{sp}(\mathbf{1}_n)^\perp$. Together, these imply that $\text{sp}(AC) \subset \text{sp}(\mathbb{K}) \cap \text{sp}(\mathbf{1}_n)^\perp$.
Next show that $\text{sp}(AC)^\perp \subset [\text{sp}(\mathbb{K}) \cap \text{sp}(\mathbb{K}\mathbf{1}_a)^\perp]^\perp$: Let $\boldsymbol{\ell} \in \text{sp}(AC)^\perp$. For any $\boldsymbol{\ell} \in \Re^n$, $\exists$ $\boldsymbol{m}$, $\boldsymbol{b}$ such that $\boldsymbol{\ell} = \mathbb{K}\boldsymbol{m} + (\text{I} - \mathbf{P}_{\mathbb{K}})\boldsymbol{b}$. If $\boldsymbol{\ell} \in \text{sp}(AC)^\perp$ then $C'A'\boldsymbol{\ell} = C'_1N^{-1}\mathbb{K}'\boldsymbol{\ell} = \mathbf{0} = C'_1N^{-1}\mathbb{K}'\mathbb{K}\boldsymbol{m} = C'_1\boldsymbol{m}$. This implies that $\boldsymbol{m} = d\mathbf{1}_a$ for some scalar d, and hence that $\mathbb{K}\boldsymbol{m} = d\mathbf{1}_n$. Therefore, $\boldsymbol{\ell} \in \text{sp}(\mathbf{1}_n) + \text{sp}(\mathbb{K})^\perp = [\text{sp}(\mathbb{K}) \cap \text{sp}(\mathbf{1}_n)^\perp]^\perp$.
It follows that $\text{sp}(AC) \subset \text{sp}(\mathbb{K}) \cap \text{sp}(\mathbf{1}_n)^\perp$ and $\text{sp}(AC) \supset \text{sp}(\mathbb{K}) \cap \text{sp}(\mathbf{1}_n)^\perp$, and therefore $\text{sp}(AC) = \text{sp}(\mathbb{K}) \cap \text{sp}(\mathbf{1}_n)^\perp$. It follows that $\mathbf{P}_{AC} = \mathbf{P}_{\mathbb{K}} - \mathbf{P}_{\mathbf{1}_n}$.

(k) $\mathbf{P}_{AC}\boldsymbol{y} = \mathbf{P}_{\mathbb{K}}\boldsymbol{y} - \mathbf{P}_{\mathbf{1}_n}\boldsymbol{y} = \mathbb{K}\bar{\boldsymbol{y}} - \bar{y}_{\cdot\cdot}\mathbf{1}_n = (\bar{y}_{i\cdot} - \bar{y}_{\cdot\cdot})$, hence

$$\begin{aligned} \boldsymbol{y}'\mathbf{P}_{AC}\boldsymbol{y} &= (\mathbf{P}_{AC}\boldsymbol{y})'(\mathbf{P}_{AC}\boldsymbol{y}) \\ &= \sum_{i=1}^{a}\sum_{j=1}^{n_i}(\bar{y}_{i\cdot} - \bar{y}_{\cdot\cdot})^2 = \sum_{i=1}^{a} n_i(\bar{y}_{i\cdot} - \bar{y}_{\cdot\cdot})^2. \end{aligned}$$

Its df is $\nu = \mathrm{tr}(\mathbf{P}_{AC}) = a - 1$. Its ncp is is δ^2 with

$$\delta^2\sigma^2 = \sum_{i=1}^{a} n_i(\eta_i - \bar{\eta})^2 = \sum_{i=1}^{a} n_i(\beta_i - \bar{\beta})^2,$$

with $\bar{\beta} = (1/a)\sum_{i=1}^{a} \beta_i$; and it is 0 iff $\beta_1 = \cdots = \beta_a$.

(l) $C'\boldsymbol{\beta} = C_1'\boldsymbol{\beta}_1 = \mathbf{0} \Longrightarrow \boldsymbol{\beta}_1 = \gamma\mathbf{1}_a$, $\boldsymbol{\beta} = (\beta_0, \gamma\mathbf{1}_a')'$, and hence $X\boldsymbol{\beta} = \mathbb{K}(\mathbf{1}_a, \mathrm{I}_a)\boldsymbol{\beta} = \mathbb{K}(\beta_0\mathbf{1}_a + \gamma\mathbf{1}_a) = (\beta_0 + \gamma)\mathbb{K}\mathbf{1}_a = (\beta_0 + \gamma)\mathbf{1}_n$, and the restricted model becomes $\mathrm{sp}(\mathbf{1}_n)$.

(m) For $\mathrm{H}_0 : C_1'\boldsymbol{\beta} = \mathbf{0}$, the full model is $\mathrm{sp}(X) = \mathrm{sp}(\mathbb{K})$ and the restricted model is $\mathrm{sp}(\mathbf{1}_n)$. Then the RMFM SS is $\boldsymbol{y}'(\mathbf{P}_{\mathbb{K}} - \mathbf{P}_{\mathbf{1}_n})\boldsymbol{y} = \boldsymbol{y}'\mathbf{P}_{AC}\boldsymbol{y}$.

(n) That $\mathrm{sp}(C_1) = \mathrm{sp}(S_a) = \mathrm{sp}(\mathbf{1}_a)^{\perp}$ was shown above. $\mathrm{sp}(\mathbf{1}_a, C_1) = \mathrm{sp}(\mathbf{1}_a) + \mathrm{sp}(\mathbf{1}_a)^{\perp} = \Re^a$. Then $\mathrm{sp}(\mathbb{K}M_C) = \mathbb{K}\{\Re^a\} = \mathrm{sp}(X)$.

(o) With the full model $\mathrm{sp}(\mathbb{K}M_C) = \mathrm{sp}(\mathbb{K}\mathbf{1}_a, \mathbb{K}C_1)$, omitting $\mathbb{K}C_1$ produces the model $\mathrm{sp}(\mathbb{K}\mathbf{1}_a) = \mathrm{sp}(\mathbf{1}_n)$, the same as the restricted model due to imposing the conditions $C'\boldsymbol{\beta} = \mathbf{0}$ on the full model.

(p) Reexpress $\boldsymbol{y}'\mathbf{P}_{AC}\boldsymbol{y} = n_1(\bar{y}_{1\cdot} - \bar{y}_{\cdot\cdot})^2 + n_2(\bar{y}_{2\cdot} - \bar{y}_{\cdot\cdot})^2$, noting that $\bar{y}_{\cdot\cdot} = (n_1\bar{y}_{1\cdot} + n_2\bar{y}_{2\cdot})/(n_1 + n_2)$.

20. (a) Follows from $\mathrm{sp}(\mathbf{P}_X) \subset \mathrm{sp}(X)$.

(b) If $\exists\ \boldsymbol{\ell}$ such that $\boldsymbol{g} = X'\boldsymbol{\ell}$, then $A\boldsymbol{g} = AX'\boldsymbol{\ell} = \mathbf{P}_X\boldsymbol{\ell} \in \mathrm{sp}(X)$.

(c) With A such that $XA' = \mathbf{P}_X$, let $A_* = \mathbf{P}_X A$, hence $\mathrm{sp}(A_*) \subset \mathrm{sp}(X)$ and $XA_*' = XA'\mathbf{P}_X = \mathbf{P}_X\mathbf{P}_X = \mathbf{P}_X$.

(d) $A\boldsymbol{g} \in \mathrm{sp}(A) \subset \mathrm{sp}(X)$.

21. Solutions are left to the reader.

A.11 Chapter 11 Proofs and Solutions

Proof of Propn. 11.1, p. 105. Suppose $\boldsymbol{b}$ and $\boldsymbol{\lambda}$ satisfy (11.1). Then $\boldsymbol{b} \in \mathrm{sp}(R)^{\perp} = \mathrm{sp}(N)$ implies that there exists $\boldsymbol{c}$ such that $\boldsymbol{b} = N\boldsymbol{c}$. And, from the first equation in (11.1),

$$N'(X'X\boldsymbol{b} + R\boldsymbol{\lambda}) = N'X'\boldsymbol{y},$$

which, because $N'R = 0$, implies that $\boldsymbol{y} - XN\boldsymbol{c}$ is in $\mathrm{sp}(XN)^{\perp}$, and therefore $XN\boldsymbol{c} = X\boldsymbol{b} = \mathbf{P}_{XN}\boldsymbol{y}$. Further, that $N'X'(\boldsymbol{y} - X\boldsymbol{b}) = \mathbf{0}$ implies that $X'(\boldsymbol{y} - X\boldsymbol{b}) \in \mathrm{sp}(N)^{\perp} = \mathrm{sp}(R)$, which implies that there exists $\boldsymbol{\lambda}$ such that $X'(\boldsymbol{y} - X\boldsymbol{b}) = R\boldsymbol{\lambda}$.

In the other direction, let $\boldsymbol{c}$ be such that $XN\boldsymbol{c} = \mathbf{P}_{XN}\boldsymbol{y}$, and let $\boldsymbol{b} = N\boldsymbol{c}$. Then $(XN)'(\boldsymbol{y} - X\boldsymbol{b}) = \mathbf{0}$ implies that $X'(\boldsymbol{y} - X\boldsymbol{b}) \in \mathrm{sp}(N)^{\perp} = \mathrm{sp}(R)$, and hence that there exists $\boldsymbol{\lambda}$ such that $X'(\boldsymbol{y} - X\boldsymbol{b}) = R\boldsymbol{\lambda}$. ■

A.17 Chapter 17 Proofs and Solutions

Proof of Propn. 17.1, p. 165. This is an alternative, step-by-step proof. Let $X_{01} = (X_0, X_1)$, so that X becomes $X = (X_{01}, X_2)$. Then N_{01} is such that

$$\begin{aligned} \text{sp}(N_{01}) &= \text{sp}(X) \cap \text{sp}(X_{01})^\perp \\ &= \text{sp}(\mathbf{P}_X - \mathbf{P}_{X_{01}}) \\ &= \text{sp}[(\mathbf{P}_X - \mathbf{P}_{X_{01}})X] \\ &= \text{sp}[(\mathbf{P}_X - \mathbf{P}_{X_{01}})X_2] \end{aligned}$$

and $X_{2*} = X_2 X_2' N_{01}$ and $X_* = (X_{01}, X_{2*})$.

Clearly $\text{sp}(X_*) \subset \text{sp}(X)$.

One way to prove that $\text{sp}(X) \subset \text{sp}(X_*)$ is to prove equivalently that $\text{sp}(X_*)^\perp \subset \text{sp}(X)^\perp$.

Let $\boldsymbol{v} \in \text{sp}(X_*)^\perp$, that is, $X_*'\boldsymbol{v} = \binom{\mathbf{0}}{\mathbf{0}} = \binom{X_{01}'\boldsymbol{v}}{X_{2*}'\boldsymbol{v}}$. This implies that $\boldsymbol{v} = (\text{I} - \mathbf{P}_{X_{01}})\boldsymbol{v}_{1*}$ for some $\boldsymbol{v}_{1*}$. Then also

$$\begin{aligned} X_{2*}'\boldsymbol{v} &= \mathbf{0} \\ &= N_{01}' X_2 X_2' (\text{I} - \mathbf{P}_{X_{01}})\boldsymbol{v}_{1*} \\ &= N_{01}' X_2 X_2' (\mathbf{P}_X - \mathbf{P}_{X_{01}})\boldsymbol{v}_{1*} \\ &= N_{01}' X_2 X_2' N_{01} \boldsymbol{v}_{1**} \end{aligned}$$

for some $\boldsymbol{v}_{1**}$ because $\text{sp}(N_{01}) = \text{sp}(\mathbf{P}_X - \mathbf{P}_{X_{01}})$. Then

$$\begin{aligned} \boldsymbol{v}_{1**}' N_{01}' X_2 X_2' N_{01} \boldsymbol{v}_{1**} &= 0 \implies \\ X_2' N_{01} \boldsymbol{v}_{1**} &= \mathbf{0} \implies \\ X_2'(\mathbf{P}_X - \mathbf{P}_{X_{01}})\boldsymbol{v}_{1*} &= \mathbf{0} \implies \\ X_2' \mathbf{P}_X (\text{I} - \mathbf{P}_{X_{01}})\boldsymbol{v}_{1*} &= \mathbf{0} \implies \\ X_2'(\text{I} - \mathbf{P}_{X_{01}})\boldsymbol{v}_{1*} &= \mathbf{0} \implies \\ X_2'\boldsymbol{v} = \mathbf{0}. \end{aligned}$$

Therefore, $X_{01}'\boldsymbol{v} = \mathbf{0}$ and $X_2'\boldsymbol{v} = \mathbf{0}$, which together imply that $\boldsymbol{v} \in \text{sp}(X)^\perp$. Therefore, $\text{sp}(X_*)^\perp \subset \text{sp}(X)^\perp$, and therefore $\text{sp}(X) \subset \text{sp}(X_*)$. Together with $\text{sp}(X_*) \subset \text{sp}(X)$, this implies that $\text{sp}(X_*) = \text{sp}(X)$.

Now prove that $\text{sp}(X_{01}) \cap \text{sp}(X_{2*}) = \{\mathbf{0}\}$. Suppose $\boldsymbol{u} \in \text{sp}(X_{01}) \cap \text{sp}(X_{2*})$, so that $\boldsymbol{u} \in \text{sp}(X_{01})$ and $\boldsymbol{u} = X_{2*}\boldsymbol{v} = X_2 X_2' N_{01}\boldsymbol{v}$ for some $\boldsymbol{v}$. Then, because $\text{sp}(N_{01}) = \text{sp}(\mathbf{P}_X - \mathbf{P}_{X_{01}})$, there exists $\boldsymbol{v}_*$ such that $N_{01}\boldsymbol{v} = (\mathbf{P}_X - \mathbf{P}_{X_{01}})\boldsymbol{v}_*$, and hence

$$\begin{aligned} (\mathbf{P}_X - \mathbf{P}_{X_{01}}) X_2 X_2' N_{01} \boldsymbol{v} &= \mathbf{0} \\ &= (\mathbf{P}_X - \mathbf{P}_{X_{01}}) X_2 X_2' (\mathbf{P}_X - \mathbf{P}_{X_{01}})\boldsymbol{v}_*, \end{aligned}$$

which implies that $X_2X_2'(\mathbf{P}_X - \mathbf{P}_{X_{01}})\boldsymbol{v}_* = X_2X_2'N_{01}\boldsymbol{v} = X_{2*}\boldsymbol{v} = \boldsymbol{u} = \mathbf{0}$. ■

Alternative proof of Proposition 17.2, p. 166. The ncp of $\boldsymbol{y}'P_3\boldsymbol{y}$ is $\boldsymbol{\delta}_3'\boldsymbol{\delta}_3$, where

$$\begin{aligned}
\boldsymbol{\delta}_3 &= (\mathbf{P}_X - \mathbf{P}_{(X_0,X_{2*})})(X_0\boldsymbol{\beta}_0 + X_1\boldsymbol{\beta}_1 + X_{2*}\boldsymbol{\beta}_2) \\
&= (\mathbf{P}_X - \mathbf{P}_{(X_0,X_{2*})})X_1\boldsymbol{\beta}_1 \\
&= (\mathbf{P}_X - \mathbf{P}_{(X_0,X_{2*})})(X_{1|0}\boldsymbol{\beta}_1 + \mathbf{P}_{X_0}X_1\boldsymbol{\beta}_1) \\
&= X_{1|0}\boldsymbol{\beta}_1 - \mathbf{P}_{(X_0,X_{2*})}X_{1|0}\boldsymbol{\beta}_1.
\end{aligned}$$

Clearly, $X_{1|0}\boldsymbol{\beta}_1 = \mathbf{0} \Longrightarrow \boldsymbol{\delta}_3 = \mathbf{0}$. There exist $\boldsymbol{z}_0$ and $\boldsymbol{z}_{2*}$ such that the last expression is $X_{1|0}\boldsymbol{\beta}_1 + X_0\boldsymbol{z}_0 + X_{2*}\boldsymbol{z}_{2*}$. Because this is a direct sum, that this expression is $\mathbf{0}$ implies that all three terms are $\mathbf{0}$, which implies that $X_{1|0}\boldsymbol{\beta}_1 = \mathbf{0}$. That is, $\boldsymbol{\delta}_3 = \mathbf{0}$ iff $X_{1|0}\boldsymbol{\beta}_1 = \mathbf{0}$, and hence $\boldsymbol{y}'P_3\boldsymbol{y}$ tests exclusively $X_{1|0}\boldsymbol{\beta}_1 = \mathbf{0}$.

Proof of Proposition 17.3, p. 167. Let R be a matrix such that $R'E_{1|0}\boldsymbol{\beta}_1$ is estimable in the model $\mathbb{K}E_*\boldsymbol{\beta}$. Then there exists a matrix L such that

$$E_*'\mathbb{K}'L = \begin{pmatrix} E_0' \\ E_{1|0}' \\ E_{2*}' \end{pmatrix}\mathbb{K}'L = \begin{pmatrix} 0 \\ E_{1|0}'R \\ 0 \end{pmatrix}. \tag{A.1}$$

It follows that $R'E_{1|0}\boldsymbol{\beta}_1 = L'\boldsymbol{\delta}_3$, because $L'\mathbb{K}E_{1|0} = R'E_{1|0}$ and $L'\mathbb{K}(E_0, E_{2*}) = 0$. ■

Proof of Proposition 17.4, p. 168. That $H_*E_0 = 0$ and $H_*E_{1|0} = E_{1|0}$ is clear from the definition of H_*. Let $\boldsymbol{j}_2 \in \mathcal{J}_2$ and $\boldsymbol{j} \in \mathcal{J}_*$. Then $\boldsymbol{j}_2 \succeq \boldsymbol{j}_* \succeq \boldsymbol{j}$ $\Longrightarrow$

$$\begin{aligned}
H_{\boldsymbol{j}}E_{\boldsymbol{j}_2}E_{\boldsymbol{j}_2}' &= \bigotimes_{i=1}^{f}\begin{cases} S_{a_i} & \text{if } j_i = 1 \text{ and } j_{2i} = 1, \\ U_{a_i} & \text{if } j_i = 0 \text{ and } j_{2i} = 1, \\ \mathbf{1}_{a_i}\mathbf{1}_{a_i}' = a_iU_{a_i} & \text{if } j_i = 0 \text{ and } j_{2i} = 0, \end{cases} \\
&= c_{\boldsymbol{j},\boldsymbol{j}_2}H_{\boldsymbol{j}}.
\end{aligned} \tag{A.2}$$

Then

$$H_{\boldsymbol{j}}E_2E_2'\mathbb{K}'N_{01} = \left(\sum_{\boldsymbol{j}_2} c_{\boldsymbol{j},\boldsymbol{j}_2}\right)H_{\boldsymbol{j}}\mathbb{K}'N_{01}. \tag{A.3}$$

Because $\text{sp}(H_{\boldsymbol{j}}) \subset \text{sp}(E_1)$, $\text{sp}(\mathbb{K}H_{\boldsymbol{j}}) \subset \text{sp}(\mathbb{K}E_1)$, and therefore $N_{01}'\mathbb{K}H_{\boldsymbol{j}} = \mathbf{0}$ for all $\boldsymbol{j} \in \mathcal{J}_*$. With $H_* = \sum\{H_{\boldsymbol{j}} : \boldsymbol{j} \in \mathcal{J}_*\}$, it follows that $H_*E_2E_2'\mathbb{K}'N_{01} = 0$. ■

B

Sampling Distributions Derived from Normally-Distributed Random Variables

B.1 Normally Distributed Random Variables.

C. R. Rao (1965) gives two equivalent definitions of random variables that follow multivariate normal distributions. The n-dimensional random variable $\boldsymbol{U}$ is said to have an n-variate normal distribution with mean vector $\boldsymbol{\mu}$ and variance-covariance matrix V if and only if:

Definition 1. For every n-vector ℓ, $\ell'\boldsymbol{U}$ has a univariate normal distribution with mean $\ell'\boldsymbol{\mu}$ and variance $\ell' V \ell$.

Definition 2. There exists an $n \times p$ matrix M, the columns of which are linearly independent, and $V = MM'$, such that $\boldsymbol{U} = \boldsymbol{\mu} + M\boldsymbol{Z}$, where $\boldsymbol{Z}$ is a p-variate random variable, the p components of which are independent standard normal random variables.

By the notation $\boldsymbol{U} \sim \mathbf{N}_n(\boldsymbol{\mu}, V)$ we shall mean that the random variable $\boldsymbol{U}$ follows an n-variate normal distribution with mean vector $\boldsymbol{\mu}$ and variance-covariance matrix V.

The standard normal density function is

$$\phi(z) = \frac{1}{\sqrt{2\pi}} e^{-z^2/2}.$$

The joint density function of the p independent standard normal components of $\boldsymbol{Z} = (Z_1, \ldots, Z_p)$ is

$$f_{\boldsymbol{z}}(\boldsymbol{z}) = \phi(z_1) \times \cdots \times \phi(z_p) = (2\pi)^{-p/2} \exp(-\boldsymbol{z}'\boldsymbol{z}/2).$$

It is useful to write down the general form of the density function of a multivariate normal random variable. As in the second definition above, let $\boldsymbol{U} = \boldsymbol{\mu} + M\boldsymbol{Z}$, and assume that $n = p$, that is, that M is a square matrix with linearly independent columns. In this case $V = MM'$ has an inverse V^{-1}. The joint density function of $\boldsymbol{U}$ is

$$f_U(\boldsymbol{u}) = (2\pi|V|)^{-n/2}\exp[-(\boldsymbol{U}-\boldsymbol{\mu})'V^{-1}(\boldsymbol{U}-\boldsymbol{\mu})/2],$$

where $|V|$ denotes the determinant of the $n \times n$ matrix V. For our purposes here, you may assume without proof that, for any $n \times n$ symmetric, pd matrix V and any n-vector $\boldsymbol{\mu}$, this function's integral over $\Re^n$ is 1. This fact is often useful in establishing other results.

It can be shown that normally-distributed random variables are independent if and only if their covariance is 0. Suppose $\boldsymbol{U}_1 = \boldsymbol{\mu}_1 + M_1\boldsymbol{Z}$ and $\boldsymbol{U}_2 = \boldsymbol{\mu}_2 + M_2\boldsymbol{Z}$, where $\boldsymbol{Z} \sim \mathbf{N}_p(0, \mathrm{I}_p)$, so that $\boldsymbol{U}_1$ and $\boldsymbol{U}_2$ are normally distributed. Then

$$\mathrm{Cov}(\boldsymbol{U}_1, \boldsymbol{U}_2) = M_1 M_2',$$

and $\boldsymbol{U}_1$ and $\boldsymbol{U}_2$ are independent if $M_1M_2' = 0$. Similarly, suppose $\boldsymbol{U} \sim \mathbf{N}_n(\boldsymbol{\mu}, V)$ and consider two linear functions $L_1'\boldsymbol{U}$ and $L_2'\boldsymbol{U}$, where L_1 and L_2 are $n \times p$ and $n \times q$ matrices, respectively. Then $\mathrm{Cov}(L_1'\boldsymbol{U}, L_2'\boldsymbol{U}) = L_1'VL_2$, and so $L_1'\boldsymbol{U}$ and $L_2'\boldsymbol{U}$ are independent if $L_1'VL_2 = 0$.

The two principal characteristics of normally-distributed random variables that we shall need as we go on are:

1. Linear combinations of normally-distributed random variables are normally distributed.

2. Normally-distributed random variables are independent if and only if their covariance is 0.

B.2 Basic Sampling Distributions Related to Normal Distributions

Chi-squared Distribution. If $Z_1, \ldots, Z_\nu$ are ν independent normal random variables, all with variance 1, then

$$X = Z_1^2 + \cdots + Z_\nu^2$$

has a chi-squared distribution with ν *degrees of freedom*. If all the Z_i have mean 0 (i.e., they are all standard normals), then X is said to have a *central* chi-squared distribution. More generally, let $\delta_i = \mathrm{E}(Z_i)$, $i = 1, \ldots, \nu$; $\delta^2 = \sum_i \delta_i^2$ is called the *non-centrality parameter*, and X is said to follow a chi-squared distribution with ν degrees of freedom and non-centrality parameter δ^2. If $\delta^2 > 0$, the distribution is said to be *non-central.*

Student's t Distribution. If $Z_0 \sim \mathbf{N}(\delta, 1)$ and $Z_1, \ldots, Z_\nu$ are standard normal random variables and $Z_0, Z_1, \ldots, Z_\nu$ are independent, then

$$T = \frac{Z_0}{\sqrt{(Z_1^2 + \cdots + Z_\nu^2)/\nu}}$$

has a Student's t distribution with ν degrees of freedom. If $\delta = 0$, the distribution is said to be *central*, while if $\delta \neq 0$ it is called a *non-central t* distribution with *non-centrality parameter* δ. A Student's t random variable is the ratio of a normal random variable with variance 1 to the square root of an independent central chi-squared random variable divided by its degrees of freedom.

***F* Distribution.** If X_1 is a chi-squared random variable with ν_1 degrees of freedom and non-centrality parameter δ^2, and X_2 is a central chi-squared random variable with ν_2 degrees of freedom, and X_1 and X_2 are independent, then

$$F = \frac{X_1/\nu_1}{X_2/\nu_2}$$

has an F distribution with ν_1 *numerator* degrees of freedom, ν_2 *denominator* degrees of freedom, and non-centrality parameter δ^2. If $\delta^2 = 0$, the distribution is said to be *central*; otherwise, it is said to be *non-central.*

When used without the 'central' or 'non-central' modifier, each name usually refers to the central distribution. However, check the context to be sure. It is the central distributions that are widely tabulated because their quantiles are needed as critical values for tests of hypotheses, construction of confidence intervals, and other inferential procedures. The non-central distributions are important for power analysis because they describe the probability distribution of the test statistic when the null hypothesis is false.

B.3 Distributions of Quadratic Forms

Suppose $\boldsymbol{Y} \sim \mathbf{N}(\boldsymbol{\mu}, \sigma^2 \mathrm{I}_n)$. Assume that the parameter set for $\boldsymbol{\mu}$ is a linear subspace $\mathcal{S} = \mathrm{sp}(X) = \{X\boldsymbol{\beta} : \boldsymbol{\beta} \in \Re^{k+1}\}$ for some given $n \times (k+1)$ matrix X. The parameter σ^2 is an unknown positive number. In this section and the next, we'll establish basic relations that will enable us to see why t-statistics and F-statistics used in inference in multiple regression actually follow Student's t and F distributions.

Let A be an $n \times n$ symmetric, idempotent matrix. Let $Q = \boldsymbol{Y}'A\boldsymbol{Y}$. We shall show that Q/σ^2 has a chi-squared distribution with $\nu = \mathrm{tr}(A)$ degrees of freedom and non-centrality parameter $\boldsymbol{\mu}'A\boldsymbol{\mu}/\sigma^2$. To do so, we must show that there exists a ν-vector $\boldsymbol{Z}$ distributed as $\mathbf{N}(\boldsymbol{\delta}, \mathrm{I}_\nu)$ such that $Q/\sigma^2 = \boldsymbol{Z}'\boldsymbol{Z} = Z_1^2 + \cdots + Z_\nu^2$ and $\boldsymbol{\delta}'\boldsymbol{\delta} = \boldsymbol{\mu}'A\boldsymbol{\mu}/\sigma^2$.

Because A is symmetric and idempotent, it is an orthogonal projection matrix, projecting onto $\mathrm{sp}(A)$. Let the ν columns of the $n \times \nu$ matrix B be an orthonormal spanning set for $\mathrm{sp}(A)$. Then, due to the uniqueness of the orthogonal projection matrix, $A = BB'$. Let $\boldsymbol{Z} = (1/\sigma)B'\boldsymbol{Y}$. Then $\mathrm{E}(\boldsymbol{Z}) = \boldsymbol{\delta} = (1/\sigma)B'\boldsymbol{\mu}$, and $\mathrm{Var}(\boldsymbol{Z}) = [(1/\sigma)B]'(\sigma^2\mathrm{I})[(1/\sigma)B] = \mathrm{I}_\nu$ because $B'B = \mathrm{I}_\nu$. Hence $\boldsymbol{Z} \sim \mathbf{N}(\boldsymbol{\delta}, \mathrm{I}_\nu)$ and

$$
\begin{aligned}
Q/\sigma^2 = \boldsymbol{Y}'A\boldsymbol{Y}/\sigma^2 &= \boldsymbol{Y}'BB'\boldsymbol{Y}/\sigma^2 \\
&= [(1/\sigma)B'\boldsymbol{Y}]'[(1/\sigma)B'\boldsymbol{Y}] \\
&= \boldsymbol{Z}'\boldsymbol{Z},
\end{aligned}
$$

the sum of squares of ν independent normal random variables, each with variance 1. Note that $\mathrm{tr}(A) = \mathrm{tr}(BB') = \mathrm{tr}(B'B) = \mathrm{tr}(\mathrm{I}_\nu) = \nu$. The non-centrality parameter of the distribution of Q/σ^2 is

$$
\begin{aligned}
\boldsymbol{\delta}'\boldsymbol{\delta} &= [(1/\sigma)B'\boldsymbol{\mu}]'[(1/\sigma)B'\boldsymbol{\mu}] \\
&= \boldsymbol{\mu}'(BB')\boldsymbol{\mu}/\sigma^2 \\
&= \boldsymbol{\mu}'A\boldsymbol{\mu}/\sigma^2.
\end{aligned}
$$

In summary, if $\boldsymbol{Y} \sim \mathbf{N}(\boldsymbol{\mu}, \sigma^2\mathrm{I}_n)$, and if A is an $n \times n$ symmetric idempotent matrix, then $\boldsymbol{Y}'A\boldsymbol{Y}/\sigma^2 \sim \chi^2_\nu(\delta^2)$, where the non-centrality parameter δ^2 is $\boldsymbol{\mu}'A\boldsymbol{\mu}/\sigma^2$ and the degrees of freedom are $\nu = \mathrm{tr}(A)$.

B.4 Independence of Linear and Quadratic Forms under Multivariate Normality

If $\boldsymbol{Y} \sim \mathbf{N}(\boldsymbol{\mu}, \sigma^2\mathrm{I})$, then linear statistics $L'\boldsymbol{Y}$ and $M'\boldsymbol{Y}$ are independent iff $L'M = 0$. We shall see now that $L'\boldsymbol{Y}$ and the quadratic form $\boldsymbol{Y}'A\boldsymbol{Y}$ are independent if $AL = 0$, and two quadratic forms $\boldsymbol{Y}'A\boldsymbol{Y}$ and $\boldsymbol{Y}'B\boldsymbol{Y}$ are independent if $AB = 0$. As part of the argument we shall use without proof this fact: if $\boldsymbol{Y}_1$ and $\boldsymbol{Y}_2$ are independent random variables (perhaps multivariate), then any functions (also perhaps multivariate) $g(\boldsymbol{Y}_1)$ and $h(\boldsymbol{Y}_2)$ are independent.

Consider independence of $\boldsymbol{Y}'A\boldsymbol{Y}$, where A is $n \times n$ symmetric, and $L'\boldsymbol{Y}$, where L is an $n \times c$ matrix. If $AL = 0$, then each column of L is orthogonal to all columns of $A' = A$, and therefore $\mathbf{P}_A L = 0$. This implies that $\mathbf{P}_A\boldsymbol{Y}$ and $L'\boldsymbol{Y}$ are independent. Further, $\boldsymbol{Y}'A\boldsymbol{Y} = (\mathbf{P}_A\boldsymbol{Y})'A(\mathbf{P}_A\boldsymbol{Y})$, a function of $\mathbf{P}_A\boldsymbol{Y}$, and therefore $L'\boldsymbol{Y}$ and $\boldsymbol{Y}'A\boldsymbol{Y}$ are independent. If $AL = 0$, then $L'\boldsymbol{Y}$ and $\boldsymbol{Y}'A\boldsymbol{Y}$ are independent.

Next consider independence of two quadratic forms, $\boldsymbol{Y}'A\boldsymbol{Y}$ and $\boldsymbol{Y}'B\boldsymbol{Y}$, where A and B are $n \times n$ symmetric matrices. If $AB = 0$, then $\mathbf{P}_A\mathbf{P}_B =$

0, and therefore $\mathbf{P}_A\boldsymbol{Y}$ and $\mathbf{P}_B\boldsymbol{Y}$ are independent. Noting that $\boldsymbol{Y}'A\boldsymbol{Y} = (\mathbf{P}_A\boldsymbol{Y})'A(\mathbf{P}_A\boldsymbol{Y})$, and similarly $\boldsymbol{Y}'B\boldsymbol{Y}$ is a function of $\mathbf{P}_B\boldsymbol{Y}$, it follows that $\boldsymbol{Y}'A\boldsymbol{Y}$ and $\boldsymbol{Y}'B\boldsymbol{Y}$ are independent. If $AB = 0$, then $\boldsymbol{Y}'A\boldsymbol{Y}$ and $\boldsymbol{Y}'B\boldsymbol{Y}$ are independent.

B.5 Exercises

For these exercises, let $\boldsymbol{Y}$ denote an n-variate normal random variable with mean vector $\boldsymbol{\mu}$ and variance-covariance matrix V. Consider $\boldsymbol{Y}$ partitioned into two parts as

$$\boldsymbol{Y} = \begin{pmatrix} \boldsymbol{Y}_1 \\ \boldsymbol{Y}_2 \end{pmatrix},$$

with $\boldsymbol{\mu}$ and V partitioned accordingly as $\boldsymbol{\mu} = (\boldsymbol{\mu}_1', \boldsymbol{\mu}_2')'$ and

$$V = \begin{pmatrix} V_{11} & V_{12} \\ V_{12}' & V_{22} \end{pmatrix}.$$

Assume that V is positive definite, which implies that V has an inverse, and that V_{11}, V_{22}, and the Schur complements $V_{11} - V_{12}V_{22}^{-1}V_{12}'$ and $V_{22} - V_{12}'V_{11}^{-1}V_{12}$ are positive definite, too.

The density function of $\boldsymbol{Y}$ is

$$f_{\boldsymbol{Y}}(\boldsymbol{y}) = (2\pi)^{-n/2}|V|^{-1/2}\exp[(-1/2)(\boldsymbol{y} - \boldsymbol{\mu})'V^{-1}(\boldsymbol{y} - \boldsymbol{\mu})].$$

For integer $m \geqslant 1$, let $\boldsymbol{Z}$ denote the multivariate standard normal, that is, the m-variate normal random variable with $\mathrm{E}(\boldsymbol{Z}) = \mathbf{0}$ and $\mathrm{Var}(\boldsymbol{Z}) = \mathrm{I}$.

1. **Completing the quadratic form.** Let A be an $n \times n$ symmetric matrix, $\boldsymbol{b}$ an n-vector, and c a scalar. Show that there exist a vector $\boldsymbol{d}$, a vector $\boldsymbol{g}$ such that $A\boldsymbol{g} = 0$, and a scalar h such that, for all $\boldsymbol{y} \in \Re^n$,

$$\boldsymbol{y}'A\boldsymbol{y} + \boldsymbol{b}'\boldsymbol{y} + c = (\boldsymbol{y} - \boldsymbol{d})'A(\boldsymbol{y} - \boldsymbol{d}) + \boldsymbol{g}'\boldsymbol{y} + h.$$

2. If A is nnd and $\boldsymbol{b} \in \mathrm{sp}(A)$ in Exercise 1, show that there exists a vector $\boldsymbol{d}$ such that

$$\boldsymbol{y}'A\boldsymbol{y} + \boldsymbol{b}'\boldsymbol{y} + c = (\boldsymbol{y} - \boldsymbol{d})'A(\boldsymbol{y} - \boldsymbol{d}) + c - \boldsymbol{d}'A\boldsymbol{d},$$

and hence that a lower bound on $\boldsymbol{y}'A\boldsymbol{y} + \boldsymbol{b}'\boldsymbol{y} + c$ is $c - \boldsymbol{d}'A\boldsymbol{d}$, and that bound is attained only for $A(\boldsymbol{y} - \boldsymbol{d}) = 0$. Further, although $\boldsymbol{d}$ giving these relations is not in general unique, this bound is.

3. Prove: If $V_{12} = 0$, then $\boldsymbol{Y}_1$ and $\boldsymbol{Y}_2$ are independent.

4. Derive the conditional pdf of $\boldsymbol{Y}_1|\boldsymbol{Y}_2 = \boldsymbol{y}_2$ to demonstrate that this conditional random variable is normally distributed. In particular, find expressions for its mean vector and variance-covariance matrix in terms of $\boldsymbol{\mu}$ and V.

5. Using either definition, show that $\boldsymbol{Y}_1 \sim \mathbf{N}(\boldsymbol{\mu}_1, V_{11})$. That is, show that marginal random variables of multivariate normal random variables are normal.

6. The moment-generating function (mgf) for $\boldsymbol{Z}$ is $m_{\boldsymbol{Z}}(\boldsymbol{t}) = \mathrm{E}(e^{\boldsymbol{t}'\boldsymbol{Z}})$, where $\boldsymbol{t} = (t_1, \ldots, t_m)' \in \Re^m$. Find $m_{\boldsymbol{Z}}(\boldsymbol{t})$.

7. Using Exercise 6, find the mgf for $\boldsymbol{Y}$. You may assume without proof that there exists a symmetric pd matrix W such that $V = WW$. Then $\boldsymbol{Y} = \boldsymbol{\mu} + W\boldsymbol{Z}$, and $m_{\boldsymbol{Y}}(\boldsymbol{t}) = e^{\boldsymbol{t}'\boldsymbol{\mu}} m_{\boldsymbol{Z}}(W\boldsymbol{t})$.

8. Let A be an $m \times m$ symmetric matrix of constants, and let $Q = \boldsymbol{Z}'A\boldsymbol{Z}$. Find the mgf of Q. If $A = \mathrm{I}$, identify the probability distribution of which this is the mgf.

9. For an $n \times n$ symmetric matrix A, find the mgf $m_A(t)$ of the quadratic form $Q_A = \boldsymbol{Y}'A\boldsymbol{Y}$.

10. For a random variable Q, $\ln m_Q(t)$ is called the *cumulant generating function* for Q. Assuming that the distribution of Q is such that it is defined, show that

$$\mathrm{E}(Q) = \left.\frac{\partial \log m_Q(t)}{\partial t}\right|_{t=0}$$

and

$$\mathrm{Var}(Q) = \left.\frac{\partial^2 \log m_Q(t)}{\partial t^2}\right|_{t=0}.$$

11. Using the cumulant generating function for $Q_A = \boldsymbol{Y}'A\boldsymbol{Y}$, find the expected value and variance of Q_A.

12. For two quadratic forms $Q_1 = \boldsymbol{Y}'A_1\boldsymbol{Y}$ and $Q_2 = \boldsymbol{Y}'A_2\boldsymbol{Y}$, where A_1 and A_2 are $n \times n$ symmetric matrices, note that you can write the joint moment generating function $m_{1,2}(t_1, t_2) = \mathrm{E}[\exp(t_1 Q_1 + t_2 Q_2)]$ simply by substituting $t_1 A_1 + t_2 A_2$ in place of tA in the expression for $m_A(t)$. Doing so, write $m_{1,2}(t_1, t_2)$. The second derivative of $\log m_{1,2}(t_1, t_2)$ with respect to t_1, t_2, evaluated at $t_1 = t_2 = 0$, gives $\mathrm{Cov}(Q_1, Q_2)$. Find $\mathrm{Cov}(Q_1, Q_2)$.

13. Show: If A is an $n \times n$ symmetric matrix and L is an $n \times c$ matrix, and if $AL = 0$, then $\mathbf{P}_A L = 0$.

14. Show: If A is an $n \times n$ symmetric matrix, then for any n-vector $\boldsymbol{y}$, $\boldsymbol{y}'A\boldsymbol{y} = (\mathbf{P}_A\boldsymbol{y})'A(\mathbf{P}_A\boldsymbol{y})$.

15. Show: If A and B are symmetric matrices such that $AB = 0$, then $\mathbf{P}_A\mathbf{P}_B = 0$.

16. Suppose $\boldsymbol{Y} \sim \mathbf{N}(\boldsymbol{\mu}, \sigma^2\mathrm{I})$, $\boldsymbol{\mu} \in \mathrm{sp}(X)$ and $\sigma^2 \in \Re_+$, where X is a known $n \times (k+1)$ matrix. Let $\boldsymbol{p}$ be a given vector in $\mathrm{sp}(X')$, that is, there exists an n-vector $\boldsymbol{\ell}$ such that $\boldsymbol{p} = X'\boldsymbol{\ell}$. For these exercises, suppose GS performed on X yields T and V such that $\mathrm{sp}(V) = \mathrm{sp}(X)$, $V'V = \mathrm{I}_m$, and $V = XT$. Let $\hat{\boldsymbol{\beta}} = TV'\boldsymbol{Y}$, so that $\hat{\boldsymbol{Y}} = X\hat{\boldsymbol{\beta}} = \mathbf{P}_X\boldsymbol{Y}$.

    (a) What is the probability distribution of $\boldsymbol{p}'\hat{\boldsymbol{\beta}}$?

    (b) What is the distribution of the vector of residuals $\boldsymbol{Y} - \hat{\boldsymbol{Y}}$?

    (c) Show that $\boldsymbol{p}'\hat{\boldsymbol{\beta}}$ and $\boldsymbol{Y} - \hat{\boldsymbol{Y}}$ are independent.

    (d) Show that residual sum of squares is $SSE = \boldsymbol{Y}'(\mathrm{I}_n - \mathbf{P}_X)\boldsymbol{Y}$.

    (e) Find $\mathrm{tr}(\mathbf{P}_X)$ and $\mathrm{tr}(\mathrm{I} - \mathbf{P}_X)$.

    (f) Find the expected value and variance of $\hat{\sigma}^2 = \boldsymbol{Y}'(\mathrm{I}_n - \mathbf{P}_X)\boldsymbol{Y}/(n-m)$.

    (g) Assuming that $\boldsymbol{\mu} \in \mathrm{sp}(X)$, what is the distribution of SSE/σ^2?

    (h) Assuming that the first column of X is a column of ones, what is the distribution of $SSR/\sigma^2 = \boldsymbol{Y}'(\mathbf{P}_X - \mathbf{P}_1)\boldsymbol{Y}/\sigma^2$?

    (i) Show that regression sum of squares and residual sum of squares, SSR and SSE, are independent (assuming that $\boldsymbol{\mu} \in \mathrm{sp}(X)$).

Bibliography

Alghanim, H., J. Antunes, D. S. B. Santos Silva, C. S. Alho, K. Balamurugan, and B. McCord (2017). Detection and evaluation of DNA methylation markers found at SCGN and KLF14 loci to estimate human age. *Forensic Science International: Genetics 31*, 81–88. [vi]

Alldredge, J. R. and N. S. Gilb (1976). Ridge regression: an annotated bibliography. *International Statistical Review 44*(3), 355–360. [128]

Allen, D. M. (1974). The relationship between variable selection and data augmentation and a method for prediction. *Technometrics 16*(1), 125–127. [109]

Anderson, R. L. and T. A. Bancroft (1952). *Statistical Theory in Research.* McGraw-Hill Book Co. [135, 189, 195, 196]

Apple (2019). Dictionary version 2.3.0. [68]

Belsley, D. A., E. Kuh, and R. E. Welsch (1980). *Regression Diagnostics.* John Wiley & Sons. [128]

Bose, R. C. (1944). The fundamental problem of linear estimation. In *Proceedings of the 31st Indian Science Congress, Part III*, pp. 2–3. [68]

Casella, G. and R. L. Berger (2002). *Statistical Inference* (2nd ed.). Pacific Grove, CA: Duxbury. [2, 61]

Chatterjee, S. and A. S. Hadi (1986). Influential observations, high leverage points, and outliers in linear regression. *Statistical Science 1*, 379–416. [128]

Christensen, R. (2010). *Plane Answers to Complex Questions: The Theory of Linear Models* (4th ed.). Springer. [1, 2]

Clarke, B. R. (2008). *Linear models: The theory and application of analysis of variance.* Hoboken, N.J.: John Wiley & Sons Inc. [1]

Cook, R. D. (1977). Letter to the editor. *Technometrics 19*, 349. [128]

Cook, R. D. and S. Weisberg (1982). *Residuals and Influence in Regression.* Chapman & Hall. [128]

Draper, N. R. and H. Smith (1998). *Applied Regression Analysis.* Probability and Statistics. New York: John Wiley & Sons, Inc. [1]

Elston, R. C. and N. Bush (1964). The hypotheses that can be tested when there are interactions in an Analysis of Variance model. *Biometrics 20*(4), 681–698. [69]

Federer, W. T. and M. Zelen (1966). Analysis of multifactor classifications with unequal numbers of observations. *Biometrics 22*(3), 525–552. [188]

Francis, I. (1973). A comparison of several analysis of variance programs. *Journal of the American Statistical Association 68*, 860–865. [135]

Freijeiro-González, L., M. Febrero-Bande, and W. González-Manteiga (2022). A critical review of lasso and its derivatives for variable selection under dependence among covariates. *International Statistical Review 90*(1), 118–145. [129]

Furnival, G. M. and R. W. Wilson (1974). Regression by leaps and bounds. *Technometrics 16*, 403–408. [127]

Gauss, C. F. (1823). *Theoria combinationis observationum erroribus minimis obnoxiae (Theory of the combination of observations least subject to error).* H. Dieterich. [63]

Gentle, J. E. (2007). *Matrix Algebra: Theory, Computations, and Applications in Statistics.* New York: Springer. [44]

Goodnight, J. H. (1976). The General Linear Models procedure. In *Proceedings of the First International SAS User's Group*, Cary, NC. SAS Institute Inc. [135, 163, 164, 174, 188]

Gram, J. P. (1883). Ueber die entwickelung reeler funtionen in reihen mittelst der methode der kleinsten quadrate. *Jrnl. für die reine und angewandte Math. 94*, 71–73. [43]

Hartley, H., J. Rao, and L. LaMotte (1978). A simple 'synthesis'-based method of variance component estimation. *Biometrics 34*, 233–242. [159]

Hector, A., S. von Felten, and B. Schmid (2010). Analysis of variance with unbalanced data: an update for ecology & evolution. *Journal of Animal Ecology 79*, 308–316. [164, 189]

Herr, D. G. (1986). On the history of ANOVA in unbalanced, factorial designs: the first 30 years. *The American Statistician 40*, 265–270. [163, 188]

Hocking, R. R. (2013). *Methods and Applications of Linear Models* (3rd ed.). John Wiley & Sons, Inc. [1, 136, 190, 196]

Hoerl, A. E. and R. W. Kennard (1970a). Ridge regression: biased estimation for nonorthogonal problems. *Technometrics 12*, 55–67. [128]

Hoerl, A. E. and R. W. Kennard (1970b). Ridge regression: biased estimation for nonorthogonal problems. *Technometrics 12*, 69–82. [128]

IBM Corporation (2021). *IBM SPSS 28 Statistics Algorithms.* IBM Corporation. [164]

Kempthorne, O. (1975). Fixed and mixed models in the analysis of variance. *Biometrics 31*, 473–486. [156, 196]

Kempthorne, O. and L. Folks (1971). *Probability, Statistics, and Data Analysis.* Ames, IA: Iowa State University Press. [78]

Khuri, A. I. (2010). *Linear Model Methodology* (1st ed.). Chapman & Hall. [1, 190]

Kuchibhotla, A. K., J. E. Kolassa, and T. A. Kuffner (2022). Post-selection inference. *Annual Review of Statistics and Its Applications 9*, 905–927. [128]

Kutner, M. H., C. J. Nachtsheim, J. Neter, and W. Li (2005). *Applied Linear Statistical Models* (5th ed.). McGraw-Hill/Irwin. [1, 2]

LaMotte, L. R. (1999). Collapsibility hypotheses and diagnostic bounds in regression analysis. *Metrika 50*, 109–119. [128]

LaMotte, L. R. (2014). The Gram-Schmidt construction as a basis for linear models. *The American Statistician 68*(1), 52–55. [43, 44]

LaMotte, L. R. (2017). Proportional subclass numbers in two-factor anova. *Statistics: A Journal of Theoretical and Applied Statistics 52*(1), 228–238. [196]

LaMotte, L. R. (2019). Corrigenda to "proportional subclass numbers in two-factor ANOVA". *Statistics: A Journal of Theoretical and Applied Statistics 53*(3), 696–698. [196]

LaMotte, L. R. (2020). A formula for Type III sums of squares. *Communications in Statistics - Theory and Methods 49*(13), 3126–3136. [164, 166]

LaMotte, L. R. (2023). Testing ANOVA effects: a resolution for unbalanced models. *Communications in Statistics – Theory and Methods 53*(22), 7860–7870. [155]

LaMotte, L. R. (2024). Essential properties of Type III* methods. *Communications in Statistics - Theory and Methods*, 1–11. [164, 166]

LaMotte, L. R. and J. Volaufova (1999). Prediction intervals via consonance intervals. *The Statistician 48*(3), 419–424. [118]

LaMotte, L. R., J. Volaufova, and S. Puntanen (2020). *On shrinkage estimators and "effective degrees of freedom"*, Chapter 14, pp. 245–254. Springer Nature Switzerland AG. [14, 109, 231]

LaMotte, L. R. and J. D. Wells (1995). Estimating maggot age from weight using inverse prediction. *Journal of Forensic Sciences 40*(4), 585–590. [120]

Langsrud, Ø. (2003). ANOVA for unbalanced data: Use Type II instead of Type III sums of squares. *Statistics and Computing 13*, 163–167. [164, 189]

Laplace, P.-S. (1816). *Théorie Analytique des Probabilités, First Supplement.* Mme Ve Courcier, Imprimeur-Libraire pour les Mathématiques, quai des Augustins, no. 57. [43]

Lehmann, E. L. (1959). *Testing Statistical Hypotheses.* New York: John Wiley & Sons, Inc. [78, 120]

Lehmann, E. L. (1986). *Testing Statistical Hypotheses* (2nd ed.). New York: Springer. [78]

Lenth, R. V. (2025). R package 'emmeans': Estimated marginal means, aka least-squares means. [171]

Macnaughton, D. B. (1998). Which sums of squares are best in unbalanced analysis of variance? MatStat Research Consulting Inc. [164, 189]

Markov, A. A. (1912). *Wahrscheinlichkeitsrechnung.* Teubner, Leipzig-Berlin, 1912. [63]

Marquardt, D. W. (1970). Generalized inverses, ridge regression, biased linear estimation, and nonlinear estimation. *Technometrics 12*, 591–612. [129]

Miller, J. (2018). Earliest known uses of some of the words of mathematics. [43]

Milliken, G. A. and D. E. Johnson (1984). *Analysis of Messy Data, Volume 1: Designed Experiments.* New York: Van Nostrand Reinhold Company. [164]

Nelder, J. A. (1977). A reformulation of linear models. *Journal of the Royal Statistical Society 140*(1), 48–77. [156]

Paik, U. B. and W. T. Federer (1974). Analysis of nonorthogonal n-way classifications. *The Annals of Statistics 2*(5), 1000–1021. [188]

Peixoto, J. L. and L. R. LaMotte (1989). Simultaneous identification of outliers and predictors using variable selection techniques. *Journal of Statistical Planning and Inference 23*, 327–343. [126, 127]

Rawlings, J. O., S. G. Pantula, and D. A. Dickey (1998). *Applied Regression Analysis: A Research Tool* (2nd ed.). Berlin: Springer. [1]

Reiersol, O. (1963). Identifiability, estimability, phenorestricting specifications, and zero Lagrange Multipliers in the Analysis of Variance. *Skandanavia Aktuarietidskr 46*, 131–142. [70]

Rencher, A. C. and G. B. Schaalje (2008). *Linear Models in Statistics.* Hoboken, NJ: John Wiley & Sons, Inc. [69]

SAS (1978). Tests of hypotheses in fixed-effects linear models. *SAS Technical Report R-101, SAS Institute, Cary, N.C.*. [163, 166, 189]

Schmidt, E. (1907). Zur theorie der linearen und nichtlinearen integralgleichungen. i. teil: Entwicklung willkürlicher funktionen nach systemen vorgeschriebener. *Mathematische Annalen 63*, 433–476. [43]

Searle, S. R. (1966). Estimable functions and testable hypotheses in linear models. Technical Report BU-213-M, Biometrics Unit, Plant Breeding Department, Cornell University. [69, 136]

Searle, S. R. (1971). *Linear Models.* New York: John Wiley & Sons, Inc. [189]

Searle, S. R. (1987). *Linear Models for Unbalanced Data.* New York: John Wiley & Sons, Inc. [189]

Searle, S. R. (1994). Analysis of variance computing package output for unbalanced data from fixed-effects models with nested factors. *The American Statistician 48*, 148–153. [189]

Searle, S. R., F. M. Speed, and H. V. Henderson (1981). Some computational and model equivalences in analysis of variance of unequal-subclass-numbers data. *The American Statistician 35*, 16–33. [189]

Seber, G. A. F. and A. J. Lee (2003). *Linear Regression Analysis* (2nd ed.). John Wiley & Sons, Inc. [1]

Seely, J. (1977). Estimability and linear hypotheses. *The American Statistician 31*(3), 121–123. [68, 70, 71, 72]

Smith, C. E. and R. Cribbie (2014). Factorial ANOVA with unbalanced data: A fresh look at the types of sums of squares. *Journal of Dairy Science 12*, 385–404. [135, 164, 190]

Snedecor, G. W. (1946). *Statistical Methods* (4th ed.). Iowa State College, Ames, Iowa: Collegiate Press, Inc. [195]

Snedecor, G. W. and W. G. Cochran (1980). *Statistical Methods* (7th ed.). The Iowa State University Press. [135, 195, 196]

Snedecor, G. W. and G. M. Cox (1935). Disproportionate subclass numbers in tables of multiple classification. Research Bulletin 180, Agricultural Experiment Station, Iowa State College of Agriculture and Mechanic Arts, Ames, IA. [135, 143, 189, 195, 196, 197]

Sokal, R. R. and F. J. Rolf (1995). *Biometry: the Principles and Practice of Statistics in Biological Research* (3rd ed.). New York: W. H. Freeman and Company. [195]

Stapleton, J. H. (1995). *Linear Statistical Models.* New York: John Wiley & Sons, Inc. [1]

Stein, C. (1956). Inadmissibility of the usual estimator for the mean of a multivariate distribution. *Proceedings of the Third Berkeley Symposium Mathematical Statistics and Probability 1*, 197–206. [128]

Stroup, W. W. (2013). *Generalized Linear Mixed Models: Modern Concepts, Methods and Applications.* Boca Raton, FL: CRC Press. [69]

Thompson, W. O. and F. B. Cady (Eds.) (1973). *The University of Kentucky Conference on Regression with a Large Number of Predictor Variables.* Department of Statistics, University of Kentucky. [5]

Tibshirani, R. (1996). Regression shrinkage and selection via the lasso. *Journal of the Royal Statistical Society Series B: Statistical Methodology 58*(1), 267–288. [129]

Venables, W. N. (2000). Exegeses on linear models. https://www.stats.ox.ac.uk/pub/MASS3/Exegeses.pdf. S-Plus User's Conference, Washington, DC, 8-9th October, 1998. [164, 170]

Wichura, M. J. (2006). *The Coordinate-Free Approach to Linear Models.* New York: Cambridge University Press. [78]

Yates, F. (1933). The principles of orthogonality and confounding in replicated experiments. *The Journal of Agricultural Science 23*, 108–145. [188]

Yates, F. (1934). The analysis of multiple classifications with unequal numbers in the different classes. *Journal of the American Statistical Association 29*(185), 51–66. [187, 188, 189, 190, 192, 195]

Index

For Product Safety Concerns and Information please contact our EU representative GPSR@taylorandfrancis.com
Taylor & Francis Verlag GmbH, Kaufingerstraße 24, 80331 München, Germany

www.ingramcontent.com/pod-product-compliance
Lightning Source LLC
Chambersburg PA
CBHW050525190326
41458CB00005B/1661
9781032981529